# LACTOFERRIN
Structure and Function

# ADVANCES IN EXPERIMENTAL MEDICINE AND BIOLOGY

## Recent Volumes in this Series

A Continuation Order Plan is available for this series. A continuation order will bring delivery of each new volume immediately upon publication. Volumes are billed only upon actual shipment. For further information please contact the publisher.

# LACTOFERRIN

## Structure and Function

Edited by

## T. William Hutchens

University of California, Davis
Davis, California

## Sylvia V. Rumball

Massey University
Palmerston North, New Zealand

and

## Bo Lönnerdal

University of California, Davis
Davis, California

SPRINGER SCIENCE+BUSINESS MEDIA, LLC

Library of Congress Cataloging-in-Publication Data

Lactoferrin : structure and function / edited by T. William Hutchens,
Sylvia V. Rumball, and Bo Lönnerdal.
        p.   cm. -- (Advances in experimental medicine and biology ;
357)
    "Proceedings of a workshop on Lactoferrin: Structure and Function,
held September 19-24, 1992, in Honolulu, Hawaii"--T.p. verso.
    Includes bibliographical references and index.
    ISBN 978-0-306-44734-1        ISBN 978-1-4615-2548-6 (eBook)
    DOI 10.1007/978-1-4615-2548-6
    1. Lactoferrins--Congresses.    I. Hutchens, T. William.
II. Rumball, Sylvia V.  III. Lönnerdal, Bo, 1938-    .  IV. Series.
QP552.L345L33  1994
574.19'245--dc20                                            94-47606
                                                                CIP

Proceedings of a workshop on Lactoferrin: Structure and Function, held September 19–24, 1992, in Honolulu, Hawaii

ISBN 978-0-306-44734-1

© 1994 Springer Science+Business Media New York
Originally published by Plenum Press New York in 1994

# PREFACE

Lactoferrin is an intriguing protein with an interesting structure and several known or suggested biological activities. We feel that attention on this protein has been too limited and diffuse, partly because it has been "hidden" among other well-known iron-binding proteins such as hemoglobin, ferritin and transferrin, but also perhaps because its biological functions are so diverse. Investigators that focus on lactoferrin represent a wide variety of medical and scientific disciplines that do not usually come together. It was our intention to improve that situation with this symposium.

In this book, experts from a variety of disciplines describe the present knowledge of the structural features of lactoferrin, its carbohydrate side chains and its capacity to bind different metal ions and anions. Several of the possible physiological functions of lactoferrin are described in detail, including the role of lactoferrin in bacterial killing, its involvement in growth and proliferation, in immune function and in iron absorption. Aspects of the molecular biology of lactoferrin and its specific interactions with different cell types are also included. Finally, as lactoferrin now has become commercially available in larger quantities, possible industrial applications are discussed. The book should give the interested reader a thorough insight into our present knowledge of lactoferrin.

The Editors

# ACKNOWLEDGMENTS

The organizers and participants thank the following sponsors of the First International Symposium on Lactoferin Structure and Function:

**Mead Johnson**

**Morinaga Milk Industries**

**New Zealand Dairy Board**

**Ross Laboratories**

**Semper/DMV/SMR**

**Snow Brand Milk Products**

**Wei-Chuan Foods**

**Wyeth-Ayerst Laboratories**

**National Institutes of Health (NICHD)**

# CONTENTS

# THREE-DIMENSIONAL STRUCTURE OF LACTOFERRIN IN VARIOUS FUNCTIONAL STATES

Edward N. Baker,* Bryan F. Anderson, Heather M. Baker,
Catherine L. Day, M. Haridas, Gillian E. Norris, Sylvia V. Rumball,
Clyde A. Smith, and David H. Thomas

Department of Chemistry and Biochemistry
Massey University
Palmerston North
New Zealand

## SUMMARY

The three-dimensional structures of various forms of lactoferrin, determined by high resolution crystallographic studies, have been compared in order to determine the relationship between structure and biological function. These comparisons include human apo and diferric lactoferrins, metal and anion substituted lactoferrins, the N-terminal half molecule of human lactoferrin, and bovine diferric lactoferrin.

The structures themselves define the nature and location of the iron binding sites and allow anti-bacterial and putative receptor-binding regions to be mapped on to the molecular surface. The structural comparisons show that small internal adjustments can allow the accommodation of different metals and anions without altering the overall molecular structure, whereas large-scale conformational changes are associated with metal binding and release, and smaller, but significant, movements accompany species variations. The results also focus on differences in flexibility between the two lobes, and on the importance of interactions in the inter-lobe region in modulating iron release from the N-lobe and in possibly enabling binding at one site to be signalled to the other.

## INTRODUCTION

Lactoferrin is of great interest because of its widespread physiological occurrence in mammals, in milk, tears, saliva and many other secretions, as well as in the neutrophilic granules of leukocytes (Brock, 1985). As a member of the transferrin family of iron binding proteins, it serves to control the levels of free iron (and possibly other metals) in body fluids. It also has pronounced antibacterial properties (Bullen *et al*, 1972; Arnold *et al*, 1977; Bellamy

---

* Address any correspondence to E.N. Baker

*Lactoferrin: Structure and Function*
Edited by T.W. Hutchens *et al.*, Plenum Press, New York, 1994

1

*et al*, 1992), as well as a number of other postulated biological roles, in the immune and inflammatory response and as a growth factor. Some of these latter activities may depend on its ability to bind to a wide variety of cells (Birgens *et al*, 1983).

The broad molecular properties of lactoferrin have been established over a number of years. It is a monomeric glycoprotein, ~80 kDa, with (for the human protein) a single polypeptide chain of 691 amino acid residues. Amino acid sequences are now known for human, bovine, mouse and porcine lactoferrins; all demonstrate the strong two-fold internal sequence homology which is characteristic of transferrins, consistent with an earlier gene duplication event (Metz-Boutigue *et al*, 1984). As for transferrin, the binding properties of lactoferrin include (1) extremely tight ($K \sim 10^{20}$) *but reversible* binding of two $Fe^{3+}$ ions per molecule, (2) an absolute requirement for a synergistically bound anion (normally $CO_3^{2-}$) with each $Fe^{3+}$ ion, and (3) the acceptance of a wide variety of other metal ions and anions in place of $Fe^{3+}$ and $CO_3^{2-}$ (Brock, 1985; Aisen and Harris, 1989). There are differences in detail between different lactoferrins, however, and between lactoferrin and transferrin, in particular the markedly greater acid stability of iron binding by lactoferrin (Mazurier and Spik. 1980). Studies of proteolytic fragments of lactoferrin have also contributed to understanding of its metal-binding, receptor-binding and anti-bacterial properties (Rochard *et al*, 1989; Legrand *et al*, 1990; Bellamy *et al* 1992).

We have undertaken a program of research into the three-dimensional structure of lactoferrin in order to determine the molecular basis of its various biological activities. This has included high resolution X-ray crystallographic analyses of various forms of the protein, including iron-bound ($Fe_2Lf$) and iron-free (apoLf) forms, metal and anion substituted lactoferrins and the recombinant N-terminal half-molecule, $Lf_N$. These studies have not only established the basic molecular structure of lactoferrin and enabled antibacterial and putative receptor-binding regions to be mapped on to the protein surface, but have also revealed the nature of the conformational changes which accompany metal binding and release and shown how it adapts to the binding of other metals and anions.

## EXPERIMENTAL

### Preparation and Crystallization

Human lactoferrin was isolated from fresh colostrum as described previously (Norris *et al*, 1989). Under these conditions, with metal ion contamination rigorously excluded during purification, the resulting protein typically had a residual iron content of 8–10% saturation. Iron-saturated or copper-saturated lactoferrins ($Fe_2Lf$ and $Cu_2Lf$) were prepared by adding two molar equivalents of ferric nitrilotriacetate or cupric chloride to solutions of the apo-protein; the preparation and characterization of these complexes and of oxalate-substituted complexes of lactoferrin, $Fe_2Ox_2Lf$ (with oxalate substituted for carbonate in both sites) and $Cu_2OxLf$ (with oxalate substituted for carbonate in one site only) are described fully elsewhere (Shongwe *et al*, 1992).

Bovine lactoferrin was prepared in essentially the same way, from fresh colostrum, although an additional step of isoelectric focusing was necessary to obtain fully-pure material. The N-terminal half-molecule of human lactoferrin, ($Lf_N$), which comprises residues 1–333, was obtained by expression of the cloned cDNA in baby hamster kidney (BHK) cells, followed by purification as described by Day *et al* (1992).

In some studies, deglycosylated lactoferrin was used, with the carbohydrate being removed using an endoglycosidase preparation from *Flavobacterium meningosepticum,* as described by Norris *et al* (1989). This preparation, which was a mixture of PNGase-F and Endo-F activities, removed the carbohydrate cleanly and completely from both sites of attachment (Asn 137 and Asn 478) on human lactoferrin. The same procedures could be used to deglycosylate the half-molecule, $Lf_N$, but not bovine lactoferrin, for which deglycosylation was incomplete. A side-effect of the method used was that iron was completely lost from the lactoferrin, perhaps because of the use of EDTA and small concentrations of the detergent nonidet in the incubation mixture. This afforded a good method for obtaining fully iron-free lactoferrin. The deglycosylated lactoferrin, once EDTA had been removed by extensive dialysis against 0.1 M $NaHCO_3$, behaved identically to native lactoferrin in iron titrations and the pH dependence of iron release.

All of the various forms of lactoferrin were crystallized using solutions of high protein concentration (50–150 mg/ml), dialyzed against buffers of low ionic strength, pH 7.8–8.0, with small amounts (typically 10% v/v) of added alcohols. (See Baker *et al*, 1991 for references).

## Structure Analyses

The three-dimensional structure of human diferric lactoferrin, $Fe_2Lf$, was determined by multiple isomorphous replacement at 3.2 Å resolution, from diffractometer data (Anderson *et al*, 1987) and subsequently refined against 2.2 Å synchrotron data (M. Haridas, B.F. Anderson and E.N. Baker, unpublished). This structure could then be used to provide molecular replacement search models for the solution of the structures of apolactoferrin, bovine lactoferrin, and the N-terminal half-molecule. Crystals of $Cu^{2+}$ and oxalate-substituted lactoferrins also proved to be isomorphous with those of $Fe_2Lf$, so that the $Fe_2Lf$ structure could be used as the starting model for their refinement (after first omitting metal ions, anions, solvent molecules, and metal ligands).

All of these structures have been refined against high resolution synchrotron data, using restrained least squares methods, with extensive rebuilding and checking (from 'omit' maps) on an interactive graphics system. The current state of each of these refinements is summarized in Table 1; with the exception of the bovine lactoferrin and $Lf_N$ structures all are now essentially complete.

## RESULTS

### General Organisation of Molecular Structure

The structure is best described in terms of that of diferric lactoferrin (Anderson et *al*, 1987, 1989). The single polypeptide chain of 691 residues is folded into two globular lobes, representing its N-terminal and C-terminal halves (the N-lobe and C-lobe). The two lobes have very similar three-dimensional structures (r.m.s. deviation 1.0 Å for 85% of $C_\alpha$ atoms after superposition), consistent with their high level of sequence similarity (~40% identity). They are joined by a short helix, residues 333–344, and the helical nature of this connecting peptide and its constant length (12 residues) appear to be features which (among others) differentiate lactoferrins from transferrins. The 'front-to-back' juxtaposition of the two lobes (Figure 1) emphasizes the asymmetry of the molecule, and the fact that although the lobes are structurally similar they are not equivalent in the context of the whole molecule.

Each lobe contains one specific metal binding site in a deep cleft between two dissimilar domains. The molecule as a whole thus comprises four domains, N1 and N2 in the N-lobe, and C1 and C2 in the C-lobe (see Figure 1). The anti-bacterial domain identified by Bellamy *et al* (1992), comprising most of helix 1 and the following β-strand (residues 13–38), is part of the surface of the N1 domain. Likewise, receptor binding studies and sequence similarities with other proteins point to the involvement of the N1 domain. and particularly its N-terminal portion, residues 1–55, in receptor binding (see below).

**Table 1.** Refinement Details for Lactoferrin Structures

| Structure | Protein atoms | Ions | Water molecules | $R^a$ | Resolution (Å) | r.m.s. dev.[b] (Å) |
|---|---|---|---|---|---|---|
| apoLf | 5342 | 1 Cl$^-$<br>1 EDTA$^{4-}$ | 387 | 0.185 | 2.0 | 0.018 |
| Fe$_2$Lf | 5322 | 2 Fe$^{3+}$<br>2 CO$_3^{2-}$ | 431 | 0.179 | 2.2 | 0.018 |
| Cu$_2$Lf | 5321 | 2 Cu$^{2+}$<br>2 CO$_3^{2-}$ | 308 | 0.196 | 2.1 | 0.018 |
| Fe$_2$Ox$_2$Lf | 5322 | 2 Fe$^{3+}$<br>2 C$_2$O$_4^{2-}$ | 122 | 0.221 | 2.3 | 0.022 |
| Cu$_2$OxLf | 5328 | 2 Cu$^{2+}$<br>1 CO$_3^{2-}$<br>1 C$_2$O$_4^{2-}$ | 307 | 0.215 | 2.1 | 0.017 |
| Lf$_N$[c] | 2591 | 1 Fe$^{3+}$<br>1 CO$_3^{2-}$ | 110 | 0.217 | 2.0 | 0.021 |
| Fe$_2$bLf[d] | 5335 | 2 Fe$^{3+}$ | — | 0.239 | 2.5 | 0.025 |

[a]Crystallographic R-factor $R = \sum ||F_o| - |F_c||/\sum |F_o|$
[b]r.m.s. dev. = root-mean-square deviation from standard bond lengths
[c]N-Terminal half-molecule of human lactoferrin
[d]Diferric bovine lactoferrin

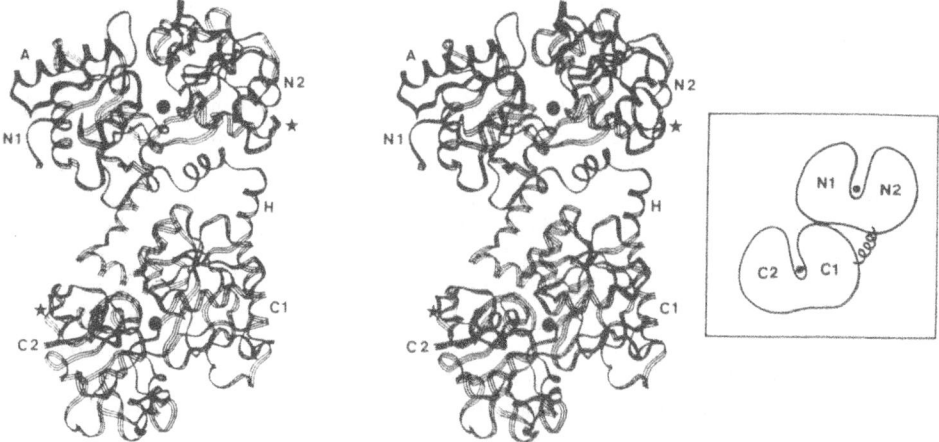

**Figure 1.** Stereo ribbon diagram showing the polypeptide chain folding for diferric human lactoferrin, Fe$_2$Lf, with the N-lobe above and the C-lobe below. The location of the four domains (N1, N2, C1, C2), the two iron sites (●), the two glycosylation sites (★), the inter-lobe connecting helix (H) and the antibacterial domain (A) of Bellamy *et al* (1992) (solid ribbon) are indicated. The front-to-back arrangement of the two lobes is shown in the inset.

## The Metal and Anion Binding Sites

The two binding sites are extremely similar. The metal is bound by four protein sidechains, 2 Tyr, 1 His and 1 Asp, and two oxygens from the bidentate $CO_3^{2-}$ anion complete a distorted octahedral geometry. As well as providing two ligands to the iron atom, the carbonate ion fits into a pocket formed by two positively-charged groups, an arginine sidechain and the N-terminus of helix 5, making good hydrogen bonds with both, and with a nearby Thr sidechain. This arrangement is illustrated schematically in Figure 2.

Other important features of the binding region are (1) a substantial water-filled cavity, beyond the Arg sidechain but within the interdomain cleft, (2) the presence of several other basic sidechains in the vicinity of the iron site (eg. Arg 210, Lys 296 and Lys 301 in the N-lobe, Lys 546 in the C-lobe) and (3) three sections of polypeptide chain which run behind each iron site, linking the two domains, two of them extended ('backbone') strands and the third an α-helix (number 11, residues 321–332). These latter three sections are intimately involved in the conformational change seen in apo-lactoferrin (below).

## Apo-lactoferrin Structure

The crystal structure of apo-lactoferrin (Anderson *et al*, 1990) reveals two striking features. Firstly, the N-lobe binding cleft is wide open, following a large-scale conformational change in this half of the molecule, and secondly the C-lobe remains closed even though no metal is bound (Figure 3). These two observations have profound implications for mechanisms of binding and release.

In the N-lobe the conformational change involves a rigid-body rotation of 54° of one domain (N2) relative to the other (N1). There are two elements to this domain movement. Firstly

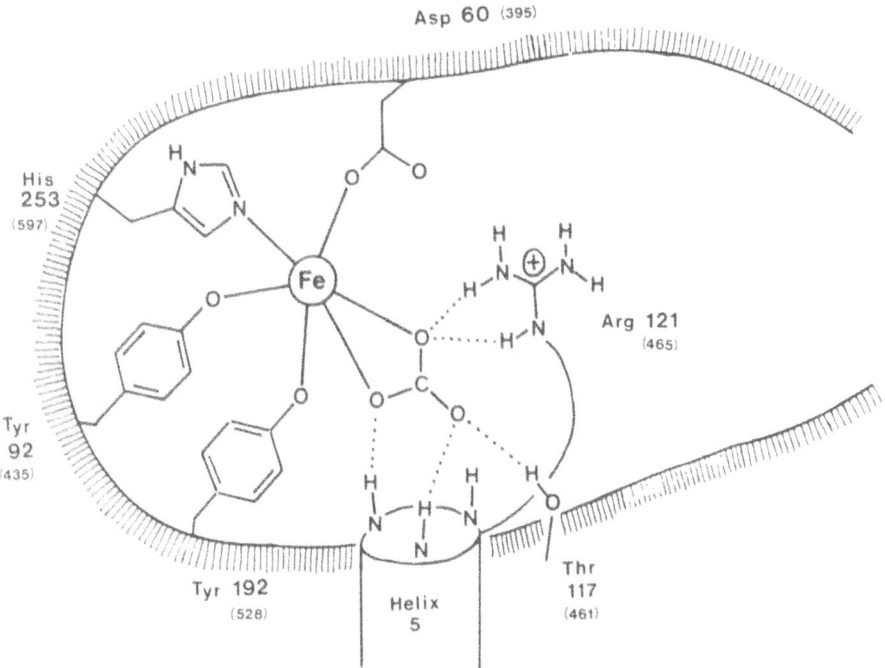

**Figure 2.** Schematic diagram showing the metal and anion binding sites for lactoferrin. Residue numbers are given for the N-lobe, with equivalent C-lobe residues in parentheses.

the two extended 'backbone' strands show an abrupt hinge at about residues Thr 90 and Val 250. This occurs without major changes in the main chain conformational angles and no change in main chain hydrogen bonding; it is simply the flexing of an antiparallel β-ribbon. Secondly a helix in domain N2 (helix 5) pivots on a second helix (helix 11) which runs behind the iron site, and this pivoting of helix 5 carries the rest of the N2 domain with it (Figure 4). Hydrogen bonds which tie the C-termini of the two helices together (the sidechain of Arg 133 hydrogen bonds to the C=O groups of residues 330 and 333) remain intact during this movement, but hydrogen bonds near their N-termini (Tyr 324 hydrogen bonding to Thr 122 and Asn 126) are broken. During this process two of the metal ligands, Tyr 92 and Tyr 192, remain with the N2 domain, close to the anion site, while the other two ligands, Asp 60 and His 253, remain associated with the N1 domain, 8–10 Å away. One other feature of the conformational change is that the relative orientations of the two lobes are also changed slightly, and the C-terminus of the final helix in the C-lobe (residues 680–691) makes a new salt bridge with Arg 249 on the N-lobe. This may provide a mechanism by which binding or release from the N-lobe is signalled to the C-lobe.

The closed C-lobe presents a most intriguing contrast, in that its structure is hardly changed from that in $Fe_2Lf$ even though no metal ion is bound. In fact the C-lobes of the two structures superimpose with an r.m.s. deviation of only 0.4 Å. There is no obvious reason why the C-lobe is not open, as the N-lobe is. A $Cl^-$ ion occupies the anion site but interacts only with one domain (C2) and there are no contacts between the lobes which could prevent the opening of both lobes at the same time. A disulphide bridge, 483–677, which has no counterpart in the N-lobe, almost certainly reduces the flexibility of the C-lobe, but should still allow it to open, albeit to a lesser extent than the N-lobe (Anderson *et al*, 1990). How then can this crystal structure be rationalised with low angle X-ray scattering studies, in solution (Grossmann *et al*, 1992), which show that *both* lobes open? The probable explanation is that in solution an equilibrium exists for the apo-protein, between open and closed states, and that crystal packing has selected one of several accessible conformations. If this is so, it also carries the implication that the open and closed states of apoLf are, at least for the C-lobe, similar in energy since the crystal packing contacts are very weak and few in number.

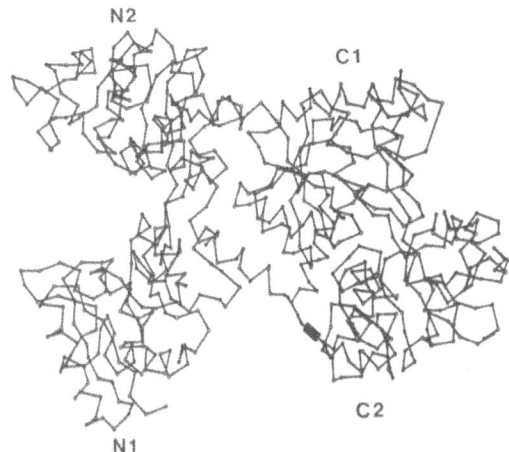

**Figure 3.** $C_\alpha$ plot for human apo-lactoferrin showing the open N-lobe (left) and the closed C-lobe (fight). The location of the disulphide bond 483–677, which may restrict the flexibility of the C-lobe is indicated (–■–).

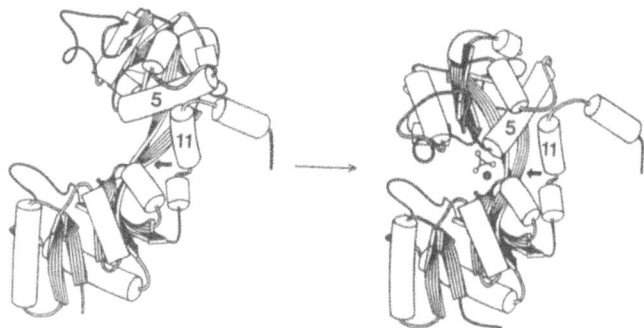

**Figure 4.** Schematic diagram showing the open to closed conformational change associated with metal binding and release by lactoferrin. The location of the hinge in the two antiparallel 'backbone' strands is indicated (←) and the two helices, 5 and 11, which pivot on one another, are labelled

## Metal and Anion Substitution

Substitution of $Cu^{2+}$ for $Fe^{3+}$, as seen in the crystal structure of $Cu_2Lf$ (Smith *et al*, 1992) produces minimal change in the protein structure. The polypeptide chains of $Fe_2Lf$ and $Cu_2Lf$ superimpose with an r.m.s. deviation of only 0.3 Å, the domain closure is essentially unchanged in each lobe, and the only significant difference appears to be a small movement of the metal ion; in the N-lobe the $Cu^{2+}$ ion is displaced 1.0 Å from the $Fe^{3+}$ position, while in the C-lobe the displacement is 0.3 Å. The effect of these displacements is to change the metal coordination slightly (Smith *et al*, 1992: Brodie *et al*, this volume).

The effects of anion substitution are seen in the crystal structures of $Cu_2OxLf$, where oxalate is substituted for carbonate in the C-lobe site, and $Fe_2(Ox)_2Lf$, where oxalate is substituted for carbonate in both sites. In each case oxalate binds to the metal in 1,2 -bidentate fashion, and fills the same anion binding pocket as carbonate, with very similar hydrogen bonding. To accommodate the larger anion, the arginine sidechain (Arg 121 in the N-lobe, Arg 465 in the C-lobe) must move aside 2–3 Å, with the consequent displacement of associated solvent molecules and (in the C-lobe) another neighbouring sidechain (Tyr 398). These changes can occur without any change to the overall protein structure, however, because of the large solvent-filled cavity within the interdomain cleft. Again the domain closure is essentially the same as in $Fe_2Lf$.

One other observation from the $Cu_2OxLf$ structure is that it is the C-lobe site which is preferentially occupied by oxalate. This highlights a paradox, that the N-lobe site offers more room for the metal ion to change its position. but the C-lobe site more readily accepts the larger oxalate ion. There are plausible explanations for this (Shongwe *et al.* 1992) but it emphasizes that predictions of detailed binding behaviour are difficult, even given high resolution structures. The results also demonstrate that differences between the two sites become more pronounced when metal ions other than $Fe^{3+}$ and anions other than carbonate are bound.

## N-terminal Half-molecule, Lf$_N$

The N-terminal half-molecule, Lf$_N$, comprises residues 1–333 of the human protein. Crystals were obtained of both glycosylated and deglycosylated Lf$_N$, but were isomorphous, implying that the structures must be essentially the same and that the carbohydrate in glycosylated Lf$_N$ is probably disordered.

With one important exception, the structure of the isolated half-molecule appears to be unchanged from that in intact lactoferrin. Although refinement is still in progress. the binding site, including the carbonate anion, is well resolved and matches that in the N-lobe of $Fe_2Lf$, even to the extent of some of the bound solvent molecules in the binding cleft. The exception concerns the helix 321–332 (helix 11) which runs behind the iron site in $Fe_2Lf$ and packs against helix 5. In Lf$_N$ residues 321–332 are no longer helical but form a third extended 'backbone' strand (Figure 5). This changes the structure of the 'hinge' region and results in the loss of hydrogen bonds at each end of helix 5 (Tyr 324 OH...Thr 122 OG1 and Tyr 324OH...Asn 126 ND2 near the N-terminus and Arg 133 NH2...333O and Arg 133 NH1...330O at the C-terminus). This change, when seen in relation to the decreased acid stability of Lf$_N$ (Day *et al.* 1992), gives important clues to the origin of the pH-dependence of iron release.

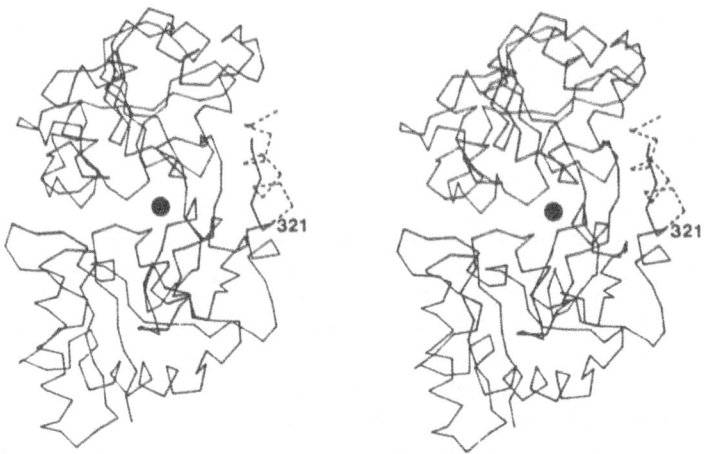

**Figure 5.** C$_\alpha$ plot for the N-terminal half-molecule of human lactoferrin showing the changed conformation following residue 321 in the recombinant half-molecule, Lf$_N$ (full lines), compared with that found in the N-lobe of intact lactoferrin (broken lines). In Lf$_N$ an extended strand is found instead of a helix, giving a 3-strand antiparallel β-sheet behind the iron site (note, however, that the conformation is only clear as far as residue 326 at present).

### Bovine Lactoferrin

The bovine lactoferrin structure analysis is at lower resolution (2.5 Å) than the others described here, and the refinement is not yet complete; thus this structure cannot be described in detail. It does, nevertheless, show some intriguing differences from the human protein. The most obvious difference is in the relative orientations of the two lobes. If the N-lobes of the human and bovine proteins are superimposed, the C-lobes do not match and an additional 12° rotation is required to bring them into register. A similar difference in lobe orientations is found when human lactoferrin and rabbit transferrin are compared (Baker and Lindley, 1992), but the movement is in a different direction.

Other features of the bovine lactoferrin structure are its different glycosylation pattern, and some potentially-significant differences in interactions in and around the binding sites. In the N-lobe, one interdomain salt bridge (Asp 217...Lys 296) which is present in human lactoferrin is lost in bovine lactoferrin, as Asp 217 is changed to Asn, and Arg 210, which interacts with the two Tyr ligands, is changed in bovine lactoferrin to Lys (as in transferrins). In the C-lobe, Tyr 415, near the back of the iron site, is changed to Arg in bovine lactoferrin and near it, at residue 447, a new glycosylation site appears. Any of these changes could account for the altered binding properties of the bovine protein, compared with human (Aisen and Leibman, 1972). Only one of the four glycosylation sites on bovine lactoferrin is also found on human lactoferrin. Also in contrast to the human protein, significant density can be seen at three of the glycosylation sites, implying that the carbohydrate is somewhat better ordered, at least near the protein surface.

## CONCLUSIONS

The structural studies presented here offer a comprehensive framework within which aspects of lactoferrin function can be interpreted. Further, many of the conclusions can be extended to other members of the transferrin family.

The binding site appears to be optimised for the binding of $Fe^{3+}$ and $CO_3^{2-}$ with respect to size, charge and stereochemistry. The 3+ charge on the metal ion is matched by the 3− charge of the Tyr and Asp ligands, and the 2− charge on the $CO_3^{2-}$ is mostly balanced by the positive charge of the Arg sidechain and the helix N-terminus. Anion binding is essential for metal binding and the reason is obvious—without it there would be a substantial net positive charge at the anion site which would destabilize metal binding. The Asp ligand also appears to be specially important in that it binds to the metal ion through one carboxylate oxygen and is hydrogen bonded to the N-termini of two helices. one in each domain, through its other carboxylate oxygen. The latter interaction must contribute substantially to the stability of the closed structure when iron is bound.

A crucial feature of the binding sites is that the metal site is formed by groups from *both* domains of a lobe, whereas the anion site is formed by only one domain (N2 or C2). This means that for strong metal binding a conformational change to close the domains over the metal is required. Anion binding, on the other hand, can occur on the apo-protein as the site is already fully formed. In fact lactoferrin has a high affinity for anions, as shown, for example, by the presence in the apo-lactoferrin crystal structure of a $Cl^-$ ion in the C-lobe anion site and a probable $EDTA^{4-}$ ion in the N-lobe site, and it may have a physiological role in the binding and transport of anions (see Brodie *et al,* this volume). Further, both the protein folding and the location and construction of the anion sites in lactoferrin suggest an evolutionary link with bacterial anion binding proteins (Baker *et al,* 1987, 1990).

The substitution of copper for iron in the Cu$_2$Lf structure shows that a metal ion of the same size can induce the same domain closure even though its charge and stereochemical preferences are different. Thus other metal ions of similar size, such as $Ga^{3+}$, $Cr^{3+}$, $Mn^{3+}$, $Co^{3+}$, $Ni^{2+}$ and $Zn^{2+}$, and probably $Al^{3+}$ should also bind with the same overall protein structure (albeit with different binding affinities, depending on how well the binding site geometry fits their stereochemical preferences). Moreover, receptor interactions of lactoferrin when these metal ions are bound should not be altered, giving support to the idea that lactoferrin could mediate the absorption of metal ions other than iron. The oxalate-substituted structures show how larger anions can be accommodated because of the *internal* flexibility within the binding cleft, again without disturbance of the overall protein structure. This leads to the idea that lactoferrin could be engineered to bind and transport other anionic compounds of possible therapeutic importance, taking advantage of the large cavity in the binding cleft.

The two-lobe, four-domain organisation of the lactoferrin molecule allows for various degrees of flexibility (Baker *et al,* 1991). The functional significance of the variations in lobe orientation between human and bovine lactoferrins, and rabbit transferrin. is unclear. They could affect receptor interactions if the receptors contact both lobes, and thus give a means of discriminating between different species. They could also, by altering the interactions between the two lobes, modulate iron binding and release, (see below).

Domain movements are clearly fundamental to the processes of binding and release by lactoferrin (and other transferrins), just as they are for other binding proteins (Quiocho, 1990). The open N-lobe and closed C-lobe in the apo-lactoferrin crystal structure are fully understandable when it is recognized that protein structures are inherently flexible, being maintained as a balance of many very weak interactions. Thus both lobes in lactoferrin must be able to open and close in the absence of a bound metal ion and the crystal structure is a 'snapshot' of one of the accessible conformations. Binding takes place when the cleft is open, with a logical series of steps being (i) initial binding of the anion in the positively charged anion pocket of the N2 (or C2) domain, (ii) binding of the metal to the anion and two Tyr ligands, all located on the N2 (or C2) domain, followed by (iii) domain closure as the metal completes its coordination by binding to the His and Asp ligands on the N1 (or C1) domain.

The structural basis of metal release from lactoferrin is less obvious, but studies of the N-terminal half-molecule (Day *et al*, 1992) offer one clue. Iron release from Lf$_N$ begins at pH 6.0 as the pH is lowered, compared with pH 4.0 for native lactoferrin. Since the iron sites appear the same, the difference in acid stability of Lf$_N$ must arise from the loss of stabilising interactions, resulting from the removal of the C-lobe and the structural change involving residues 321–333 at the back of the iron site. These changes are close to the hinge region of the N-lobe and may thus alter the relative stabilities of the open and closed forms of this lobe. It is interesting to note that the characteristic difference between lactoferrins and transferrins is the more facile (higher pH) release of iron from transferrins. This is primarily a difference in the *N-lobes,* however, since the C-lobe of transferrin releases iron at a lower pH, more akin to lactoferrin (pH ~ 4.5); moreover the pH dependence of iron release from Lf$_N$ is very similar to that of the N-lobe of transferrin (Day *et al.* 1992). Thus we suggest that the difference in iron release from lactoferrins. compared with transferrins, derives, at least in part, from stabilising interactions between the two lobes which are different in the two proteins.

The apo-lactoferrin structure also offers a hint that the C-lobe is less flexible than the N-lobe. It is already known that the C-lobe of transferrin binds iron more strongly and releases it more slowly and at lower pH; the lesser flexibility of the C-lobe may explain this (if the results from lactoferrin can be translated to transferrin) but also raises again an earlier idea that the two lobes may have different functional roles. This idea receives further support from a recent study on receptor-mediated iron release from transferrin (Bali and Aisen, 1991), which shows that the receptor acts to 'pull' iron from the C-lobe and suggests that iron release from the N-lobe (mediated by pH) is less significant. This has important implications for lactoferrin, too. One of the most serious doubts about suggestions that lactoferrin functions in iron delivery is the question of how iron could be released from it, given that release *in vitro* occurs 2 pH units lower than for transferrin. If pH-dependent release is less important and receptor action all-important, however. lactoferrin, too, may be able to release iron to cells, and the focus shifts to the need to characterise lactoferrin receptors and their properties.

Other functional properties of lactoferrin are likely to depend on its surface structure. The antibacterial peptide isolated by Bellamy *et al* (1992) from both human and bovine lactoferrins comprises the first α-helix in the molecule (residues 14–31) and the following β-strand (residues 32–39). These residues form part of the surface of the N1 domain. remote from the metal binding site. The helix is amphipathic. with hydrophobic groups protruding into the core of the molecule and hydrophilic groups projecting into solution—most prominent of the latter are a number of basic sidechains, Arg 24, Lys 28, Arg 29, Lys 30, Lys 38 and Arg 39, which together with the nearby N-terminus, with its 3 Arg residues (Arg 2, Arg 3 and Arg 4), could be involved in interactions with cells. This is the most basic region on the whole molecule. If this region determines the bactericidal activity of intact lactoferrin, however, why is the iron-loaded protein inactive, given that the region appears equally exposed in both iron-bound and iron-free forms? Perhaps the explanation lies in the greater flexibility of the apo-protein, which may allow better contact with the bacterial surface.

The anti-bacterial region described above has also been implicated as part of a possible receptor-binding surface on both lactoferrin and transferrin. Most of the evidence is indirect, based on (i) weak sequence similarities between residues 1–55 and several lymphoma transforming proteins, (ii) the fact that this region forms a single exon (number 2) of the transferrin gene and is a structurally discrete unit on the molecular surface, and (iii) structural similarities between this region and the receptor binding region of apolipoprotein E, to whose receptor lactoferrin also binds. In addition to the forgoing, however, there is also direct evidence from binding studies with human lymphocyte receptors for lactoferrin which implicates the N1 domain as a primary receptor binding site (Rochard *et al,* 1989). It should now be possible to test these suggestions by mutagenesis.

Finally, the crystallographic studies have revealed little about the structure of the carbohydrate on lactoferrin, since it is apparently disordered, giving rise to little or no interpretable density. This does, however, argue against any structural role of the carbohydrate, since it is not involved in specific interactions with the protein. For the same reason it seems unlikely to have a role in metal binding, and our own studies, comparing native and deglycosylated lactoferrin, have shown that iron binding and release, pH stability and heat stability, and resistance to proteolysis are all largely unaffected by the presence or absence of the carbohydrate.


## ACKNOWLEDGMENTS

This work is supported by the U.S. National Institute of Child Health and Human Development (grant HD-20859), the Wellcome Trust, the Health Research Council of New Zealand and the New Zealand Dairy Research Institute. C.L.D. acknowledges receipt of a Postgraduate Scholarship.


## REFERENCES

Aisen, P. and Harris, D.C. (1989). In *Iron Carriers and Iron Proteins* (Loehr, T.M., ed. pp. 241–351, VCH Publishers, New York.

Aisen P. and Leibman, A. (1972). *Biochim. Biophys. Acta* 257, 314–323.

Anderson, B.F., Baker, H.M., Dodson, E.J., Norris, G.E., Rumball, S.V., Waters, J.M. and Baker, E.N. (1987). *Proc. Natl. Acad. Sci. USA* 84, 1769–1773.

Anderson, B.F., Baker, H.M., Norris, G.E., Rice, D.W. and Baker, E.N. (1989). *J. Mol. Biol.* 209, 711–734.

Anderson, B.F., Baker, H.M., Norris, G.E., Rumball, S.V. and Baker, E.N. (1990). *Nature (London)* 344, 784–787.

Arnold, R.R., Cole, M.F. and McGhee, J.R. (1977). *Science* 197, 263–265.

Baker, E.N. and Lindley, P.F. (1992). *J. Inorg. Biochem.* 47, 147–160.

Baker, E.N., Rumball, S.V. and Anderson, B.F. (1987) *Trends in Biachem Sci.* 12, 350–353.

Baker, E.N., Anderson, B.F., Baker, H.M., Haridas, M., Norris, G.E., Rumball, S.V. and Smith, C.A. (1990). *Pure Appl. Chem.* 62, 1067–1070.

Baker, E.N., Anderson, B.F., Baker, H.M., Haridas, M., Jameson, G.B., Norris, G.E., Rumball, S.V. and Smith, C.A. (1991) *Int. J. Biol. Macromol.* 13, 122–129.

Bali, P.K. and Aisen. P. (1991). *Biochemistry* 30, 9947–9952.

Bellamy, W., Takase, M., Yamauchi, K., Wakabayashi, H., Kawase, K. and Tomita. M. (1992). *Biochim. Biophys. Acta* 1121, 130–136.

Birgens, H.S., Hansen, N.E., Karle, H. and Ostergaard Kristensen, L. (1983). *Br. J. Haematol* 54, 383–391.

Brock, J.H. (1985). In *Metalloproteins* (Harrison, P.M., ed.). Part 2, pp. 183–262, Macmillan, London.

Bullen, J.J., Rogers, H.J., and Leigh, L. (1972). *Brit Med.* J. 3, 69–75.

Day, C.L., Stowell, K.M., Baker, E.N. and Tweedie, J.W. (1992). *J. Biol. Chem.* 267, 13857–13862.

Grossmann, J.G., Neu, M., Pantos, E., Schwab, F.J., Evans, R.W., Townes-Andrews. E., Lindley, P.F., Appel, H., Thies, W-G and Hasnain, S.S. (1992). *J. Mol. Biol* 225, 811–819.

Legrand, D., Mazurier, J., Colavizza, D., Montreuil, J. and Spik, G. (1990). *Biochem. J.* 266, 575–581.

Mazurier, J. and Spik, G. (1980). *Biochim. Biophys. Acta* 629, 399–408.

Metz-Boutigue, M-H., Jolles, J., Mazurier, J., Schoentgen, F., Legrand, D., Spik, G. and Jolles, P. (1984). *Eur. J. Biochem.* 145, 659–676.

Norris, G.E., Baker, H.M. and Baker, E.N. (1989). *J. Mol. Biol.* 209, 329–331.

Quiocho, F.A. (1990). *Phil. Trans. Roy. Soc. Lond.* B326, 341–351.

Rochard, E., Legrand, D., Mazurier, J., Montreuil, J. and Spik, G. (1989). *FEBS Lett.* 255, 201–204.

Shongwe, M.S., Smith, C.A., Ainscough, E.W., Baker, H.M., Brodie, A.M. and Baker. E.N. (1992). *Biochemistry.* 31, 4451–4458.

Smith, C.A., Anderson, B.F., Baker, H.M. and Baker. E.N. (1992). *Biochemistry* 31, 4527–4533.

# CHARACTERIZATION OF TWO KINDS OF LACTOTRANSFERRIN (LACTOFERRIN) RECEPTORS ON DIFFERENT TARGET CELLS

Geneviève Spik, Dominique Legrand, Béatrice Leveugle, Joël Mazurier, Takashi Mikogami, Jean Montreuil, Annick Pierce, and Elisabeth Rochard

Université des Sciences et Technologies de Lille
Laboratoire de Chimie Biologique (Unité Mixte de Recherche n°111 du CNRS)
59655 Villeneuve d'Ascq Cedex, France

## ABSTRACT

Lactotransferrin (Lf), an iron-binding glycoprotein present as a major component in the specific granules of human neutrophilic granulocytes is released in the blood during the acute phase of infection and participates in the regulation of the host-defence mechanisms. Our previous observations (Mazurier et al., 1989) showing i) that the activation by PHA of T-lymphocytes induces the appearance at the cell surface of Lf-receptors which are absent from the membrane of resting lymphocytes and ii) that Lf becomes a growth factor for the activated lymphocytes, led us to undertake a series of researches on the presence of Lf receptors at the surface of different blood cells.

Characterization of Lf receptors was performed by flow cytofluorimetry using either Lf labelled on its glycan moiety with fluorescein or purified anti-lymphocyte Lf receptor antibodies.

High affinity receptors for Lf were characterized only at the surface of human activated lymphocytes and of non-activated platelets. These two receptors possess common physico-chemical properties and antigenic epitopes.

Low affinity receptors for Lf were characterized on monocytes, eosinophils and neutrophils. These receptors are immunologically different from those found on activated lymphocytes and on non-activated platelets.

Cell-lines of human lymphocyte T and megakaryocyte possess lactotransferrin receptors whose properties are similar to those found on peripheral blood cells.

The soluble form of the receptor identified in the lymphocytes T culture medium possesses a molecular mass close to that of the membrane receptor suggesting that the cytoplasmic tail of the receptor should be very short.

## INTRODUCTION

Lactotransferrin (Lf), also called lactoferrin, a member of the transferrin family, is biosynthesized either by the epithelial secretory cells [1] or by the early cells of the neutrophilic blood series [2]. The Lf secreted by the epithelial cells is found at the surface of the mucosae or in the secretory biological fluids such as milk [3]. The Lf biosynthesized by the precursors of the neutrophils is stored in the secondary granules of the mature cells in association with other cationic proteins such as lysozyme and alkaline phosphatase [4]. During infection and after stimulation of the neutrophils, 60 to 70% of the secondary granules are discharged from the cell and the amount of Lf liberated in the blood may be about 30 g per day [5]. This huge amount of Lf is rapidly cleared by the liver and the plasma concentration remains very low [6]. An important problem to be solved concerns the biological significance of this plasma Lf. One hypothesis is that the liberated Lf interacts with the blood cells and increases the mechanisms of immunological defence of the organism.

Several studies have been performed in the literature concerning the binding of Lf to the blood cells. In particular, it was shown that no binding of Lf occurs to red-blood cells [7] and to platelets [8] but that Lf binds to monocytes [9, 10], to a subpopulation of lymphocytes B [9], to leukaemic cells [8] and to phytohemagglutinin activated lymphocytes [11]. Although the potential ability of lactotransferrin to bind to different blood cells has been recognized, the nature of the interactions between the iron-binding protein and the cells has not yet been clearly elucidated. In fact, in addition to the interaction between Lf and a specific receptor characterized in the pHA-activated lymphocytes [11], binding of the lactotransferrin to a membrane lectin which recognizes fucose residues has been reported [12, 13]. The increased binding of Lf to the monocyte/macrophage differentiated HL60 cells was found to be due to interactions between the Lf-lipopolysaccharide complex and the lipopolysaccharide-receptors [14]. Moreover, ionic interactions between Lf and cell membrane DNA have been reported [10, 15].

In order to define the nature of the different blood cells which bind Lf and the nature of the interactions, we have undertaken two types of binding assays. The first ones were performed using radiolabelled Lf and the binding parameters were defined by Scatchard plot analysis. The second ones were realized by flow cytometry using fluorescent-Lf.

Binding of iodine radiolabelled-Lf to PHA-activated or anti-CD3-activated lymphocytes led to the conclusion that the binding which is saturable, reversible and specific must be mediated by a receptor [11]. The dissociation constants of iron-free and iron-saturated Lf were found to be identical [11]. An optimal pH binding of 7.2 was found for the two forms of Lf [11]. These results which are quite different from those described for human serum transferrin [16] suggest that the fate of Lf and of its two iron ions inside the cell should be quite different.

By flow cytometry, the binding of Lf to the blood cells was, in a first step, analyzed by using FITC-labelled Lf. In these conditions, we were unable to detect a binding of the fluorescent Lf to the PHA-activated lymphocytes [17]. This inhibition was further explained by the fact that fluorescein was coupled to a lysine residue which is located in the domain N1 corresponding to the receptor-binding site [18, 19]. To overcome this inhibition we have coupled the Lf to the sugar part of the molecule [20] and used this fluorescent-Lf for flow cytometry experiments.

To define the structural relationships between the Lf receptors characterized on different blood cells, the receptor present on activated lymphocytes was isolated by affinity chromatography [11] and polyclonal antibodies against the purified receptor were raised in the rabbits.

In the present data, we describe the characterization by flow-cytometry of two types of Lf receptors at the surface of blood cells. Moreover, we demonstrate that the physicochemical and immunological properties of the receptors found on activated lymphocytes and on non-activated platelets are very similar to those of a soluble receptor isolated from the culture medium of a lymphocytic cell-line.

## MATERIAL AND METHODS

### Cell Preparation

Leucocyte fraction was obtained from anti-coagulated human blood from healthy volunteers after lysis of the red-blood cells using FACS lysing solution. Lymphocytes and monocytes were isolated by Ficoll-metrizamide [21]. Platelets were isolated according [22] and eosinophils were obtained from human blood of hyper-eosinophilic subjects.

The human T lymphoblastic leukemia cell line (Jurkat cells) and the megakaryoblastic cell line were routinely cultured in RPMI medium supplemented with fetal calf serum. Cell-free cultured media were concentrated using a Filtron ultrafiltration apparatus.

### Binding Experiments

Binding experiments were performed by the direct method using fluorescent-Lf or by the indirect method using unlabelled-Lf revealed by unlabelled primary antibodies and secondary FITC-anti-rabbit antibodies. In some experiments the Lf receptors were revealed by unlabelled anti-lymphocyte-Lf-receptor antibodies and FITC-anti-rabbit antibodies.

Fluorescent human lactotransferrin (HyF-Lf) was obtained by coupling the carbohydrazinoamino-fluorescein (HyF) to aldehyde groups obtained after mild periodic oxidation of the N-acetylneuraminic acid residues [20]. Binding of Hyf-Lf to blood cells was achieved by incubating each cell fraction at a density of $5 \times 10^5$ with HyF-Lf at a concentration ranging from $1 \times 10^{-8}$ to $1 \times 10^{-7}$ M for high affinity receptors and from $1 \times 10^{-6}$ to $5 \times 10^{-6}$ M for low affinity receptors. Non-specific binding was measured in presence of 100 molar excess of unlabelled lactotransferrin. The data were collected on a Becton-Dickinson FACscan cytofluorimeter.

### Physicochemical Properties of the Receptors

The Lf receptors characterized on activated lymphocytes, on platelets and in the culture medium were isolated by immunoaffinity-chromatography on anti-lactotransferrin antibodies immobilized on Sepharose 4B [11]. SDS polyacrylamide gel electrophoresis of the purified receptors were performed according to [23].

## RESULTS

### Screening of the Human Blood Cells

In a first step, the binding of Lf to human activated lymphocytes and to platelets was measured using both HyF-Lf and unlabelled-Lf detected by fluorescent antibodies. As shown in Fig. 1 similar patterns were obtained suggesting that the binding affinity constants and the number of binding sites should be the same for the two kinds of cells. For both cells, the saturation of the binding occurred at a protein concentration of $12.5 \times 10^{-8}$ M. The binding was inhibited in the presence of a 100 fold molar excess of unlabelled Lf demonstrating the reversibility of the binding.

When a concentration of 0.1 μM was used to analyze the binding of Lf to the other blood cells, no binding occurs to red blood cells, neutrophils, eosinophils and resting lymphocytes

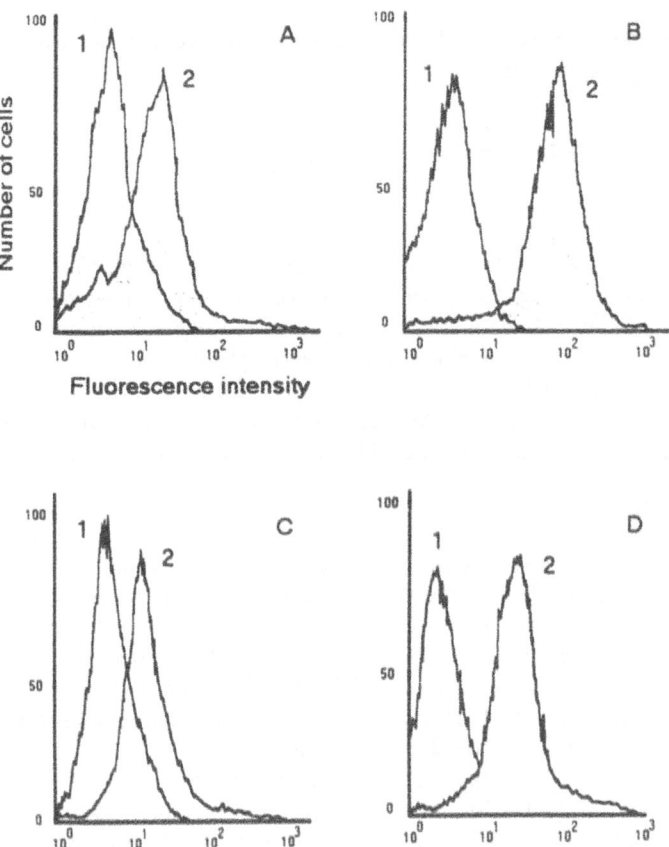

**Figure 1.** Analysis by flow cytofluorimetry of the binding: first, of fluorescent lactotransferrin (HyF-Lf) to human activated lymphocytes (A) and to human non-activated platelets (C) and, second, of unlabelled lactotransferrin, to human activated lymphocytes (B) and to human non-activated platelets (D) as revealed by fluorescent antibodies. Peak 1 ~ corresponds to the unspecific binding of lactotransferrin to the cells; peak 2 ~ corresponds to the total binding capacity of lactotransferrin to the cells.

(Fig. 2A) These results suggest that high affinity receptors for human Lf are found only on activated lymphocytes and on platelets. When the concentrations of HyF-Lf and unlabelled Lf were 50 fold increased, presence of Lf receptors was revealed on neutrophils, eosinophils and monocytes (Fig. 2B). These receptors are called low affinity Lf receptors.

## Structural Analysis of the Lf Receptors

Isolation of Lf receptors from activated lymphocytes and platelets was performed by anti-ligand-affinity chromatography using immobilized antilactotransferrin IgG. Both purified receptors analyzed by SDS-polyacrylamide gel electrophoresis showed a single protein band possessing a molecular mass of 105 kDa (Fig. 3, lanes A 1 and A 2). The electrophoretic migration of these two protein bands was not modified in reductive conditions (Fig. 3, lanes B 1 and B 2 ) indicating that both receptors are constituted of a single polypeptide chain.

However, electrophoretic migration of the bands was modified after endoglycosidase (N-glycanase) treatment suggesting that the Lf receptors are glycosylated. From these

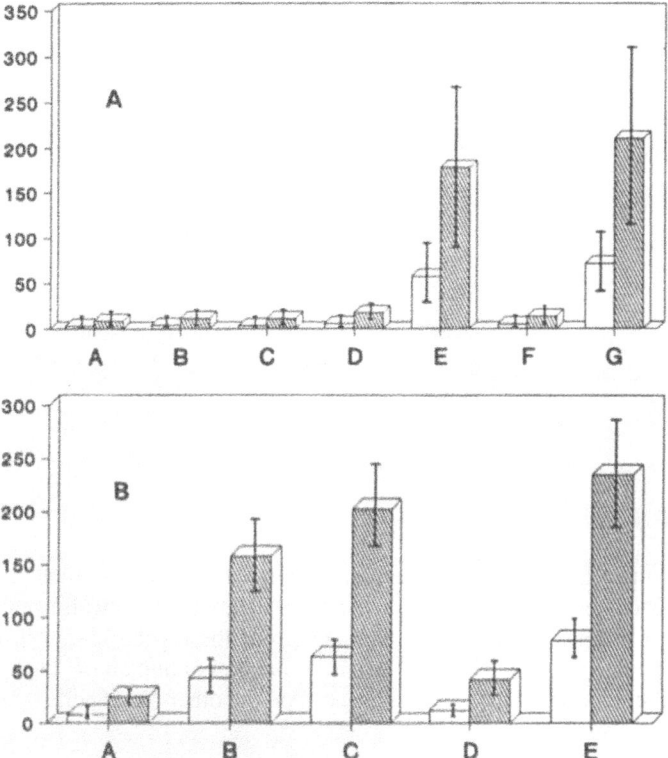

**Figure 2.** A: Binding of fluorescent lactotransferrin (HyF-Lf) (0.1 μM) [⎯⎯⎯⎯] and of unlabelled lactotransferrin (0.1 μM) as revealed by fluorescent antibodies [▨▨▨▨] to (a) red blood cells, (b) neutrophils, (c) eosinophils, (d) resting lymphocytes, (e) activated lymphocytes, (f) monocytes/macrophages, and (g) platelets. B: Binding of fluorescent lactotransferrin (HyF-Lf) (5 μM)[⎯⎯⎯⎯] and of unlabelled lactotransferrin (5 μM) as revealed by fluorescent antibodies [▨▨▨▨] to (a) red blood cells, (b) neutrophils, (c) eosinophils, (d) resting lymphocytes, and (e) monocytes/macrophages.

results it was concluded that the Lf receptors found on activated lymphocytes and on platelets possess very similar physicochemical properties. These properties were similar to those of the membrane receptors isolated from human T lymphoblastic leukemia cells (Jurkat cells) and from megakaryoblastic cells. However a difference in the molecular mass of about 10 kDa was observed when we compared the electrophoretic migration of the activated lymphocyte Lf receptor to that of the solubilized receptor isolated from the culture medium of these cells indicating that the cytoplasmic tail of the receptor should be short.

In order to complete the comparative studies realized on the different blood cell receptors, polyclonal antibodies prepared against the activated lymphocytes Lf receptor were used to detect by flow cytometry the presence of immunologically related Lf receptors. The results obtained indicate that these polyclonal antibodies bind to platelets but do not bind to neutrophils, eosinophils and monocytes suggesting that no immunological relationships exist between high and low affinity Lf receptors.

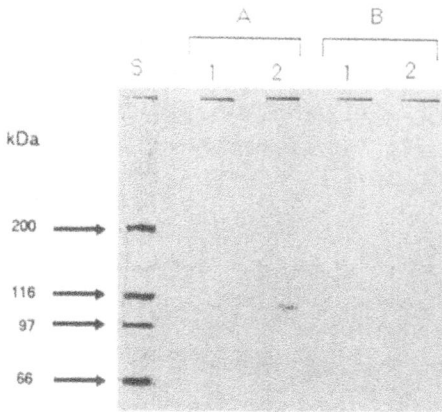

**Figure 3.** SDS-Polyacrylamide gel electrophoresis in non-reductive conditions (A) and in reductive conditions (B) of lactotransferrin receptors isolated from human activated-lymphocytes (lane 1) and human non-activated platelets (lane 2).

## CONCLUSION

High affinity Lf receptors have been characterized for the first time at the surface of activated-lymphocytes [11] and of non-activated-platelets [20], as well as at the surface of corresponding cell-lines and in the culture medium of these cells [24]. All these receptors recognize a binding site located in the N-terminal domain I of human Lf. They differ from the low-affinity receptors which have been evidenced on the other blood cells.

## ACKNOWLEDGMENTS

This work was supported in part by the Université des Sciences et Technologies de Lille, the Ministère de l'Education Nationale and the Centre National de la Recherche Scientifique (Unité Mixte de Recherche n°111 du CNRS, Director: Prof. André Verbert).

## REFERENCES

1. Masson PC, Heremans JF.(1966) Studies on lactoferrin, the iron-binding protein of secretions. Protides Biol. Fluids 14:115–123.

2. Rado TA, Bollekens J, St Laurent G, Parker L, Benz EJ.(1984) Lactoferrin biosynthesis during granulopoiesis. Blood 64:1103–1109.

3. Montreuil J, Mullet S.(1960) Isolement d'une lactosidérophiline du lait de Femme. C.R. Acad. Sc. Paris 250:1736–1737 ; Montreuil J, Tonnelat J, Mullet S.(1960) Préparation et propriétés de la lactosiderophiline (lactotransferrine) du lait de Femme. Biochim. Biophys. Acta 45:413–421.

4. Leffel MS, Spitznagel JK.(1972) Association of human lactoferrin with lysozyme in granules of polymorphonuclear leukocytes. Infect. Immun. 6:761–765.

5. Sawatzki G.(1987) The role of iron-binding proteins in bacterial infections. in: "Iron Transport in microbes, plants and animals" (Winkelmann G., van der Helm D., Neilands J.B. Eds), VCH Verlagsgesellschaft mbH, Weinheim, Federal Republic of Germany, 477–489.

6. Bennett RM, Mohla C.(1976) A solid phase radioimmunoassay for the measurement of lactoferrin in human plasma: variations with age, sex and disease. J. Lab. Clin. Med. 88:156–165.

7. Zapolski EJ, Princiotto JV.(1976) Failure of rabbit reticulocytes to incorporate conalbumin or lactoferrin iron. Biochim. Biophys. Acta 421:80–86.

8. Birgens HS, Karle H, Hansen NE, Kristensen L.(1984) Lactoferrin receptors in normal and leukaemic blood cells. Scand. J. Haematol. 33:275–280.

9. Bennett RM, Davis J.(1981) Lactoferrin binding to human peripheral blood cells: an interaction with a B-enriched population of lymphocytes and a subpopulation of adherent mononuclear cells. J. Immunol. 127:1211–1216.

10. Moguilevsky N, Courtoy PJ, Masson PL.(1985) Study of lactoferrin binding sites at the surface of blood monocytes. In Proteins of Iron Storage and Transport. ( Spik G, Montreuil J, Crichton R, Mazurier J. Eds) Elsevier, 199–202.

11. Mazurier J, Legrand D, Hu WL, Montreuil J, Spik G.(1989) Expression of human lactotransferrin receptors in phytohemagglutinin-stimulated human peripheral blood lymphocytes. Eur. J. Biochem. 179:481–487.

12. Prieels JP, Pizzo SV, Glasgow LR, Paulson JC,Hill RL.(1978) Hepatic receptor that specifically binds oligosaccharides containing fucosyl N-acetylglucosamine linkages. Proc. Natl. Acad. Sci. USA 75:2215–2219.

13. Goavec M, Mazurier J, Montreuil J, Spik G. (1985) Rôle des glycannes dans la fixation de la sérotransferrine et de la lactotransferrine sur les macrophages alvéolaires humains. C.R. Acad. Sc. Paris 16: 689–694.

14. Miyazawa K, Mantel C, Lu L, Morrison DC, Broxmeyer HE. (1991) Lactoferrin-lipopolysaccharide interactions. Effect on lactoferrin binding to monocyte/macrophage differentiated HL60 cells. J. Immunol. 146:723–729.

15. Bennett RM, Davis J, Cambell S, Portnoff S. (1983) Lactoferrin binds to cell membrane DNA. J. Clin. Invest. 71:611–618.

16. Dautry-Varsat A, Ciechanover A, Lodish HF. (1983) pH and recycling of transferrin during receptor mediated endocytosis. Proc. Natl. Acad. Sci. USA 80: 2258–2262.

17. Legrand D, Mazurier J, Maes P, Rochard E, Montreuil J, Spik G. (1991) Inhibition of the specific binding of human lactotransferrin to human peripheral blood PHA stimulated lymphocytes by fluorescein labelling and location of the binding site. Biochem. J. 276:733–738.

18. Rochard E, Legrand D, Mazurier J, Montreuil J, Spik G. (1989) The N-terminal domain I of human lactotransferrin binds specifically to phytohemagglutinin-stimulated peripheral blood human lymphocyte receptors. FEBS-Lett. 255:201–204.

19. Legrand D, Mazurier J, Elass A, Rochard E, Maes P, Montreuil J, Spik G.(1992) Molecular interactions between human lactotransferrin and the phytohemagglutinin-activated human lymphocyte lactotransferrin receptor lie in two loop-containing regions of the N-terminal domain I of human lactotransferrin. Biochemistry 31:9243–9251.

20. Leveugle B, Mazurier J, Legrand D, Mazurier C, Montreuil J, Spik G. (1993) Lactoferrin binding to its platelet receptor inhibits platelet aggregation. Eur. J. Biochem. 213:1205–1211.

21. Boyum A. Isolation of mononuclear cells and granulocytes from human blood. Scand. J. Clin. Invest. 97l 77–89.

22. Patscheke H, Wörner P. (1978) Platelet activation detected by turbidometric shape-change analysis. Differential influence of cytochalasin B and prostaglandin E. Thrombos. Res. 12: 485–496.

23. Laemmli UK.(1970) Cleavage of structural proteins during assembly of the head of bacteriophage T4. Nature 227:680–685.

24. Bao YB, Leveugle B, Liu JL, Collard A, Coppe P, Roche AC, Nillesse N, Spik G, Mazurier J. Immunolocalization of the lactotransferrin receptor on the human T lymphoblastic cell line Jurkat. Eur. J. Cell Biol. (submitted for publication)..

# PRIMARY AND THREE-DIMENSIONAL STRUCTURE OF LACTOTRANSFERRIN (LACTOFERRIN) GLYCANS

Geneviève Spik, Bernadette Coddeville, Joël Mazurier, Yves Bourne ,[†] Christian Cambillaut, and Jean Montreuil

Université des Sciences et Technologies de Lille
Laboratoire de Chimie Biologique (Unité Mixte de Recherche n° 111 du
    CNRS)
59655 Villeneuve d'Ascq Cedex, France

[†]Faculté de Médecine, Unité de Recherche Associée n°1296 du CNRS
Boulevard Pierre Dramard
13326 Marseille Cedex 15, France

## SUMMARY

In order to establish relationships between glycan structure and biological activity, the authors undertook a comparative study of the glycan primary structure of different transferrins from several species. By associating permethylation-mass spectrometry and [1]H-NMR spectroscopy, the primary structure of the human, bovine, caprine, murine and porcine lactotransferrin glycans were determined. Using the same methods, the glycan structure of 9 serotransferrins was determined. The results obtained led to the conclusion that glycans are specific for each transferrin and, for a given transferrin, specific to the species. No relationship could be established between primary structure and function of transferrin glycans. Glycan molecular modelling, molecular dynamics simulations and X-ray diffraction studies of free glycans confirm the mobility in space of antennae. In contrast, the glycan associated with a protein is immobilized into only one conformation, as in the case of glycan-lectin associations or of "internal" glycan-protein interactions, like in rabbit serotransferrin, in which the glycan forms a bridge between the two lobes of the peptide chain, and maintains the protein in a biologically active conformation. In the case of human sero- and lactotransferrins, the glycans are in an external position on the molecules and could play a role of recognition signals.

## INTRODUCTION

Transferrins constitute a homogeneous group of proteins which possess well known common physicochemical and biological properties : high degree of homology, single polypeptide chain organized in two lobes, reversible binding capacity of two ferric ions, iron transport and recognition by specific membrane receptors. In addition, transferrins are glycosylated except those of some fishes. Consequently, they represent a remarkable model for answering the question "Are glycans markers of Evolution?", on the one hand, and, on the other hand, for establishing relationships between glycan structure and biological functions and, thus, for answering the following second question: "Why are transferrins glycosylated?"

The knowledge and the understanding of the biological activity of any kind of molecule passes through the knowledge of its 3D-structure which depends itself on the knowledge of the primary structure of the molecule.

These are the reasons why we undertook a comparative study on glycan primary structure of transferrins (sero-, ovo- and lactotransferrins) of different tissues and species [for general review, see ref. 21].

## SOME BASIC DATA ON GLYCOCONJUGATES

### Structure

Glycoconjugates result from the covalent linkage of a carbohydrate called glycan with a protein (glycoproteins) [for reviews, see Ref. 1, 14–17] or a lipid (glycolipids).

In glycoproteins, glycans are conjugated to peptide chains through two types of primary covalent linkages: N-glycosyl and O-glycosyl linkages leading to the definition of two classes of glycoproteins: N-glycosyl and O-glycosyl-proteins. Up to now, no O-glycosidic linkage has been found in transferrins.

The only N-glycosidic bond characterized so far in glycoproteins is the N-acetylglu-cosaminyl-asparagine: GlcNAc($\beta$1-N)Asn. On the contrary, the O-glycosidic type offers a wide variety of linkages since all hydroxylated amino-acids : Ser, Thr, OH-Lys, OH-Pro and Tyr are implicated in the association glycan-protein.

All of the N-glycosylproteins, whatever their origin is: plants, fungi, microorganisms or viruses, possess in common a mannotriosido-di-N-acetylchitobiose "inner-core" (Fig. 1b) which is, consequently, invariant and non-specific [14].

The N-glycans derive from the substitution of the inner-core by a wide variety of oligosaccharidic structures which carry the specificity, bear the variable fraction of the glycans and are called *antennae* [14].

The N-glycosylproteins are divided into three families on the basis of the primary structure of the antennae [14]. In the first family, the inner-core is substituted by mannose residues only, leading to the *oligomannosidic* or *high-mannose* type. The second one, called the *N-acetyllac-tosaminic* or *complex* type, is derived from the substitution of the core by a variable number of N-acetyllactosamine Gal($\beta$1-4)GlcNAc residues. In the third family, the glycans contain both oligomannosidic and N-acetyllactosaminic type structures, giving the *hybrid* type, not yet found in transferrins.

The same protein can bear different glycan structures and the same site of glycosylation can be occupied by a wide variety of glycans. At the moment, we do not know the origin and the significance of the so-called microheterogeneity of glycans.

### Biological Role of Glycans

Thanks to a series of discoveries and observations, it is now well known that glycans act i) in the folding and in the maintenance of protein conformation, ii) in the protection of the peptide chain against proteolytic attack, iii) in the decrease in immunogenicity of proteins, iv) in the recognition and association of viruses, microorganisms and fungi with cell membrane lectins, v) in cell recognition and adhesion and in cell-contact inhibition.

## PRIMARY STRUCTURE OF LACTOTRANSFERRIN GLYCANS

Since the discovery of lactotransferrins in human [18] and bovine [8] milks, active research on their glycan moieties has been carried out in our laboratory and extended to the lactotransferrins of other species.

The primary structure of human (hLf), bovine (bLf), caprine (cLf), and porcine (pLf) lactotransferrin glycans was determined by associating methylation analysis and ¹H-NMR spectroscopy.

### Results

1. All of the milk Lfs (Table 1) contain biantennary glycans of the N-acetyllactosaminic type, α-1,6-fucosylated on the N-acetyl-glucosamine residue linked to the peptide chain. Only the human Lf contains α-1,3-fucosylated "peripheral" N-acetylglucosamine residues (Fig. 1). Interestingly, human leucocyte Lf is not fucosylated and possesses the same structure as the biantennary glycans of human serotransferrin (Fig. 2). Consequently, the results obtained by using human milk Lf in experiments on non-mucosal cells must be carefully interpreted and must take into account the presence of fucose in the glycans.

2. Bovine, caprine and porcine Lf contain additional glycans of the oligomannosidic type (Fig. 3). In this regard, it is worthwhile noting that the relative proportions of glycans of oligomannosidic and N-acetyllactosaminic type vary with the period of lactation suggesting that glycan biosynthesis could be controlled by hormones.

3. Only human Lf possesses poly-N-acetyllactosaminic glycans (Fig. 4).

4. Bovine Lf is characterized by glycans possessing α-1,3-linked Gal residues in the terminal non-reducing position (Fig. 5) and, as well as porcine Lf, by GalNAc residues replacing Gal residues (Fig. 6).

5. Glycans are not distributed at random along the peptide chain as shown in Figure 7. In bovine Lf, Asn-233 and 545 are substituted by glycans of the oligomannosidic type only,

**Table 1.** Characteristics of Lactotransferrin Glycans from Different Species

| Origin | Sugar content | Number of glycans | Type of glycans | | Fucosylation | | GalNAc |
|--------|---------------|-------------------|------|--------|-------|-------|--------|
| | | | M | L | α-1,3 | α-1,6 | |
| hLf [a] | 6.40 | 2 | no | bi | + | + | no |
| hLf [b] | 5.90 | 2 | no | bi | no | no | no |
| mLf | 3.50 | 1 | no | bi | no | + | no |
| cLf | 11.00 | 4 | 2 | bi (2) | no | + | no |
| bLf | 11.20 | 4 | 2-4 | bi (1-2) | no | + | + |
| pLf | 3.40 | 1 | 0 | bi | no | + | + |

M: Oligomannosidic glycans

L: N-acetyllactosaminic glycans

[a]Human lactotransferrin from milk

[b]Human lactotransferrin from polymorphonuclear leucocytes

mLf, cLf, bLf, and pLf: murine, caprine, bovine, and porcine milk lactotransferrins, respectively

bi: Biantennary structure

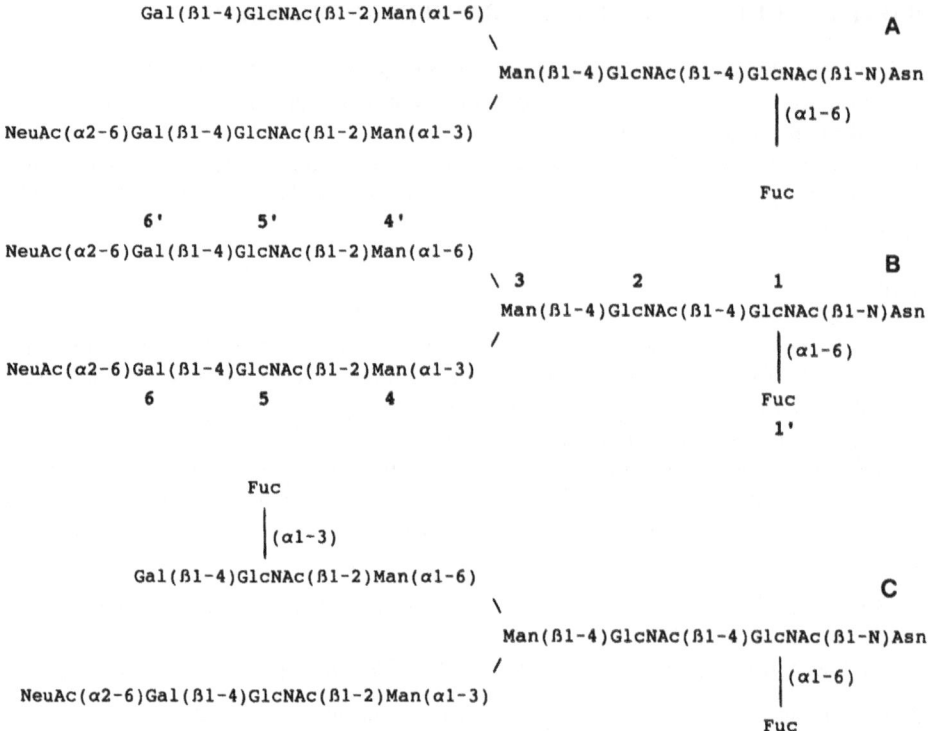

```
Gal(ß1-4)GlcNAc(ß1-2)Man(α1-6)                                              A
                                \
                                 Man(ß1-4)GlcNAc(ß1-4)GlcNAc(ß1-N)Asn
                                /                          |(α1-6)
NeuAc(α2-6)Gal(ß1-4)GlcNAc(ß1-2)Man(α1-3)
                                                          Fuc

           6'       5'       4'
NeuAc(α2-6)Gal(ß1-4)GlcNAc(ß1-2)Man(α1-6)                                    B
                                \ 3       2        1
                                 Man(ß1-4)GlcNAc(ß1-4)GlcNAc(ß1-N)Asn
                                /                          |(α1-6)
NeuAc(α2-6)Gal(ß1-4)GlcNAc(ß1-2)Man(α1-3)                Fuc
           6        5        4                            1'


           Fuc
            |(α1-3)
Gal(ß1-4)GlcNAc(ß1-2)Man(α1-6)                                              C
                                \
                                 Man(ß1-4)GlcNAc(ß1-4)GlcNAc(ß1-N)Asn
                                /                          |(α1-6)
NeuAc(α2-6)Gal(ß1-4)GlcNAc(ß1-2)Man(α1-3)                Fuc
```

**Figure 1.** Primary structure of human (A, B, C) [15, 20, 22–24], bovine [20] caprine (A, B) [20] and murine (B) [10] milk Lf glycans. Numbers from 1 to 4,4′ refer to the inner-core common to all of the N-glycosylproteins. NeuAc: N-acetylneuraminic acid; Gal: galactose; GlcNAc: N-acetylglucosamine; Man: mannose; Fuc: L-fucose.

and Asn-368 and 476 by glycans of both types. In addition, the glycan containing GalNAc is located at the Asn-476 only.

## Conclusion

The comparative study of the primary structure of Lf originating from 5 different species, as well as that of 10 serotransferrins and 2 ovotransferrins [21], leads to the conclusion that the carbohydrate moieties of transferrins are specific to each of them on the basis of the primary structure of glycans and/or of their number and of their position in the peptide chain. This observation allows us to give a positive answer to the aforementioned first question: "Are the glycans markers of Evolution?".

```
NeuAc(α2-6)Gal(ß1-4)GlcNAc(ß1-2)Man(α1-6)
                                \
                                 Man(ß1-4)GlcNAc(ß1-4)GlcNAc(ß1-N)Asn
                                /
NeuAc(α2-6)Gal(ß1-4)GlcNAc(ß1-2)Man(α1-3)
```

**Figure 2.** Primary structure of human polymorphonuclear leukocyte Lf [7].

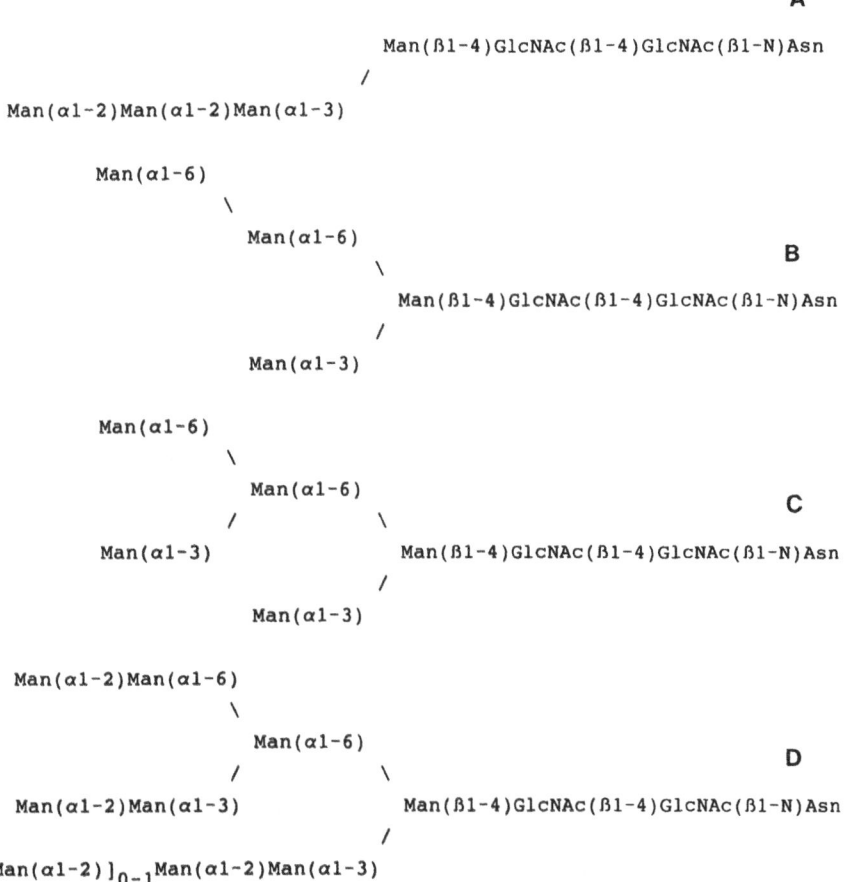

A

Man(ß1-4)GlcNAc(ß1-4)GlcNAc(ß1-N)Asn
/
Man(α1-2)Man(α1-2)Man(α1-3)

Man(α1-6)
\
Man(α1-6)
\                                    B
Man(ß1-4)GlcNAc(ß1-4)GlcNAc(ß1-N)Asn
/
Man(α1-3)

Man(α1-6)
\
Man(α1-6)
/                \                                    C
Man(α1-3)        Man(ß1-4)GlcNAc(ß1-4)GlcNAc(ß1-N)Asn
/
Man(α1-3)

Man(α1-2)Man(α1-6)
\
Man(α1-6)
/                \                                    D
Man(α1-2)Man(α1-3)    Man(ß1-4)GlcNAc(ß1-4)GlcNAc(ß1-N)Asn
/
$[Man(α1-2)]_{0-1}$Man(α1-2)Man(α1-3)

**Figure 3.** Primary structure of the glycans of the oligomannosidic type from bovine [9] and caprine [unpublished results] milk Lf.

In contrast, we must confess that, on the basis of the determination of the glycan primary structure of transferrins in general, we are unable to establish relationships between primary structure and function of the glycans and to answer the aforementioned second question: "Why are transferrins glycosylated?" and also to the following one: "How to explain the great microheterogeneity of their glycan moieties?". At the moment, the mystery thickens since, as observed in our laboratory and in others, the partial or complete deglycosylation of transferrins does not seem to affect the reversible iron fixation, the recognition and association with the reticulocyte membrane, or the iron transfer into the cell.

Could the determination of glycan 3D-structure bring the answer to these important questions?

## TRIDIMENSIONAL STRUCTURE OF LACTOTRANSFERRIN GLYCAN

Initially, the only view that we had of the 3D structure of glycans was mainly speculative. In fact, it resulted from molecular building concerning biantennary glycan from human sero-

```
            Fuc
             |
             |(α1-3)
Gal(ß1-4)GlcNAc(ß1-3)Gal(ß1-4)GlcNAc(ß1-2)Man(α1-6)
                                                    \
                                                     Man(ß1-4)GlcNAc(ß1-4)GlcNAc(ß1-N)Asn
                                                    /
                    Gal(ß1-4)GlcNAc(ß1-2)Man(α1-3)            |(α1-6)
                                                             Fuc
```

```
           Fuc
            |
            |(α1-3)
       Gal(ß1-4)GlcNAc(ß1-2)Man(α1-6)
                                     \
                                      Man(ß1-4)GlcNAc(ß1-4)GlcNAc(ß1-N)Asn
                                     /
Gal(ß1-4)GlcNAc(ß1-3)Gal(ß1-4)GlcNAc(ß1-2)Man(α1-3)         |(α1-6)
  |(α2-6)                                                  Fuc
NeuAc
```

**Figure 4.** Primary structure of the poly-N-acetyllactosaminic glycans from human Lf [12].

```
      Gal(α1-3)Gal(ß1-4)GlcNAc(ß1-2)Man(α1-6)
                                             \
                                              Man(ß1-4)GlcNAc(ß1-4)GlcNAc(ß1-N)Asn
                                             /
      NeuAc(α2-6)Gal(ß1-4)GlcNAc(ß1-2)Man(α1-3)       | (α1-6)
                                                     Fuc
```

**Figure 5.** Primary structure of a biantennary glycan from bovine milk Lf with an α-1,3-Gal residue in the terminal non-reducing position [5].

```
      Gal(ß1-4)GlcNAc(ß1-2)Man(α1-6)                              A
                                    \
                                     Man(ß1-4)GlcNAc(ß1-4)GlcNAc(ß1-N)Asn
                                    /
NeuAc(α2-6)GalNAc(ß1-4)GlcNAc(ß1-2)Man(α1-3)
```

```
[NeuAc(α2-6)]_{0-1}GalNAc(ß1-4)GlcNAc(ß1-2)Man(α1-6)              B
                                                   \
                                                    Man(ß1-4)GlcNAc(ß1-4)GlcNAc(ß1-N)Asn
                                                   /
      NeuAc(α2-6)Gal(ß1-4)GlcNAc(ß1-2)Man(α1-3)            |(α1-6)
                                                        [Fuc]_{0-1}
```

**Figure 6.** Primary structure of a biantennary glycan from bovine and porcine Lf with a GalNAc residue replacing a Gal residue [5].

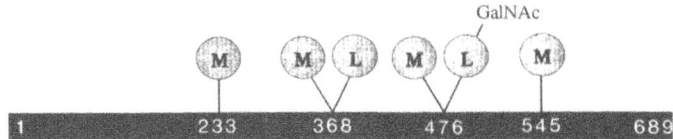

**Figure 7.** Assignment of the different glycans to their glycosylation sites on the bovine Lf peptide chain [unpublished results]. M, L : oligomannosidic and N-acetyllactosaminic type glycans, respectively.

and lactotransferrins whose glycan primary structure had just been determined [19]. This approach led to the description of the Y-, T-, bird- and broken wing-conformation (Fig. 8) [for review, see refs. 14–17]. This situation was due to the difficulties encountered in the crystallization of glycoproteins. Only the conformation of the glycan moiety of the Fc fragment of the human IgG has been fully resolved. Concerning other crystallized glycoproteins, only the 3D-structure of the first monosaccharides in the terminal reducing position is generally resolved. Application of sophisticated techniques like NMR, RPE and neutron scattering led to results which favoured the emerging concept of conformation interconversions due to the mobility of antennae. This concept was recently verified on the basis of molecular modelling and X-ray diffraction data.

## Molecular Modelling and Molecular Dynamics Simulation of Sero-and Lactotransferrins

**Experimental Procedures.**    Procedures are described in detail in refs. 6 and 13. Briefly, all calculations were carried out on an Evans and Sutherland PS 350 graphic station and on a Vax G320 host-computer. The energy calculations and the molecular dynamics simulations were performed using the Tripos 5.4 force fields of the Sybyl-software. The three dimensional Connolly representation of the surface of the molecular models was obtained from Hydra.

The atomic coordinates of rabbit serotransferrin as determined at 3.3 Å resolution were a generous gift from Dr. Lindley. The interactions between the glycan moiety and the amino-acid side chains were analyzed using Manosk software.

## Results

1.  On the basis of the computer calculations, we demonstrate that, in the absence of any interaction with a protein, a high number of glycan conformations exists due to the mobility of antennae of which the concept is thus confirmed. The different conformations can be classified into five basic conformations, four of which having already been described. In fact, in addition to the Y-, T-, bird- and broken wing-conformations, a "back-folded wing" conformation is energetically feasible [13].

    These results have been confirmed by molecular dynamics simulations from 0 to 200 ps [6]. In fact, starting from a bird- conformation, the 3D-structure evolved through successive transitional states to a new, compact and energetically favorable conformation which had not been described previously. In this conformation, both antennae are organized in two coplanar loops rolled in a contrary direction and oriented perpendicularly to the plane of the di-N-acetyl-chitobiose residue leading to a so-called "lobster conformation" (Fig. 9).

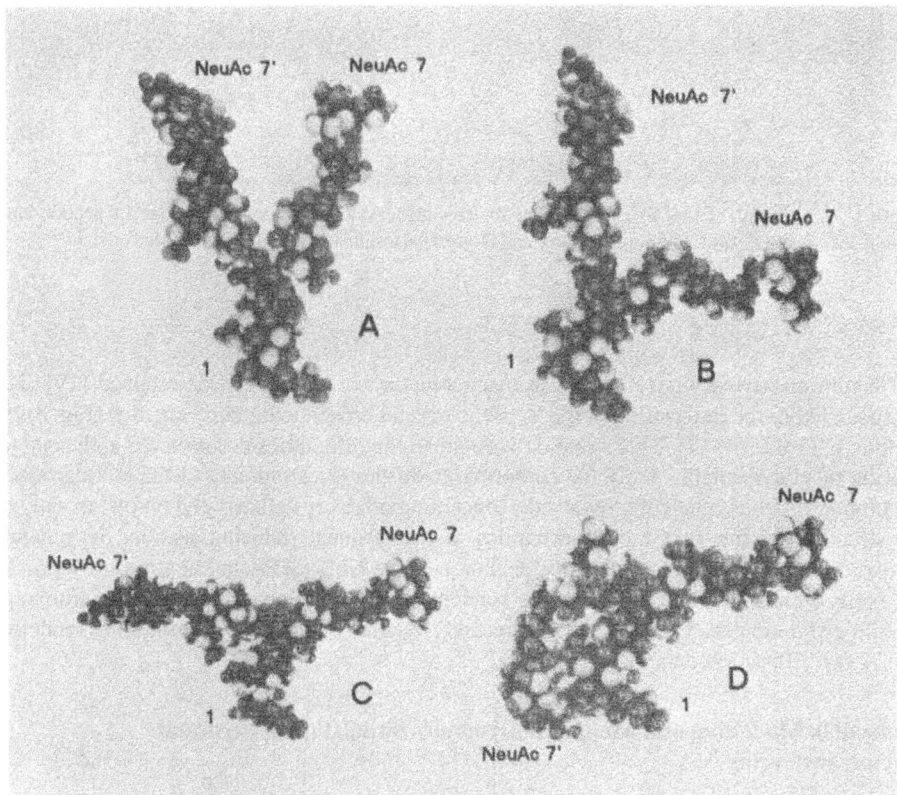

**Figure 8.** Spatial conformation of a biantennary glycan of the N-acetyllactosaminic type [14–17]. A, B, C, D: Y-, T-, bird- and broken wing conformations, respectively. Numbers correspond to the numbering used in structure B of Figure 1.

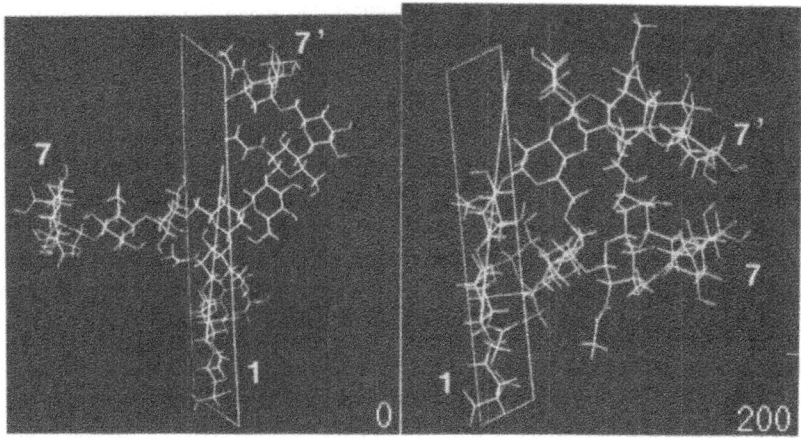

**Figure 9.** Three-dimensional representation of selected (0 and 200 ps) conformations of the disialylated mono-fucosylated biantennary glycan of human Lf [6]. 0 ps: bird-conformation; 200 ps: "lobster-conformation". Numbers correspond to the numbering used in structure B of Figure 1.

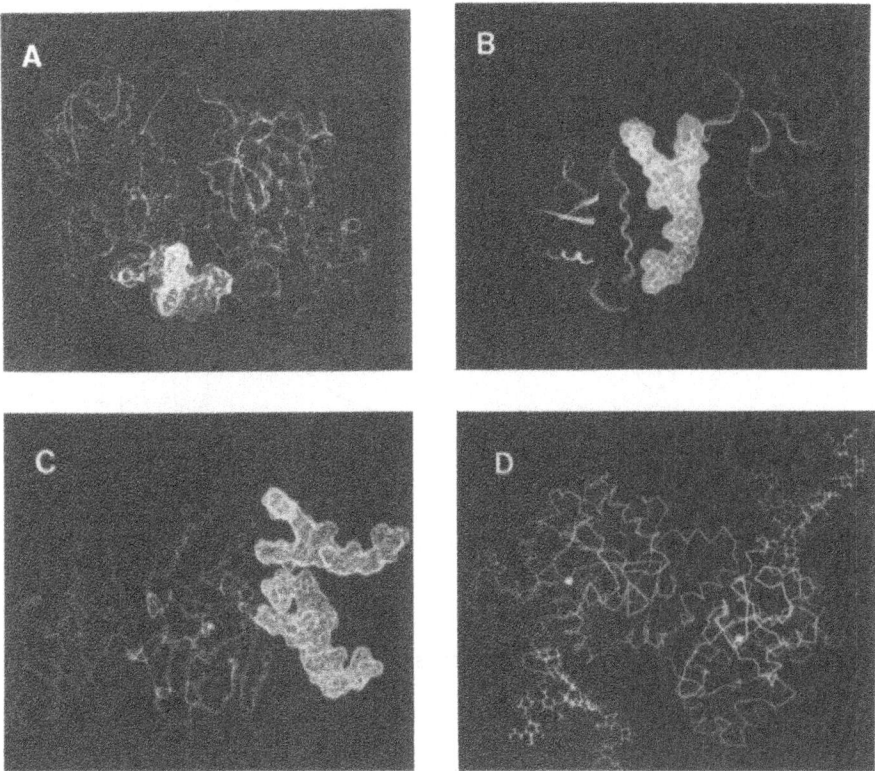

**Figure 10.** Molecular modelling of rabbit serotransferrin (A and B) glycan [13] and of human serotransferrin (C) and lactotransferrin (D) [unpublished results].

2. In contrast, the glycan linked to the protein, in the rabbit serotransferrin molecule for example, is immobilized into only one conformation: the broken wing-conformation. In fact, the only data [2] delivered by X-ray diffraction of the glycan of this transferrin was the localization of one NeuAc residue in the vicinity of α-helix n°9 (residues 254–271). The molecular modelling we performed shows that only the broken wing-conformation fits the X-ray diffraction. The NeuAc residue of the α-l,3-antenna is in the vicinity of the side chain of the following amino-acids: Ser-254, Asp-257, Glu-271 (hydrogen-bonds) and Phe-272 (hydrophobic interactions) (Figs. 10A and B). These results strongly support the hypothesis that, in the case of rabbit serotransferrin, the glycan reinforces the association of the two lobes and contributes to maintain the protein moiety in a biologically active 3D-conformation. On the contrary, in human sero- and lactotransferrins, the glycans do not seem to interact with the peptide chain (Figs. 10C and D). They are free in space and could thus play a role of recognition signals.

**Conclusion.** The computer data we obtained underline the flexibility and mobility of the antennae and their conformational adaptability and capacity for adopting only one 3D-conformation when the glycan interacts with an other molecule. This view has been recently confirmed by X-ray diffraction studies.

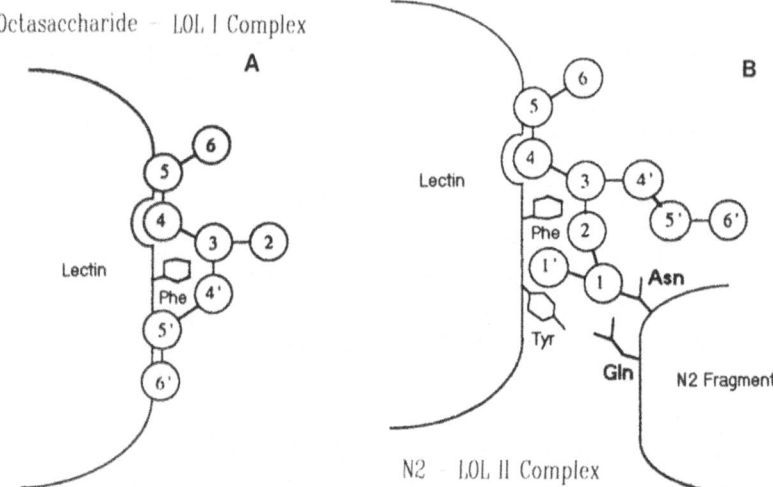

**Figure 11.** Spatial conformation of a free glycan (A) and of the N2-glycopeptide from human lactotransferrin (B) determined by X-ray diffraction [from refs. 3 and 4]. Numbers correspond to the numbering used in structure B of Figure 1.

### X-ray Diffraction of Glycan Lectin Complexes

Recently, Cambillau *et al.* have devised an elegant method of co-crystallization of the iso-lectins I and II of *Lathyrus ochrus* (LOL I and II) with glycans or glycoproteins [3, 4]. Human Lf glycan and N2-glycopeptide [11] were co-crystallized with LOL and X-ray diffraction data from single crystals were collected at 2.3 Å or 3.5 Å resolution.

**Results.** The obtained results are illustrated in Figure 11 which shows that the 3D-conformation of the free glycan (bird-conformation) and of the N2-conjugated glycan (broken wing-conformation) are different.

**Conclusion.** The concept of flexibility of the antennae proposed years ago is now firmly established on experimental data. The mobility of glycan antennae allows them to adapt their configuration to a receptor molecule. The concept of the antennae changing their configuration according to the environment fits perfectly to the concept of glycans acting as recognition signals.

### ACKNOWLEDGEMENTS

This work was supported in part, by the Université des Sciences et Technologies de Lille, the Ministère de l'Education Nationale and the Centre National de la Recherche Scientifique (Unité Mixte de Recherche n°111 du CNRS, Director : Prof. André VERBERT). Molecular modelling and molecular dynamics simulation have been carried out in collaboration with Prof. Gérard VERGOTEN and Dr. Manuel DAUCHEZ (Université des Sciences et Technologies de Lille).

### REFERENCES

1. Allen HJ, Kisailus EC, (1992) Glycoconjugates. Composition, structure and function. Marcel Dekker, New York.

2. Bailey S , Evans RE, Garrat RC, Gorinski B, Hasnain S, Horsburgh C, Thuti H, Lindley PF, Mydin A, Sarra R, Watson JL, (1988) Molecular structure of serum transferrin at 3.3 angstroms resolution. Biochemistry, 27:5804–5812.

3. Bourne Y, Nésa MP, Rougé P, Mazurier J, Legrand D, Spik G, Montreuil J, Cambillau C, (1992) Crystallization and preliminary X-ray diffraction study of Lathyrus ochrus isolectin II complexed to the human lactotransferrin N2 fragment. J Mol Biol, 227:938–941.

4. Bourne Y, Rougé P, Cambillau C, (1992) X-ray structure of a biantennary octosaccharide-lectin complex refined at 2.3 angströms resolution. J Biol Chem, 267:197–203.

5. Coddeville B, Strecker G, Wieruszeski JM, Spik G, Vliegenthart JFG, van Halbeek H, Peter-Katalinic J, Egge H, (1992) Heterogeneity of bovine lactotransferrin glycans. Characterization of α-D-Galp-(1→3)-β-D-Gal and α-NeuAc-(2→6)-β-D-Galp-NAc-(1→4)-β-D-GlcNAc substituted N-linked glycans. Carbohydr Res, 236:145–164.

6. Dauchez M, Mazurier J, Montreuil J, Spik G, Vergoten G, (1992) Molecular dynamics simulations of a monofucosylated biantennary glycan of the N-acetyllactosamine type: the human lactotransferrin glycan. Biochimie, 74:63–74.

7. Derisbourg P, Wieruszeski JM, Montreuil J, Spik G, (1990) Primary structure of glycans isolated from human lactotransferrin. Absence of fucose residues questions the proposed mechanism of hyposideremia. Biochem J, 269:821–825.

8. Groves ML, (1960) The isolation of a red protein from milk. J Am Chem Soc, 82:3345–3350.

9. van Halbeek H, Dorland L, Vliegenthart JFG, Spik G, Chéron A, Montreuil J, (1981) Structure determination of two oligomannoside-type glycopeptides obtained from bovine lactotransferrin by 500-MHz [1]H-NMR spectroscopy. Biochim Biophys Acta, 675:293–296.

10. Leclerq Y, Sawatzki G, Wieruszeski JM, Montreuil J, Spik G, (1987) Primary structure of the glycans from mouse serum and milk transferrins. Biochem J, 247:571–578.

11. Legrand D, Mazurier J, Metz-Boutigue MH, Jollès J, Jollès P, Montreuil J, Spik G, (1984) Characterization and localization of an iron-binding 18 kDa glycopeptide isolated from the N-terminal half of human lactotransferrin. Biochim Biophys Acta, 787:90–96.

12. Matsumoto A, Yoshima H, Takasaki S, Kobata A (1982) Structural study of the sugar chains of human lactotransferrins: finding of four novel complex-type asparagine-linked sugar chains. J Biochem (Japan) 91:143–152.

13. Mazurier J, Dauchez M, Vergoten G, Montreuil J, Spik G, (1991) Modélisation moleculaire des glycannes: structure tridimensionnelle et interaction avec la fraction protéique. L'exemple de la sérotransferrine de Lapin. CR Acad Sci Paris, 313 Série III:7–14.

14. Montreuil J, (1974) Recent data on the structure of the carbohydrate moiety of glycoproteins. Metabolic and biological implications. Pure Appl Chem, 42:431–477.

15. Montreuil J, (1980) Primary structure of glycoprotein glycans. Basis for the molecular biology of glycoproteins. Adv Carbohydr Chem Biochem, 37:157–223.

16. Montreuil J, (1982) Glycoproteins. In Comprehensive Biochemistry (Newberger A, van Deenen LLM eds) Elsevier, Amsterdam, 19B/II:115–132.

17. Montreuil J, (1984) Spatial conformation of glycans and glycoproteins. Biol. Cell, 51:115–132.

18. Montreuil J, Mullet S, (1960) Isolement d'une lactosidérophiline du lait de Femme. CR Acad Sci Paris, 250:1736–1737. Montreuil J, Tonnelat J, Mullet S, (1960) Préparation et propriétés de la lactosidérophiline (lactotransferrine du lait de Femme). Biochim Biophys Acta, 45:413–421.

19. Spik G, Bayard B, Fournet B, Strecker G, Bouquelet S, Montreuil J, (1975) Complete structure of two carbohydrate units of human serotransferrin. FEBS Lett, 50:296–299.

20. Spik G, Coddeville B, Legrand D, Mazurier J, Léger D, Goavec M, Montreuil J, (1985) A comparative study of the primary structure of glycans from various sero-, lacto- and ovotransferrins. In Proteins of Iron Storage and Transport (Spik G, Montreuil J, Crichton RR, Mazurier J eds.) Elsevier, Amsterdam, 47–51.

21. Spik G, Coddeville B, Montreuil J, (1988) Comparative study of the primary structures of sero-, lacto- and ovotransferrin glycans. Biochimie, 70:1459–1569.

22. Spik G, Mazurier J, (1977) Comparative structural and conformational studies of polypeptide chain, carbohydrate moiety and binding sites of human serotransferrin and lactotransferrin. In Proteins of Iron Metabolism (Brown EB, Aisen P, Fielding J, Crichton RR eds.) Grune and Stratton, New York, 143–151.

23. Spik G, Strecker G, Fournet B, Bouquelet S, Montreuil J, Dorland L, van Halbeek H, Vliegenthart JFG, (1982) Primary structure of the glycan from human lactotransferrin. Eur J Biochem, 121:413–419.

24. Spik G, Vandersyppe R, Fournet B, Bayard B, Charet P, Bouquelet S, Strecker G, Montreuil J, (1974) Structure of glycopeptides isolated from human serotransferrin and lactotransferrin. Proc Intern Symp on Glycoconjugates, Villeneuve d'Ascq, 1973 (Montreuil J, ed.) Editions du CNRS, Paris, 1:483–499.

# SYNERGISM AND SUBSTITUTION IN THE LACTOFERRINS

Andrew M. Brodie, Eric W. Ainscough, Edward N. Baker,
Heather M. Baker, Musa S. Shongwe, and Clyde A. Smith

Department of Chemistry and Biochemistry
Massey University
Palmerston North, New Zealand

## SUMMARY

The anion binding properties of human lactoferrin (Lf), with $Fe^{3+}$ or $Cu^{2+}$ as the associated metal ion, highlight differences between the two sites, and in the anion binding behaviour when different metals are bound. Carbonate, oxalate and hybrid carbonate-oxalate complexes have been prepared and their characteristic electronic and EPR spectra recorded. Oxalate can displace carbonate from either one or both anion sites of $Cu_2(CO_3)_2Lf$, depending on the oxalate concentration, but no such displacement occurs for $Fe_2(CO_3)_2Lf$ although it does for the bovine analogue. Addition of oxalate and the appropriate metal ion to apoLf under carbonate-free conditions gives dioxalate complexes with both $Fe^{3+}$ and $Cu^{2+}$. The anion sites as determined from the crystal structures of $Fe_2(CO_3)_2Lf$, $Fe_2(C_2O_4)_2Lf$, $Cu_2(CO_3)_2Lf$, and $Cu_2(CO_3)(C_2O_4)Lf$ have been compared. Both the carbonate and oxalate ions bind in bidentate fashion to the metal, except that the carbonate ion in the N-lobe site of dicupric lactoferrin is monodentate. The hybrid copper lactoferrin complex shows that the oxalate ion binds preferentially in the C-lobe site in a bidentate mode. A series of complexes containing the synergistic anion O,N-chelates with increasing substitution on the N atom (glycinate, iminodiacetate and nitrilotriacetate) have been prepared with iron bovine lactoferrin for comparison with the O,O-chelate oxalate. Overall these observations lead to a generalised model for synergistic anion binding by transferrins and allow comparisons to be made with nonsynergistic anions such as citrate and succinate.

## INTRODUCTION

Lactoferrin is a member of the family of iron binding proteins known as the transferrins (Harris & Aisen, 1989). A key feature of the binding properties of transferrins is the synergistic relationship between metal and anion binding; two $Fe^{3+}$ ions are bound in the specific sites, but only if two $CO_3^{2-}$ (or other suitable) anions are concomitantly bound. Just as other di-, tri- and tetravalent metal ions can substitute for $Fe^{3+}$, so other anions can substitute for $CO_3^{2-}$. In a study of the binding of over thirty anions to transferrin, Schlabach and Bates (1975) concluded that synergistic anions must possess a carboxylate group and a proximal electron donor group,

*Lactoferrin: Structure and Function*
Edited by T.W. Hutchens *et al.*, Plenum Press, New York, 1994

and obey certain steric limits. This then led to their "interlocking sites" model, whereby the anion was simultaneously bound to the metal ion (through the electron donor group) and a cationic group on the protein (through the carboxyl group).

Numerous spectroscopic studies have given evidence of direct binding of the anion to the metal (Harris & Aisen, 1989). Some of these studies also suggested the presence of a neighbouring cationic protein group (Zweier, 1980; Zweier et al., 1981), in accord with the "interlocking sites" model, and chemical modification experiments implicated an essential arginine sidechain (Rogers et al., 1978).

Nevertheless in spite of the spectroscopic studies important questions remained unanswered. The bound carbonate ion plays a key role in metal binding and release and direct crystallographic evidence of the metal-anion interaction was needed. This came with the refinement of the structure of the diferric lactoferrin complex, $Fe_2(CO_3)_2Lf$, which proved bidentate coordination of the carbonate anion to the iron (Anderson et al., 1989).

The acceptance of a wide variety of other anions in place of carbonate, raises questions of how the proteins adapt to sterically and chemically different ions. The extension of bidentate carbonate coordination to a more general model, covering other anions, and its relationship with the "interlocking sites" model, is not straightforward. Spectral results have been interpreted in terms of both bidentate and monodentate oxalate coordination (Eaton et al., 1989; Bertini et al., 1986a; Sola, 1990).

There is also evidence of differences between the anion sites, both within a given transferrin and between different transferrins. These differences appear much more pronounced for anions other than carbonate (Zweier & Aisen, 1977; Ainscough et al., 1983; Zweier, 1980; Bertini et al., 1986b).

In this paper we discuss the recently reported details of the binding of both carbonate and oxalate ions to human lactoferrin (Lf), from carbonate complexes of diferric lactoferrin and dicupric lactoferrin, $Fe_2(CO_3)_2Lf$ and $Cu_2(CO_3)_2Lf$, and a hybrid carbonate-oxalate complex of dicupric lactoferrin, $Cu_2(CO_3)(C_2O_4)Lf$ (Shongwe et al., 1992) as well as some new results for the dioxalate complex, $Fe_2(C_2O_4)_2Lf$. The oxalate binding seen in the latter two complexes allows us to elaborate on the original "interlocking sites" model and to advance structural explanations for some of the effects noted in previous solution studies. At the same time we present further solution studies of oxalate binding, complementary to the crystallographic studies, which confirm the inequivalence of the two sites in lactoferrin and demonstrate the quite profound differences in binding behaviour which can result from the use of different metal ions. Where appropriate results involving bovine lactoferrin are included and comparisons are made with nonsynergistic anions.

## EXPERIMENTAL

All glassware was soaked in nitric acid and thoroughly rinsed with doubly distilled water prior to use. All buffers were passed through a column of a chelating resin (Bio-Rad Chelex 100, 100–200 mesh, sodium form) before use to minimise contamination by adventitious metal ions. Isotopically pure $^{65}Cu$ was obtained as $^{65}CuO$ from Oak Ridge National Laboratory and dissolved in concentrated HCl. All other reagents were of the highest purity commercially available.

Electron paramagnetic resonance (EPR) spectra were recorded at 110 K, with protein concentrations in the range $1.25 \times 10^{-4}$ to $3 \times 10^{-4}$ mol/L, by using a Varian E-I04A spectrometer equipped with a Varian E-257 variable temperature accessory. Spectral g-values were calibrated with (diphenylpicryl)hydrazyl (DPPH) as a standard.

Experiments requiring rigorous exclusion of carbon-dioxide or carbonate ions were manipulated under the flow of argon using standard Schlenk-line techniques (Shriver, 1969).

Human and bovine lactoferrins were prepared as described previously (Norris et al., 1989; Norris et al., 1986).

The carbonate complexes of human lactoferrin with $Fe^{3+}$ or $Cu^{2+}$ were prepared by a modification of previously-reported methods (Ainscough et al., 1979). Stoichiometric amounts of ferric nitrilotriacetate or cupric chloride, respectively, were added to solutions of apolactoferrin (10 to 16 mg/ml) in 0.03 M Hepes buffer, pH 7.8, containing 0.01 M $NaHCO_3$ and 0.1 M NaCl. The protein solutions were then passed down a previously equilibrated Chelex column to remove any non-specifically bound metal ions. Oxalate complexes were prepared either by displacement of carbonate from $M_2(CO_3)_2Lf$ complexes or by addition of oxalate and metal ion to carbonate-free solutions of apolactoferrin. In the displacement method, solid sodium oxalate was added, in a 30-, 50- or 100-fold molar excess, to solutions of the appropriate $M_2(CO_3)_2Lf$ complex (10 to 16 mg/ml) in 0.03 M Hepes buffer, pH 7.8, containing 0. 1 M NaCl. In the second method, a 50-fold molar excess of solid sodium oxalate was added to the solution of apolactoferrin (10 to 16 mg/ml) in 0.03 M Hepes, pH 7.8, containing 0. 1 M NaCl. The protein solution was transferred into a Schlenk tube and its pH was lowered to 4.1 by addition of 0. 1 M HCl. The solution was then made $CO_2$-free by continual evacuation and flushing with argon over a period of 1 to 2 hours. At the end of this time the pH was restored to 7.8 by addition of $CO_2$-free ammonia. Metal binding was then carried out by the addition of aliquots of ferric nitrilotriacetate or cupric chloride, as before; these and all subsequent manipulations were carried out under an atmosphere of argon.

Bovine lactoferrin complexes with anions other than carbonate were prepared from bovine apolactoferrin under carbonate-free conditions as described for the oxalate human lactoferrin complexes.

The crystal structures of $Fe_2(CO_3)_2Lf$, $Cu_2(CO_3)_2Lf$ and $Cu_2(CO_3)(C_2O_4)Lf$ have been refined to 2.2, 2.1 and 2.2 Å, respectively (Shongwe et al., 1992). Full details of the refinements of these structures and that of $Fe_2(C_2O_4)_2Lf$ will be published elsewhere.


# RESULTS AND DISCUSSION

## Iron Carbonate Lactoferrin

Human lactoferrin, when isolated from fresh colostrum, is 8–10% saturated with iron. Addition of further iron, either as iron(II), (viz., ferrous ammonium sulphate) or as iron(III) (viz., ferric nitrilotriacetate) rapidly saturates the two specific iron binding sites as all the apolactoferrin is converted to diferric bis(carbonate)lactoferrin, $Fe_2(CO_3)_2Lf$. The orange-red colour of the iron-saturated protein is a result of the intense visible absorption band at 466 nm which is assigned as a phenolate ($\pi$) to iron ($d_{\pi}$) type charge transfer absorption and identifies iron-tyrosine binding (Ainscough et al., 1980). The EPR spectrum, with its g' 4.3 resonance, found for all the ferric carbonate transferrins, points to a unique distorted six coordinate environment for the ferric ion, the detail of which can be understood from the crystallographic results. The ferric ion is bound to four protein ligands—two phenolate oxygens, one imidazole nitrogen and a carboxylate oxygen from an aspartate. The remaining two sites are occupied by the synergistic carbonate ion, which binds in a bidentate mode (Figure 1a). Angles around the $Fe^{3+}$ ion are expected to be distorted from regular octahedral values of 90°, especially the O-Fe-O angle imposed by the small bite of the bidentate carbonate ion. The exquisite fit of the anion between the metal and the protein explains the high specificity for $Fe^{3+}$ and carbonate in the transferrins. The full hydrogen bonding potential of the carbonate is used as it is locked into place by interactions with positively charged groups of the N-terminus of helix 5 and the side chain of an arginine residue.

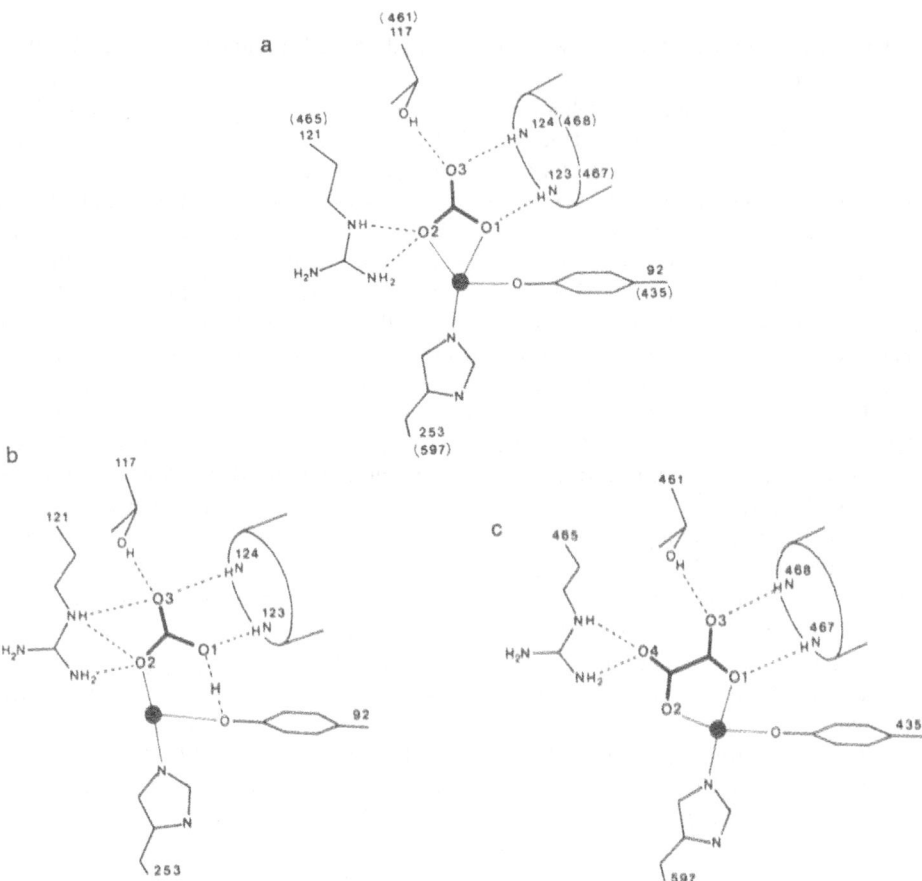

**Figure 1.** Schematic diagrams showing the anion interactions with the protein and the metal ion in (a) the bidentate carbonate site in each lobe of $Fe_2(CO_3)_2Lf$ and the C-lobe of $Cu_2(CO_3)_2Lf$; (b) the monodentate (bi)carbonate site in the N-lobes of $Cu_2(CO_3)_2Lf$ and $Cu_2(CO_3)(C_2O_4)Lf$; and (c) the oxalate site in each lobe of $Fe_2(C_2O_4)_2Lf$ and the C-lobe of $Cu_2(CO_3)(C_2O_4)Lf$.

## Iron Oxalate Lactoferrin Complexes

Replacing the physiologically important carbonate anion with oxalate allows subtle site differences between human and bovine lactoferrin and between lactoferrin and serum trans-ferrin to be revealed. The dioxalate complex of human lactoferrin, $Fe_2(C_2O_4)_2Lf$ is prepared from apolactoferrin by addition of 2 mol of ferric citrate to a carbonate-free solution of the protein containing a 50-fold molar excess of oxalate at pH 7.8. This complex is purple-red ($\lambda_{max}$ 482 nm) compared with the orange-red carbonate complex ($\lambda_{max}$ 466 nm). Spectro-photometric titrations carried out in the absence of $CO_3^{2-}$ clearly indicate a required stoichiometry of 2 mol of Fe(III) and 2 mol of $C_2O_4^{2-}$ per mole of protein. The EPR spectrum (Figure 2) is markedly different from that of $Fe_2(CO_3)_2Lf$ but similar to that reported for the serum transferrin complex $Fe_2(C_2O_4)_2Tf$ (Aisen et al., 1972; Dubach et al., 1991). However there is a more pronounced splitting of the g' ~ 6 resonance observed for the transferrin complex which implies small site differences with an increase in the E/D EPR zero field splitting parameter from ~0.04 to nearer 0.1 (Dubach et al., 1991). A mixed-anion complex $Fe_2(CO_3)(C_2O_4)Lf$ can also be prepared by exposure of a carbonate-free solution of $Fe_2(C_2O_4)_2Lf$ to atmospheric $CO_2$ for several weeks. The EPR of this complex comprises two resonances, centred around $g' \sim 4.3$ and $g' \sim 6$ (Figure 2), characteristic of carbonate and oxalate binding respectively. The complex $Fe_2(C_2O_4)_2Lf$ is sufficiently stable in air to be crystallized and the X-ray structural results show the oxalate ion to be bound in a 1,2-bidentate mode in both C- and N-lobe sites (Figure 1c). The larger bite angle of the oxalate ion as compared with $CO_3^{2-}$, allows for a more regular octahedral geometry for the $Fe^{3+}$ ions in line with the EPR spectral differences observed between $Fe_2(C_2O_4)_2Lf$ and $Fe_2(CO_3)_2Lf$. The hydrogen bonding interactions of $C_2O_4^{2-}$ with the protein are similar to those of the carbonate ion and involve both the helix 5 N-terminus and the arginine residues. However, the larger size of the oxalate compared with carbonate causes a local rearrangement beyond the anion binding site as the side chains of the arginine residues are pushed away.

It is not possible to replace the synergistic carbonate anion from the human lactoferrin complex, $Fe_2(CO_3)_2Lf$, by substitution reactions. For example even up to a 200 molar excess of oxalate does not cause a shift in the visible absorption band at 466 nm to the longer wavelengths expected for oxalate substitution. In contrast, the carbonate ion in diferric bovine lactoferrin is replaced by excess oxalate over a period of two weeks at 4°C to give $Fe_2(CO_4)_2bLf$, with identical spectroscopic properties to the complex prepared in the absence of carbonate and to the human lactoferrin analogue. These results show that while the ligand

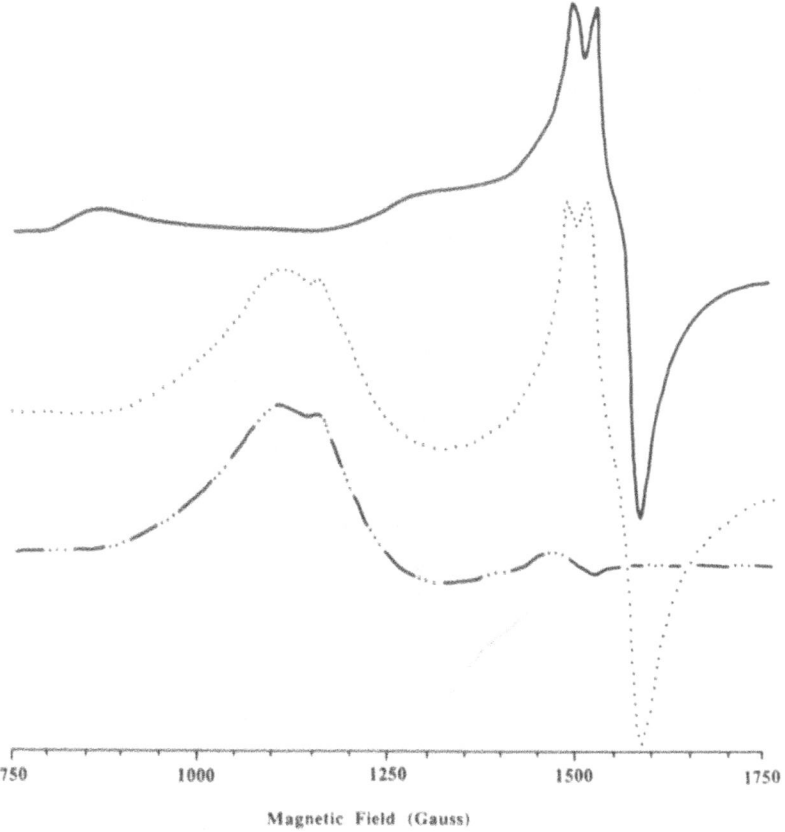

Magnetic Field (Gauss)

**Figure 2.** EPR spectra of carbonate and oxalate complexes of diferric human lactoferrin. Solutions in 0.3 M Hepes buffer, pH 7.8, containing 0.1 M NaCl. Spectra are $Fe_2(CO_3)_2Lf$ (—); $Fe_2(CO_3)(C_2O_4)Lf$ (·····); and $Fe_2(C_2O_4)Lf$ (—···—).

coordination to the iron is basically the same for the dioxalate complexes of human and bovine lactoferrin, and serum transferrin, there must be differences beyond the immediate coordination sphere with bovine lactoferrin being more flexible than the human protein.

Although the X-ray crystallographic refinement of bovine diferric bis(carbonate)lactoferrin is not yet complete (Baker et al., 1991 ) differences between the bovine and human proteins, which could account for these observations, can be noted. In the N-lobe in human lactoferrin an aspartic acid residue (Asp217) is not present in the bovine protein but is replaced by an asparagine. The result is the loss of a salt bridge (Lys296...Asp217) in bovine lactoferrin linking the two domains which close up over the metal as it binds (Baker et al., 1991). A second difference is the switch of an arginine residue (Arg210) in human lactoferrin to a lysine in the bovine protein. The arginine is hydrogen bonded to two tyrosines (Tyr92 and Tyr82), one of which is bound directly to $Fe^{3+}$ (Tyr 92). In the bovine protein the replacement lysine no longer interacts directly with the tyrosines, but indirectly via an intervening water molecule. It is stressed that such species differences do not alter the main features of the iron coordination environment but they could provide an explanation for the fact that oxalate ions will displace the synergistic carbonate ions from bovine diferric lactoferrin but not from its human analogue. The loss of an interdomain salt bridge and specific tyrosine- hydrogen bond interactions could increase the flexibility of the closed structure and explain why the carbonate ion is not locked so firmly in place between the $Fe^{3+}$ and the polypeptide chain in bovine lactoferrin. It is of interest to note that rabbit difernc transferrin is similar to bovine diferric lactoferrin in that it also has a lysine residue in place of Arg210 in the human protein (Baker & Lindley, 1992).

### Iron O,N-anion Bovine Lactoferrin Complexes

The metal-anion-protein interaction can be further probed with anions which will cause both electronic and steric perturbations. Such a series is neatly provided by glycinate (GLY), iminodiacetate (IDA) and nitrilotriacetate (NTA). These anions provide a change from the O,O donor set of carbonate and oxalate to an O,N donor set and as well provide for an increased steric bulk along the series on the N-substituent [$-NH_2$ to $-N(CH_2CO_2H)H$ to $-N(CH_2CO_2H)_2$ respectively]. Under carbonate free conditions at pH 7.8 the bovine lactoferrin complexes $Fe_2(GLY)_2bLf$, $Fe_2(IDA)_2bLf$ and $Fe_2(NTA)_2bLf$ can be prepared in a manner similar to that described for the oxalate complex $Fe_2(C_2O_4)_2Lf$ with spectrophotometric titrations (Figure 3) confirming the stoichiometry. As for both the human and bovine lactoferrin oxalate complexes the visible absorption bands move to longer wavelengths on carbonate replacement (Table 1) but again it is the EPR spectra that point to differences in iron-anion-protein interactions. The EPR

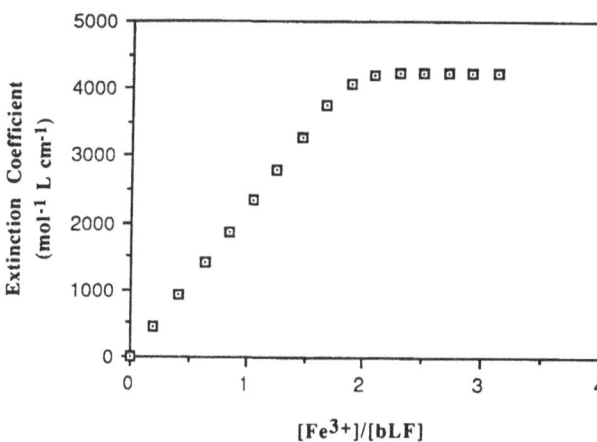

**Figure 3.** Titration of ferric citrate with carbonate-free bovine apolactoferrin in the presence of excess glycinate (0.025 M Tris buffer, pH 7.8).

spectrum of $Fe_2(GLY)_2bLf$ is virtually identical to that of the serum transferrin oxalate complex $Fe_2(C_2O_4)_2Tf$ (Aisen et al., 1972; Dubach et al., 1991 ) with the pronounced splitting of the $g' \sim 6$ resonance. This firmly establishes O,N chelation of the glycinate anion to $Fe^{3+}$ and implies the $Fe^{3+}$-GLY-bLf and $Fe^{3+}$-$C_2O_4^{2-}$-Tf interactions must be similar. In contrast, the more bulky IDA and NTA anions show a strong resonance at $g' \sim 4.3$ (although it is not split as in the $CO_3^{2-}$ case). Chelation of the anion is still assumed but the results point to a perturbation being forced on the iron coordination environment by interaction of the non-coordinated-$CH_2CO_2H$ anion substituents with the surrounding protein structure. The side chain of the arginine residue may be pushed back even further than for oxalate substitution.

**Table 1.** Charge Transfer Absorption Maxima for Human and Bovine Lactoferrin Complexes.

| Complex | $\lambda_{max}$ (nm) |
| --- | --- |
| $Fe_2(CO_3)_2Lf$ | 466 |
| $Fe_2(C_2O_4)(CO_3)Lf$ | 476 |
| $Fe_2(C_2O_4)_2Lf$ | 482 |
| $Fe_2(CO_3)_2bLf$ | 465 |
| $Fe_2(C_2O_4)_2bLf$ | 482 |
| $Fe_2(GLY)_2bLf$ | 490 |
| $Fe_2(IDA)_2bLf$ | 490 |
| $Fe_2(NTA)_2bLf$ | 478 |
| $Cu_2(CO_3)_2Lf$ | 434 |
| $Cu_2(C_2O_4)(CO_3)Lf$ | 424 |
| $Cu_2(C_2O_4)_2Lf$ | 420 |

## Copper Carbonate Lactoferrin

Differences in the two metal binding sites of lactoferrin become more pronounced when metal ions other than iron are bound. This is clearly revealed when copper(II) is substituted for iron(III). The dicupric bis(carbonate)lactoferrin complex has a characteristic visible maximum at 434 nm giving it a yellow colour. The EPR shows a typically axial Cu(II) spectrum with $g_{\parallel} = 2.314$, $g_{\perp} = 2.060$ and $A_{\parallel} = 150$ G (Figure 4). In dicupric lactoferrin, $Cu_2(CO_3)_2Lf$, the structural results show the carbonate binding is generally similar to that in $Fe_2(CO_3)_2Lf$. However, there are differences in detail between the N- and C-terminal sites and in the metal-carbonate interactions. In the C-terminal site the carbonate ion makes the same hydrogen bonds with the protein as in the C-lobe of the diferric protein. Again the $CO_3^{2-}$ anion coordination is bidentate. In contrast, in the N-terminal site, the carbonate is monodentate and the copper coordination is approximately square pyramidal (Figure 1b). There appears to be a resulting hydrogen bonding change and the form of the anion may be $HCO_3^-$ rather than $CO_3^{2-}$ with a hydrogen bond linking it to tyrosine 92.

## Copper Oxalate Lactoferrin Complexes

The reactions to form the dicupric lactoferrin complexes are summarized in Figure 5. Titration of apolactoferrin with $Cu^{2+}$, in carbonate-free conditions, in the presence of a 50-fold molar excess of oxalate, at pH 7.8, showed clearly that 2 moles of $Cu^{2+}$ are bound per mole of protein. The resulting yellow complex had an intense visible absorption maximum at 420 nm. The EPR spectrum (Figure 4) with $g_{\parallel} = 2.345$, $g_{\perp} = 2.070$, $A_{\parallel} = 142$ G, is similar to but not identical to that of $Cu_2(CO_3)_2Lf$. Titration of apolactoferrin with oxalate, in carbonate-free conditions, in the presence of two equivalents of $Cu^{2+}$, at pH 7.8, showed that 2 moles of $C_2O_4^{2-}$ are bound per mole of protein, giving the same complex and establishing the formulation as $Cu_2(C_2O_4)_2Lf$.

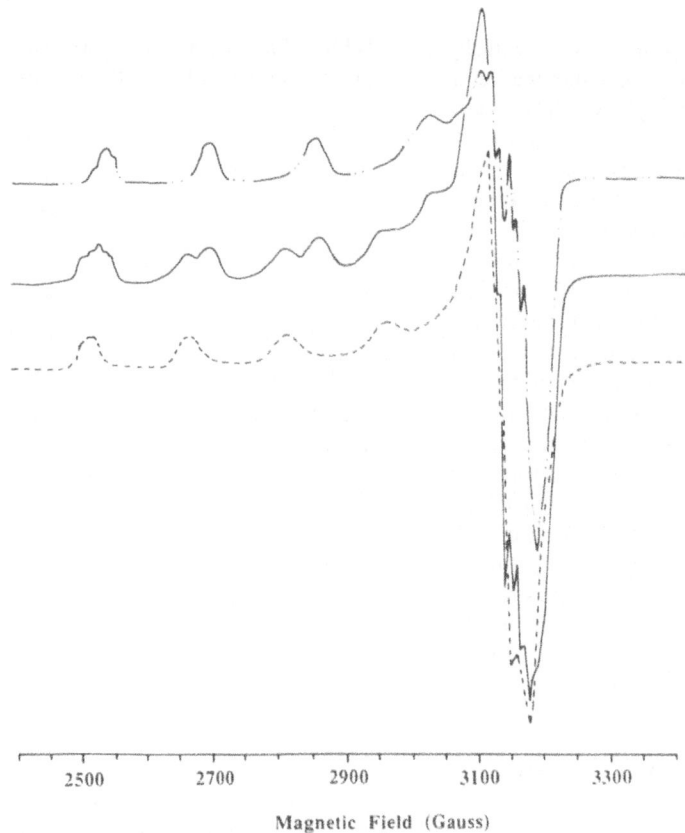

**Figure 4.** EPR spectra of carbonate and oxalate complexes of dicupric human lactoferrin. Spectra are $Cu_2(CO_3)_2Lf$ (— ·· —); $Cu_2(CO_3)(C_2O_4)Lf$ (—); $Cu_2(C_2O_4)_2Lf$ (- - -).

Addition of a 50-fold molar excess of oxalate to the carbonate complex, $Cu_2(CO_3)_2Lf$, caused the displacement of carbonate from one of the two specific sites, giving the hybrid species $Cu_2(CO_3)(C_2O_4)Lf$. It showed a shift in the visible absorption maximum to 424 nm and the EPR spectrum is easily recognisable as an overlap of the spectra of $Cu_2(CO_3)_2Lf$ and $Cu_2(C_2O_4)_2Lf$ (Figure 4), in which the superhyperfine structure on the low field line comprises two independent triplets superimposed to form a five-line resonance.

The mixed-anion complex could also be prepared by displacement of oxalate from $Cu_2(C_2O_4)_2Lf$. Thus a solution of $Cu_2(C_2O_4)_2Lf$, if exposed to atmospheric $CO_2$ for approxi-

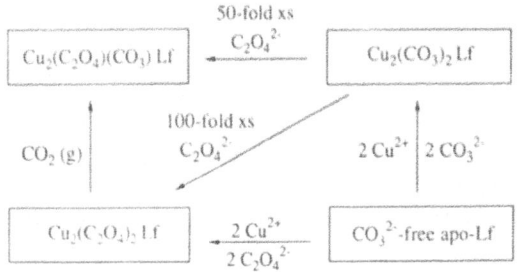

**Figure 5.** Reaction scheme for preparation and interconversion of carbonate, oxalate and hydrbid carbonate-oxalate complexes of dicupric lactoferrin.

mately 4 weeks, shows changes in both the electronic and EPR spectra characteristic of $Cu_2(CO_3)(C_2O_4)Lf$, e.g., $\lambda_{max}$ shifted from 420 to 424 nm. The hybrid complex is sufficiently stable under ambient conditions to be crystallized.

In the hybrid complex the oxalate ion occupies the C-terminal site where, as in $Fe_2(C_2O_4)_2Lf$, it is bound to the metal in a 1,2-bidentate mode (Figure 1c). Again the side chain of the arginine residue is displaced (by ~ 2 Å) away from the anion. In the N-terminal site of this hybrid complex the carbonate ion is bound in monodentate fashion exactly as in the N-terminal site of $Cu_2(CO_3)_2Lf$.

The reasons for the C-terminal site binding oxalate preferentially when $Cu^{2+}$ is the metal ion are not immediately obvious. There is certainly as much room for the oxalate in the N-lobe site as in the C-lobe. The most likely explanation is that it is the precise detail of the metal-anion-protein complex at each site which matters. The combination of $Fe^{3+}$ and $CO_3^{2-}$ is highly favourable at both sites both sterically and electronically and the anion is not easily substituted by oxalate. When $Cu^{2+}$ is the metal ion the metal-carbonate-protein interactions change. As noted earlier, in the N-lobe the $CO_3^{2-}$ ion is monodentate and may in fact be $HCO_3^-$ with a hydrogen bond linking it to tyrosine 92. Although the Tyr92 ligand is in an almost identical position to its counterpart in $Fe_2(CO_3)_2Lf$, the $Cu^{2+}$ has moved away causing an elongation of the Cu-O bond. In the C-lobe, the $CO_3^{2-}$ does not bind symmetrically to the $Cu^{2+}$ ion as it does to $Fe^{3+}$ in $Fe_2(CO_3)_2Lf$, but in an asymmetric bidentate fashion. These observations emphasize the unique character of the $Fe^{3+}$-$CO_3^{2-}$-lactoferrin interaction. The N-lobe site provides a more favourable geometry (square pyramidal) than the C-lobe and perhaps for this reason the C-lobe is more easily substituted by oxalate.

The oxalate binding preferences for lactoferrin mirror those for copper-substituted ovotransferrin, to which oxalate binds preferentially in the C-terminal site, and for which di-oxalate complexes are obtained using carbonate-free apo-ovotransferrin but mono-oxalate complexes when prepared in air (Zweier, 1980). In contrast, serum transferrin binds only one oxalate with $Cu^{2+}$, in the N-terminal site (Zweier & Aisen, 1977).

## Nonsynergistic Anion Binding to Iron Carbonate Bovine Lactoferrin

Synergistic anions, such as carbonate, oxalate and glycinate, are unique in that they are essential for specific metal binding to the lactoferrins. They bind in a special anion binding cavity and link the metal ion to the protein. However, other anions known as nonsynergistic anions, such as perchlorate and chloride, interact with serum transferrin and affect the structure and reactivity of the specific metal binding sites (Harris & Aisen, 1989). Little is known about the interaction of nonsynergistic anions with the lactoferrins. Anions such as citrate and succinate do not act as synergistic anions towards bovine lactoferrin. In experiments with these anions under carbonate free conditions, similar to those described for the preparation of $Fe_2(C_2O_4)_2bLf$, the development of a visible absorption band is not observed which implies the lack of specific site binding for iron. Exposure to ambient carbon dioxide causes the immediate development of the characteristic red colour as $Fe_2(CO_3)_2bLf$ forms. Citrate, however, perturbs the $Fe_2(CO_3)_2bLf$ complex. When it is added in a 100-fold molar excess, the $g' \sim 4.3$ EPR resonance is affected. The relative intensity of the low field doublet peaks change and as further citrate is added the doublet nature of the signal is lost (Figure 6). Despite this change in the EPR spectrum the electronic and CD spectra are virtually unchanged. The effect appears to be species specific as the EPR of human diferric bis(carbonate)lactoferrin is unchanged by similar concentrations of citrate. It also appears that citrate binds to bovine iron lactoferrin in the milk *in vivo*. For instance, during the isolation of bovine lactoferrin from colostrum, when the crude

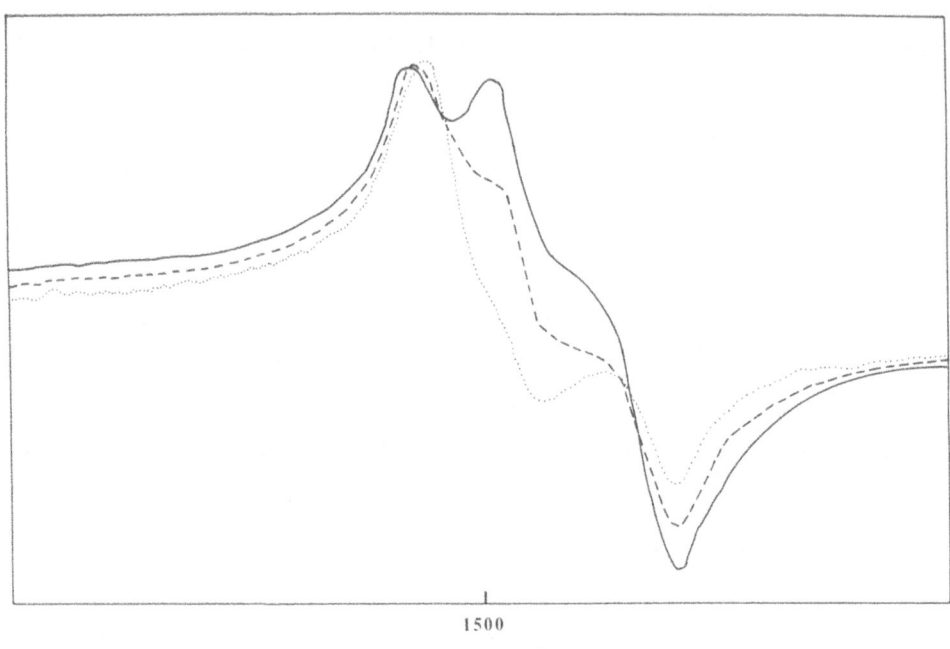

1500

Magnetic Field (Gauss)

**Figure 6.** The perturbation of the EPR spectrum of bovine diferric lactoferrin by citrate. Spectra are $Fe_2(CO_3)_2bLf$ (————); $Fe_2(CO_3)_2bLf$ after the first isolation step from bovine colostrum (- - - -); and $Fe_2(CO_3)_2bLf$ plus excess citrate (.....).

extract is loaded directly onto a CM-Sephadex column without prior dialysis, the EPR spectrum of the isolated iron bovine lactoferrin shows it to have citrate bound (Figure 6).

Preliminary results indicate other anions such as succinate also act in a nonsynergistic fashion but EPR evidence shows glycinate which also can be a synergistic anion towards bovine lactoferrin can also bind elsewhere. In these latter cases, the effect on the EPR spectrum is similar to that observed when ions such as chloride bind nonsynergistically to serum transferrin (Folajtar & Chasteen, 1982), viz., the $g' \sim 9$ signal intensity is enhanced and a broad shoulder appears on the low field side of the $g' \sim 4.3$ resonance. Nonsynergistic anion binding of anions, such as succinate and glycinate to lactoferrin and chloride and phosphate to transferrins, requires at least one site near to the specific metal binding site. Since the positive arginine 121 residue, which is linked to the synergistic carbonate ion by hydrogen bonds, has beyond it a cavity occupied by water molecules, it could provide one binding site for such secondary anions close enough to $Fe^{3+}$ to perturb its EPR spectrum. It is also possible that in the apoprotein, when $Fe^{3+}$ is not present in the specific site, nonsynergistic anions bind in the synergistic anion cavity and are displaced when specific metal binding occurs. Evidence for this comes from the x-ray structure of apolactoferrin which shows a chloride ion located in the C-lobe synergistic anion cavity (Baker et al., 1991) and the fact that it is now generally accepted (Harris & Stenback, 1988; Zweier et al., 1981; Kojima & Bates, 1981) that synergistic anion binding precedes specific metal binding to the transferrins.

The nonsynergistic binding of citrate to bovine lactoferrin appears to be unique and indicates specific interactions with amino acid residues near the iron binding site which are not possible in human lactoferrin. A possible site in the C-lobe for citrate binding could be near the "hinge" region behind the iron site. A change from Tyr415 in human lactoferrin to an arginine residue in $Fe_2(CO_3)_2bLf$ introduces a second positive charge near to the positive lysine 546 which is already present.

## CONCLUSIONS

Comparison of the synergistic anion binding in the lactoferrin complexes leads to a reappraisal of the classic Schlabach and Bates (1975) "interlocking sites" model for anion binding. The anion does indeed bridge between the metal ion and positively charged groups on the protein, as in that model. However, the common feature of all the synergistic anions identified by Schlabach and Bates is that all have both a carboxyl group and, on a proximal carbon atom, a potential electron donor ligand X. By analogy with the carbonate and oxalate sites described here, we suggest that the primary feature of anion binding to transferrins is the hydrogen bonding of the carboxyl group of the anion to the N-terminus of helix 5. The anion then coordinates the metal through X and one carboxylate oxygen (Figure 7).

Anions larger than $CO_3^{2-}$ force the arginine sidechain to move away (this is possible because of the solvent-filled interdomain cavity beyond). Whether or not the arginine hydrogen bonds to an anion, as for carbonate and oxalate, will depend on the nature of the substituents on the proximal carbon. This is probably a major factor in the strength of anion binding and the favourable interactions of oxalate with the arginine sidechain probably account for the fact that it is the only anion whose binding affinity approaches that of carbonate.

Whether or not metal or anion substituted lactoferrins other than iron carbonate lactoferrin have a physiological importance is uncertain at present although the facile binding of copper could mean lactoferrin has a role as a copper carrier. The fact that the overall protein structure is the same for $Fe_2(CO_3)_2Lf$ and $Cu_2(CO_3)_2Lf$, irrespective of differences in the metal-carbonate-protein interactions, implies copper could be transported and delivered to cells as well as iron. Studies on complexes other than the physiologically important iron carbonate lactoferrins do however point to site and species differences which in themselves may have a significance. For instance the behaviour of bovine diferric laetoferrin is more akin to that of human dicupric than human diferric lactoferrin when it comes to carbonate displacement by oxalate signifying that explanations for species differences must be sought beyond the immediate metal coordination environment. Furthermore the observation of a monodentate copper carbonate interaction in the copper lactoferrins could provide a pointer to metal ion release if protonation of the anion is an important step in this process.

Preliminary studies with nonsynergistic anions (e.g., citrate) also point to species differences and suggest that the physiological importance of nonsynergistic anion binding to lactoferrins should not be overlooked. Milk is not a uniform body fluid but a mammary gland

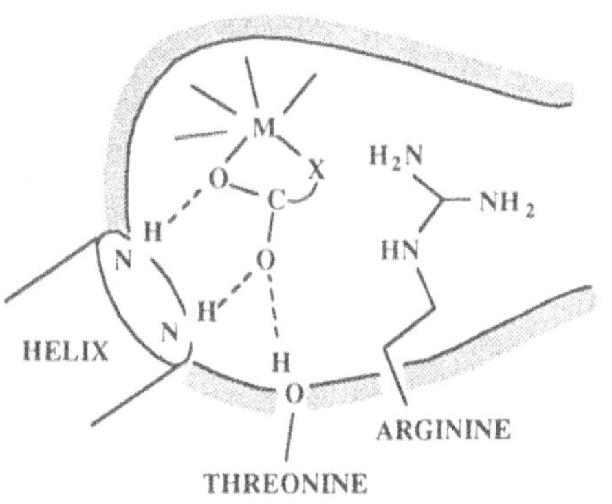

**Figure 7.** Generalised model for synergistic anion binding to transferrins.

secretory product which contains many anionic species. The unique behaviour of bovine lactoferrin towards citrate may be related to the higher citrate concentration in bovine milk (4–8 mM) than in human milk (2–3 mM). Citrate has been reported to inhibit the bacteriostic properties of bovine lactoferrin (Reiter, 1983) presumably by influencing iron uptake by the protein. Finally it is suggested that the nonsynergistic anion binding properties of the transferrins could be exploited to carry anionic drugs, e.g., antitumour ruthenium (III) agents (Kratz et al., 1992).

## REFERENCES

Ainscough, E.W., Brodie, A.M., & Plowman, J.E. (1979). Inorg. Chim. Acta 33:149–153.

Ainscough, E.W., Brodie, A.M., Plowman, J.E., Bloor, S.J., Loehr, J.S., & Loehr, T.M. (1980). Biochemistry 19:4072–4079.

Ainscough, E.W., Brodie, A.M., McLachlan, S.J., & Ritchie, V.S. (1983). J. Inorg. Biochem. 18:103–112.

Aisen, P., Pinkowitz, R.A., & Leibman, A. (1972). Ann. NY Acad. Sci. 222:337–346.

Anderson, B.F., Baker, H.M., Norris, G.E., Rice, D.W., & Baker, E.N. (1989). J. Mol. Biol. 209:711–734.

Baker, E.N., Anderson, B.F., Baker, H.M., Handas, M., Jameson, G.B., Norris, G.E., Rumball, S.V., & Smith, C.A. (1991). Int. J. Biol. Macromol. 13:122–129.

Baker, E.N., & Lindley, P.F. (1992). J. Inorg. Biochem., 47:147–160.

Bertini, I., Luchinat, C., Messori, L., Scozzafava, A., Pellacani, G., & Sola, M. (1986a). Inorg. Chem. 25:1782–1786.

Bertini, I., Luchinat, C., Messori, L., Monnanni, R., & Scozzafava, A. (1986b). J. Biol. Chem. 261:1139–1146.

Dubach, J., Gaffney, B.J., More, K., Eaton, G.R., & Eaton, S.S. (1991). Biophys. J. 59:1091–1100.

Eaton, S.S., Dubach, J., More, K.M., Eaton, G.R., Thurman, G., & Ambruso, D.R. (1989). J. Biol. Chem. 264:4776–4781.

Folajtar, D.A., & Chasteen, N.D. (1982). J. Am. Chem. Soc. 104:5775–5780.

Harris, D.C., & Aisen, P. (1989). in Physical Bioinorganic Chemistry (Loehr, T.M., Ed.) Vol. 5, pp 239–351, VCH Publishers, New York.

Harris, W.R., & Stenback, J.Z. (1988). J. Inorg. Biochem. 33, 211-223.

Kojima, N., & Bates, G.W. (1981). J. Biol. Chem. 256:12034–12039.

Kratz, F., Mulinacci, N., Messori, L., Bertini, I., & Keppler, B.K. (1992). Metal Ions in Biology and Medicine 2:69–74.

Norris, G.E., Anderson, B.F., Baker, E.N., Gärtner, A.L., Ward, J., & Rumball, S.V. (1986). J. Mol. Biol. 191:143–145.

Norris, G.E., Baker, H.M., & Baker, E.N. (1989). J. Mol. Biol. 209, 329-331.

Reiter, B. (1983). Int. J. Tiss. Reac. 5:87–96.

Rogers, T.B., Borresen, T., & Feeney, R.E. (1978). Biochemistry 17:1105–1109.

Schlabach, M.R., & Bates, G.W. (1975). J. Biol. Chem. 250:2182–2188.

Shongwe, M., Smith, C.A., Ainscough, E.W., Baker, H.M., Brodie, A.M., & Baker, E.N. (1992). Biochemistry 31:4451–4457.

Shriver, D.F. (1969). The Manipulation of Air Sensitive Compounds, McGraw-Hill, New York.

Sola, M. (1990). Eur. J. Biochem. 194:349–353.

Zweier, J.L. (1980). J. Biol. Chem. 255:2782–2789.

Zweier, J.L., & Aisen, P. (1977). J. Biol. Chem. 252:6090–6096.

Zweier, J.L., Wooten, J.B., & Cohen, J.S. (1981). Biochemistry 20:3505–3510.

# SALT EFFECTS ON THE PHYSICAL PROPERTIES OF THE TRANSFERRINS

N. Dennis Chasteen,[†] John K. Grady,[†] Robert C. Woodworth,[‡] and
Anne B. Mason[‡]

†Department of Chemistry
University of New Hampshire
Durham, New Hampshire 03824

‡Department of Biochemistry
University of Vermont
Burlington, Vermont 05405

## SUMMARY

Salts are known to have a pronounced effect on the spectroscopic, thermodynamic and kinetic properties of human serum transferrin. The present study was undertaken to examine the effect of NaCl on the related proteins ovotransferrin and lactoferrin. EPR difference spectroscopy was used to probe changes in the metal site of these proteins. Sodium chloride was found to perturb the $g' = 4.3$ EPR spectra of both ovotransferrin and lactoferrin but in different ways. The spectrum of ovotransferrin is reduced in amplitude with a broad feature appearing at $g' = 4.8$ whereas there is a loss of resolution of the doublet feature at the peak of the EPR derivative spectrum for lactoferrin. The increase in the amplitude of the ovotransferrin EPR difference spectrum (spectrum without NaCl minus spectrum with NaCl) as a function of NaCl concentration is suggestive of saturation binding. A Hill plot binding isotherm gave $n = 1.87 \pm 0.32$ and $\log K = 1.49 \pm 0.03$ for ovotransferrin, where n is the number of $Cl^-$ ions binding to either one or both iron containing lobes of the protein and K is the overall association constant. Preliminary measurements with lactoferrin gave $n = 1.95 \pm 0.34$ and $\log K = 1.41 \pm 0.06$. These results are similar to those previously reported for serum transferrin and suggest that $Cl^-$ binds to all the transferrins with strong pairwise cooperativity. This binding may reflect a functional role for chloride and other physiological anions in the uptake and release of iron by the transferrins.

## INTRODUCTION

Twenty years ago Price and Gibson [1] reported that the spectroscopic properties of the iron centers of serum transferrin and ovotransferrin are changed by the presence of $NaClO_4$. Sodium perchlorate at a concentration of 0.1–0.8 M caused a reduction in amplitude and resolution of the characteristic $g' = 4.3$ electron paramagnetic resonance (EPR) signal of

*Lactoferrin: Structure and Function*
Edited by T.W. Hutchens *et al.*, Plenum Press, New York, 1994

transferrin. These workers attributed this effect to a perchlorate-induced conformational change(s) in the protein. The $VO^{2+}$ EPR signal of vanadyl transferrin was subsequently shown to be affected by perchlorate also [2], demonstrating that the effect is not metal ion specific.

Further studies have shown that the thermodynamic properties of the protein are influenced also by sodium perchlorate and other salts [3]. At pH 7.5, the presence of salt increases the thermodynamic stability of iron(III) binding to the N-terminal site relative to the C-terminal site; this effect is enhanced as the pH is increased in the range pH 7– 8.4 [3, 4]. Similar salt effects on the kinetic stability of the two sites have been observed [4, 5]. Thus, increasing salt concentration increases the rate of iron removal from the C-terminal site but decreases the rate for the N-terminal site [4, 5]. The magnitude of the effect depends on the chelator, e.g., citrate, EDTA, NTA and pyrophosphate, and the salt, e.g., NaF, NaBr, NaI, $NaNO_3$, $Na_2SO_4$, and $NaClO_4$, employed. In contrast, studies of the rate of iron removal from N-terminal and C-terminal monoferric transferrins by the chelator 3,4-LICAMS (N, N′,N″-tris(5-sulfo-2,3-di-hydroxybenzoyl)-1,5,10-triazadecane) have shown that the rate of iron removal increases for both sites as a function of KCl concentration [6]. Most significant, however, is the fact that the rates extrapolate to zero at zero ionic strength of the medium, an observation strongly suggesting that a salt-induced conformational change is a prerequisite for iron release from the protein.

A number of mechanisms have been postulated to account for the observed salt effects on the kinetics of iron removal from transferrin [6 ,7 ,8]. All of the proposed mechanisms involve binding of nonsynergistic anions such as $Cl^-$ and $ClO_4^-$ to functionally important cationic sites on the protein. The presence of modifier or allosteric sites to which nonsynergistic anions bind was originally proposed by Chasteen and coworkers on the basis of EPR spectral measurements and kinetic data [1, 4 , 9–11].

Folajtar and Chasteen [10] carried out EPR spectrometric titrations of diferric transferrin and of the N- and C-terminal monoferric transferrins as a function of anion concentrations of the salts NaSCN, NaClO4, Na3HP2O7, Na3ATP, NaCl, Na2HPO4, Na2AMP, Na2SO4, NaF, and NaBF4. The change in the amplitude of the EPR difference spectrum as a function of added salt was interpreted by a model in which anion binding to the protein induces a conformational change which is reflected by the change in the $Fe^{3+}$ g′ = 4.3 EPR spectrum. Binding isotherms gave the number of anions bound, n (2 per lobe) and the overall apparent binding constant, K.

To date the locations of the nonsynergistic anion binding sites are unknown. We have therefore undertaken EPR spectrometric titrations with ovotransferrin and lactoferrin to determine whether these proteins also bind nonsynergistic anions in a fashion similar to human serum transferrin. This data in conjunction with the amino acid sequences for ovotransferrin [12], lactoferrin [13], and serum transferrin [14] and the x-ray structures of lactoferrin and transferrin [15, 16, 17] will be useful in beginning to understand the origin of the salt effect on the transferrins and in locating possible nonsynergistic anion binding sites on these proteins. We report here our findings for diferric ovotransferrin and some preliminary results for diferric lactoferrin.

## EXPERIMENTAL

Ovotransferrin was purchased from Sigma Chemical Company. Diferric ovotransferrin was prepared by the addition of 5 mM Fe(III):nitrilotriacetate, 1:2 metal/ligand, to the 0.25 mM apoprotein in 0.1 M Hepes/Na, 10 mM $NaHCO_3$ buffer, pH 7.5. After incubation overnight, the nitrilotriacetate was removed by dialysis against 0.1 M $NaClO_4$, 10 mM Hepes, pH 7.5 for 15 hr followed by successive dialyses against 10 mM Hepes to remove the $NaClO_4$. The protein was concentrated to 0.22 mM on an Amicon ultrafiltration cell with 0.1 M Hepes, 10 mM $NaHCO_3$ buffer, pH 5.

EPR spectrometric titrations were carried out at 77 K on a Varian E-4 X-band EPR spectrometer with the following instrument settings: power = 20 mW, 100 KHz peak-to-peak modulation amplitude = 10 G, microwave frequency = 9.371 GHz, time constant = 0.3 s, and scan rate = 2000 G/8 min. Measurements at 77 K were made using a quartz dewar inserted in the rectangular TE 102 cavity. The protein sample (400 μL) was loaded in 4 mm o.d., 3 mm i.d. quartz EPR sample tube and frozen in liquid nitrogen. Successive microliter additions of 5 M NaCl were made to the same protein sample by thawing the tube between EPR measurements. Difference spectra and double integrals were calculated on a PC Compatible computer using the software EPRWare from Scientific Software Services. The EPR signal amplitude was corrected for sample dilution from the added NaCl. For example the corrections were 2.9% and 10.9% for the 0.16 and 0.50 M NaCl samples, respectively.

## RESULTS

Figure 1 shows the EPR spectrum of diferric ovotransferrin in 0.1 M Hepes, 10 mM NaHCO$_3$ with and without 0.5 M NaCl. The amplitude of the spectrum is reduced by 36% in NaCl; in addition a broad hump at g' ≈ 4.8 appears to low field of the g' = 4.3 signal located at 1600 G. Thus it is evident that salt induces changes in the metal site coordination, presumably

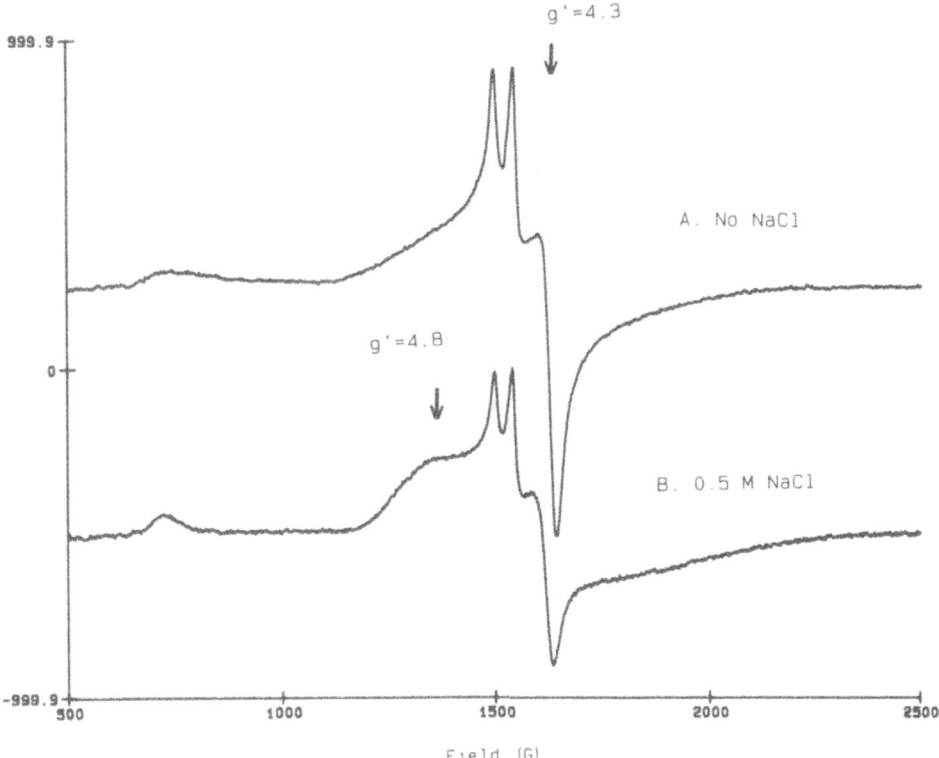

**Figure 1.** Frozen solution EPR Spectra of diferric ovotransferrin (A) without and (B) with 0.5 M NaCl. Conditions: 0.22 mM protein in 0.1 M Hepes/Na, 10 mM NaHCO$_3$, pH 7.5 with or without 0.5 M NaCl, 77 K. See Experimental for instruments settings.

because the conformational state of the protein has been altered. This effect was examined in further detail.

Figure 2 illustrates the reduction in the peak-to-trough amplitude of the $g' = 4.3$ signal as a function of added NaCl. The curve is biphasic. There is a relatively pronounced reduction in EPR signal amplitude up to a salt concentration of approximately 0.15 M (designated by an arrow in Fig. 2) followed by a more gradual reduction to 0.5 M NaCl where the experiment was terminated. We attribute the first phase to relatively strong binding of $Cl^-$ to specific modifier sites on the protein (*vide infra*) and the second phase to weaker binding at other sites. Double integration of the $g' = 4.3$ EPR signal demonstrated that the EPR spectrum accounted for essentially all of the $Fe^{3+}$ regardless of the concentration of NaCl; therefore the reduction in EPR amplitude was not due to the loss of $Fe^{3+}$ binding to the protein.

Figure 3 shows a series of EPR difference spectra obtained by subtracting the spectra in the presence of salt from that in the absence of salt. The amplitude (h) of the EPR difference spectrum as a function of NaCl concentration is indicative of saturation binding (Fig. 4). To model the data we assume the following simple equilibrium:

$$\text{Fe-Lobe} + n\text{Cl}^- \leftrightarrow \text{Fe-Lobe-Cl}_n \tag{1}$$

Here Fe-Lobe represents iron bound in either the N- or C-terminal lobe of the protein. Fe-Lobe exhibits the EPR spectrum of the native protein and Fe-Lobe-Cl$_n$ exhibits the perturbed spectrum in the presence of NaCl (Fig. 1). Here we do not concern ourselves with the small differences in EPR signal of the two iron binding sites. The equilibrium constant is given by

$$K = \frac{[\text{Fe} - \text{Lobe} - \text{Cl}_n]}{[\text{Fe} - \text{Lobe}][\text{Cl}^-]^n} = \frac{\theta}{(1 - \theta)[\text{Cl}^-]^n} \tag{2}$$

Where the degree of saturation of the binding sites $\theta$ is given by $\theta = h/h_{max}$. Here h is the height of the EPR difference spectrum and $h_{max}$ is the maximum height, i.e. the value at binding site saturation. Equation 2 in logarithmic form becomes:

$$-\log(\theta/1 - \theta) = -n\log[\text{Cl}^-] - \log K \tag{3}$$

Figure 5 illustrates the Hill plot for ovotransferrin. A straight line is observed (correlation coefficient = 0.95), as predicted by equation 3 from which values of $n = 1.87 \pm 0.32$ and $\log K = 1.49 \pm 0.07$ (standard errors) are obtained, corresponding to the binding of two $Cl^-$ ions to either one or both lobes of the protein. These values are in good agreement with the previously reported values of $n = 2.00 \pm 0.11$ and $\log K = 1.29 \pm 0.02$ for ovotransferrin [10].

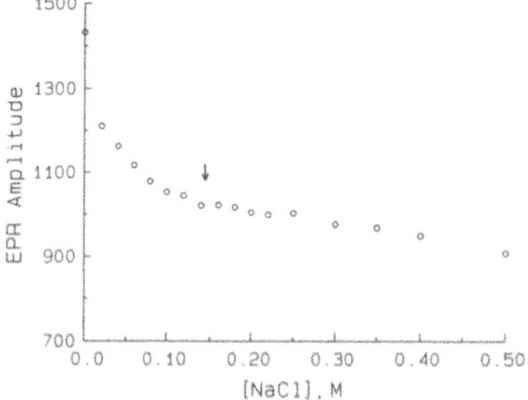

**Figure 2.** Amplitude of the $g' = 4.3$. EPR signal in Figure 1 as a function of NaCl concentration.

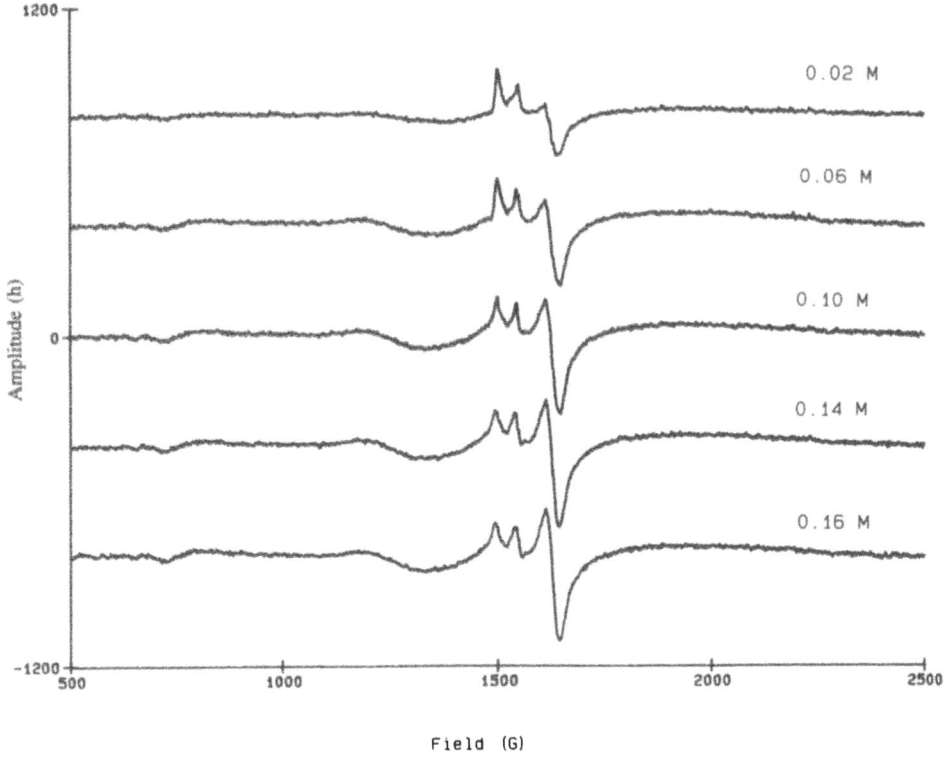

**Figure 3.** g′ = 4.3. EPR difference spectra of ovotransferrin at various NaCl concentrations. The spectra in the presence of NaCl were subtracted from the spectrum in the absence of NaCl.

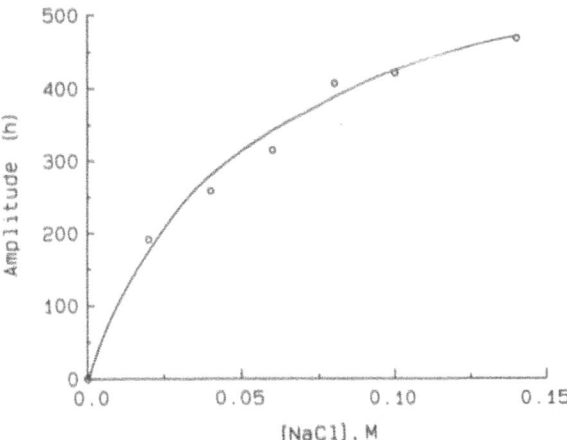

**Figure 4.** Amplitude (h) of the EPR difference spectrum versus NaCl concentration.

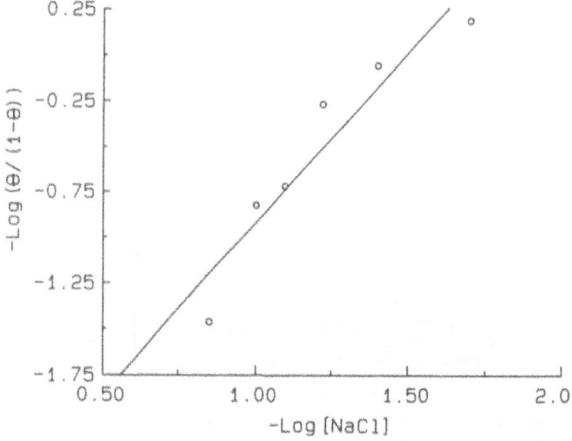

**Figure 5.** Hill plot of the data in Figure 4. The least-squares line is shown. The slope gives the value of n and the intercept of the abscissa gives log K. Note that the origin of the graph is not (0,0).

Preliminary measurements with lactoferrin showed that NaCl causes a loss in resolution of doublet feature at the peak of the $g' = 4.3$ EPR signal but no reduction in overall amplitude of the signal (data not shown). The difference spectra were used to construct a Hill plot in the concentration range 0–0.15 M NaCl from which values of n = 1.95 ± 0.34 and log K = 1.41 ± 0.06 similar to those of ovotransferrin were obtained.

## CONCLUSIONS

The Hill plots for ovotransferrin and lactoferrin have a slope of n = 2. This result suggests pairwise binding of $Cl^-$ in a highly cooperative fashion where $K_2 \gg K_1$, $K_1$ and $K_2$ being the binding constants for the first and second chlorides. Positive cooperativity in chloride binding has been observed for serum transferrin also [10]. Table 1 summarizes chloride binding data for diferric ovotransferrin, diferric lactoferrin, diferric transferrin, the C-terminal monoferric transferrin, the N-terminal monoferric transferrin and the recombinant N-terminal half-molecule of transferrin. In every instance n ≈ 2 cooperativity is observed (Table 1). In the case of transferrin it is known that two functional groups undergo proton dissociation at pH ≈ 7.5 with strong positive cooperativity [3], suggesting that proton dissociation/association and the cooperativity in chloride binding may be related in some way.

**Table 1.** Stoichiometries and Equilibrium Constants for Chloride Binding

| Protein | n[a] | log K[a] | Reference |
|---------|------|----------|-----------|
| Diferric ovotransferrin | 1.87 (0.17) | 2.79 (0.40) | This work |
| Diferric lactoferrin | 1.95 (0.34) | 2.75 (0.41) | This work |
| Diferric transferrin | 2.15 (0.07) | 1.97 (0.03) | 10 |
| C-terminal monoferric transferrin | 2.13 (0.13) | 2.11 (0.04) | 10 |
| N-terminal monoferric transferrin | 2.02 (0.20) | 2.75 (0.12) | 10 |
| Recombinant N-terminal half molecule transferrin[b] | 2.20 (0.32) | 1.41 (0.05) | 18 |

[a]Standard errors are given in parentheses from linear least squares fit to the Hill plot (equation 3).
[b]Reference 21.

The log K value of 1.49 for ovotransferrin corresponds to an overall association constant of $K = K_1 K_2 = 31$ or a root-mean-square average association constant per $Cl^-$ of $K^{1/2} = 5.6$ $M^{-1}$, a value typical of anion binding to cationic groups on proteins [4 & 10 and references therein]. Both diferric ovotransferrin (log K = 1.49) and diferric lactoferrin (log K = 1.41) bind chloride more weakly than diferric transferrin (log K = 1.97) but not appreciably so.

The data in Table 1 show that pairwise cooperativity of $Cl^-$ binding occurs in both N- and C-terminal lobes of transferrin; $Cl^-$ binding is strongest in the N-lobe, log K = 2.75 versus log K = 2.11 for the C-lobe. In the experiments reported here with ovotransferrin and lactoferrin, only the diferric proteins were used. Therefore it is unclear whether only one or both lobes of these proteins are involved in the binding of chloride with strong pairwise positive cooperativity. Computer models show that pairwise binding in either one or both lobes will give n = 2 Hill plots [10].

Comparison of the sequences of lactoferrin, ovotransferrin and serum transferrin reveals a number of cationic groups which are conserved among both domains of all three proteins [13]. Baker and coworkers [16a] have examined the structure of lactoferrin and the sequences of the various transferrins with regard to conserved basic residues located in the interdomain cleft where the iron binds. They have identified a number of residues which are either conserved among all of the transferrins or replaced by another basic residue when mutated. They suggest that Arg-121 (Arg-465 in the C-lobe) and Arg-210 (Lys-546 in the C-lobe) (lactoferrin numbering) are likely candidates for nonsynergistic anion binding, both being within 6.5 Å of the iron. Since basic residues are found at these positions in both lobes of all three protein (e.g., these residues are Arg-124 and Lys-206 in the C-lobe of serum transferrin), additional $Cl^-$ binding studies with N- and C-terminal monoferric ovotransferrin and lactoferrin should aid in establishing whether these residues are the ones involved in nonsynergistic anion binding. It should be noted, however, that while chloride binds in both lobes of human transferrin, it binds the strongest in the N-lobe. In contrast, only minor spectral changes from perchlorate are seen in the N-lobe, the perchlorate effect being very pronounced on the EPR spectrum of the C-lobe [11, 19, 20]. Thus it is possible that the binding sites and origin of the salt effects may not be the same for $Cl^-$ and $ClO_4^-$.

Modification of either lysines or arginines on diferric transferrin renders the protein unresponsive to perchlorate [19]; however, lysine modification causes spectral changes in the C-lobe which are essentially identical to those seen by addition of perchlorate to the native protein [11]. In addition, lysine modification accelerates iron loss from the C-lobe but stabilizes iron in the N-lobe; a similar phenomenon is observed by addition of perchlorate to the native protein [11]. Thus, it is evident that the molecular origin of the perchlorate effect is complex, being different in the N- and C-lobes, and furthermore may involve both lysine and arginine residues.

## ACKNOWLEDGMENTS

Supported by grants GM 20194 and DK 21739 of the National Institutes of Health.

## REFERENCES

1. Price, E.M. and Gibson, J.F. (1972) Electron paramagnetic resonance evidence for a distinction between the two iron-binding sites in transferrin and in conalbumin. J. Biol. Chem. 247:8031–8035.

2. Cannon, J.C. and Chasteen, N.D. (1975) Nonequivalence of the metal binding sites in vanadyl-labeled human serum transferrin. Biochemistry 14:4573–4577.

3. Chasteen, N.D. and Williams, J. (1981) The influence of pH or the equilibrium distribution of iron between the metal-binding sites of human transferrin. Biochem. J. 193:717–727.

4. Williams, J., Chasteen, N.D., and Morton, K. (1982) The effect of salt concentration on the iron binding properties of human transferrin. Biochem. J. 201: 527–532.

5. Baldwin, D.A. and de Sousa, D.M.R. (1981) The effect of salts on the kinetics of iron release from N-terminal and C- terminal monoferric transferrins. Biochem. Biophys. Res. Commun. 99:1101–1107.

6. Kretchmar, S.A. and Raymond, K.N. (1988) Effects of ionic strength on iron removal from the monoferric transferrins. Inorg Chem. 27:1436–1441.

7. Bertini, I., Hirose, J., Luchinat, C., Messori, L., Piccioli, M. and Scozzafava, A. (1988) Kinetic studies of metal removal from transferrins by pyrophosphate. Investigation of iron(III) and manganese(III) derivatives. Inorg. Chem. 27:2405-2409.

8. Harris, W.R. and Bali, P.K. (1988) Effects of anions on the removal of iron from transferrin by phosphoric acids and pyrophosphate. Inorg. Chem. 27:2687–2691.

9. Chasteen, N.D. (1983) Transferrin: A perspective. in Advances in Inorganic Biochemistry (Theil, E.C., Eichhorn, G.L., and Marzilli, L.G., eds.), Elsevier, New York, Vol. 5: 201-232.

10. Folajtar, D.A. and Chasteen, N.D. (1982) Measurement of nonsynergistic anion binding to transferrin by EPR difference spectroscopy. J. Am. Chem. Soc. 104:5775–5780.

11. Thompson, C.P., McCarty, B.M., and Chasteen, N.D. (1986) The effect of salts and amino group modification on the iron binding domains of transferrin. Biochem. Biophys. Acta. 870:530–537.

12. (a) Jeltsch, J.-M. and Chambon, P. (1982) The complete nucleotide sequence of chicken ovotransferrin mRNA. Eur. J. Biochem. 122: 291-295; (b) Williams, J., Elleman, T.C., Kingston, I.B., Wilkins, A.G., and Kuhn, K.A. (1982) The primary sequence of hen ovotransferrin. Eur. J. Biochem. 122:297–303.

13. Metz-Boutigue, M.-H., Jolles, Mazurier, J., Schoentgen, F., Legrand, D., Spik, G., Montreuil, J., and Jolles, P. (1984) Human lactotransferrin: Amino acid sequence and structural comparisons with other transferrins. Eur. J. Bioochem. 145:659–676.

14. MacGillivray, R.T.A., Mendez, E., Sinha, S.K., Sutton, M. R., Lineback-Zins, J. and Brew, K. (1982) The complete amino acid sequence of human serum transferrin. Proc. Natl. Acad. Sci. USA 79:2504–2508.

15. Anderson, B.F., Baker, H.M., Dodson, E.J., Norris, G.E., Rumball, S.V., Waters, J.M. (1987) Structure of lactoferrin at 3.2 Å. Proc. Natl. Acad. Sci. USA 84:1769–1773.

16. (a) Anderson, B.F., Baker, H.M., Norris, G.E., Rice, D. W., and Baker, E.N. (1989) Structure of human lactoferrin: crystallographic structure analysis at 2.8 Å resolution. J. Mol. Biol. 209:711–734; (b) Shongwe, M.S., Smith, C.A., Ainscough, E.W., Baker, H.M.,. Brodie, A.M., and Baker, E.N. (1992) Anion binding by human lactoferrin: results from crystallographic and physicochemical studies. Biochemistry 31:4451–4458.

17. (a) Bailey, S., Evans, R.W., Garratt, R.C., Gorinsky, B., Hasnain, S., Horsburgh, C., Jhoti, H., Lindeley, P.F., Mydin, A., Sarra, R., and Watson, J.L. (1988) Molecular structure of serum transferrin at 3.3 Å. Biochemistry 27:5804–5812; (b) Sarra R., Garratt, R., Gorinsky, B., Jhoti, H. and Lindley, P. (1990) High-resolution x-ray studies of rabbit serum transferrin: preliminary structure analysis of the N-terminal half-molecule at 2.3 Å resolution. Acta Cryst. B46:763–771.

18. Chasteen, N.D., Grady, J.K., Woodworth, R.C. and Mason, A., unpublished data.

19. Chasteen, N.D., Thompson, C.P. and Rines, J.P. (1983) Spectroscopic and chemical modification studies of anion-transferrin interactions. In Structure and function of iron storage and transport proteins (Urushizaki, I., Aisen, P., Listowsky, I. and Drysdale, J.W., eds.) Elsevier, New York, pp 241–246.

20. Baldwin, D.A., de Sousa, D.M.R. and Ford, G. (1982) The effects of salts and detergents on the relative lability and structure of the N-terminal and C-terminal monoferric transferrins. In The Biochemistry and Physiology of Iron (Saltman, P. and Hegenauer, J. eds.), Elsevier Biomedical, Amsterdam, pp 57–65.

21. Woodworth, R.C., Mason, A.B., Funk, W.D., and McGillivray R.T.A. (1991) Expression and initial characterization of five site-directed mutants of the N-terminal half-molecule of human transferrin. Biochemistry 30, 10824–10829.

# INTERACTION OF LACTOFERRIN WITH SEQUESTERED TRANSITION METAL IONS

Tai-Tung Yip and T. William Hutchens

Department of Food Science and Technology
University of California
Davis, California

## INTRODUCTION

All living organisms require several different transition metal ions to develop, grow, and function as designed. Since many of these metal ions, like copper, zinc and iron are so reactive that the biochemical reactions dependent on these essential elements typically involve the formation of specific protein-metal ion complexes or chelates. Indeed, the absorption, transport, cellular accumulation, and steady-state molecular distribution of these transition metal ions within and between intracellular compartments would appear to be determined, in part, by a series of directional metal ion transfer events between macromolecules.

To understand how specific proteins impact on metal ion bioavailability, we must understand the structural determinants of selective metal ion sequestration and directional transfer by these proteins. Well-defined model systems and improved analytical capabilities are required. Toward this end, we have formulated a model system in detail (1) and recently developed new technologies for the identification and investigation of sequence-specific protein- and peptide-metal ion interactions (2–8). We have used this information to prepare model bioactive protein surface domains by immobilizing synthetic peptides representing authentic protein surface metal-binding sites (1, 9). The purpose of this investigation was to use this model system to study the interaction of human milk lactoferrin with surface-bound, more biochemically-relevant form of, transition metal ions, subsequent biospecific metal ion transfer, and explore the possibililty of lactoferrin playing a physiological role in transition metal ion sequestration and transport.

## RESULTS AND DISCUSSION

### I. Interaction of Lactoferrin with Immobilized Iminodiacetate-Cu(II)

Iminodiacetate (IDA) has a structure similar to half of an EDTA molecule. IDA chelates metal ions effectively with two carboxylate groups and an imino group, all of which are very common ligands in protein system for forming co-ordinate bonds with transition metal ions. Therefore the immobilized IDA system serves as a good starting point to test the ability of lactoferrin to interact with sequestered metal ion.

*Lactoferrin: Structure and Function*
Edited by T.W. Hutchens *et al.*, Plenum Press, New York, 1994

Human lactoferrin bound strongly to immobilized IDA-Cu(II) at neutral pH, and we have studied the affinity and mechanism of interaction quite extensively (1, 10). These past findings are summarized as follows: a) lactoferrin interacted with immobilized copper via its surface histidine residues rather than by the two specific metal-binding sites, b) occupancy of the specific metal-binding sites by different metal ions resulted in protein surface structural changes that can be detected by the immobilized copper ion, c) there was a ready transfer of copper ion from immobilized IDA-Cu(II) to apo-lactoferrin.

## II. Interaction of Lactoferrin with Immobilized IDA-Zn(II)

Lactoferrin bound to immobilized IDA-Zn(II) marginally at neutral pH; prolonged washing with the same pH 7.4 buffer would disrupt this interaction (Figure 1). The occupancy of specific metal-binding sites of lactoferrin with iron or copper did not decrease the affinity of binding. However, in contrast to immobilized IDA-Cu(II), the immobilized IDA-Zn(II) could not differentiate the subtle structural changes caused by metal ion occupancy as described above (Figure 1).

Furthermore, there was no detectable transfer of zinc from the immobilized IDA-Zn(II) to the apo-lactoferrin under the present experimental conditions.

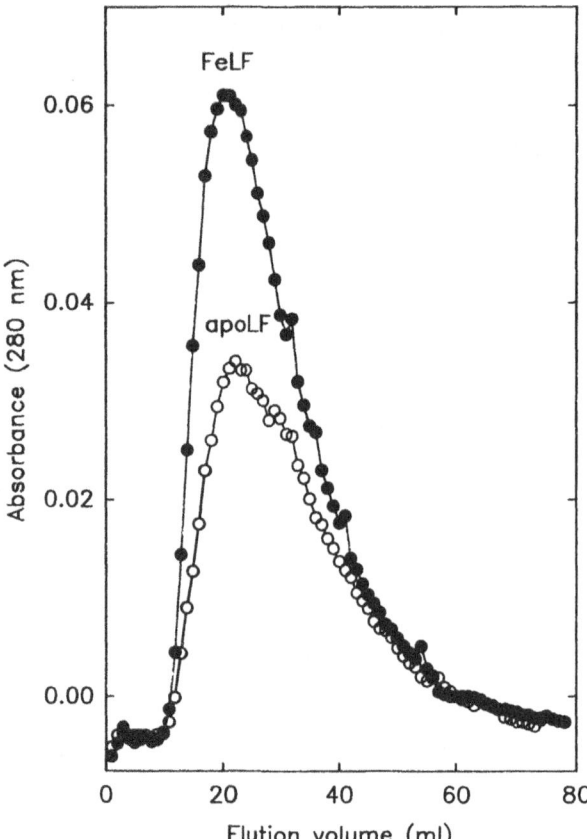

**Figure 1.** Interaction of apo- and Fe(III)-lactoferrin with immobilized IDA-Zn(II). Apo- or Fe(III)-saturated lactoferrin was applied to a Chelating Sepharose (1 × 5 cm, Fast Flow, Pharmacia) loaded with Zn(II) and equilibrated in 20 mM sodium phosphate 0.5 M NaCl, pH 7.4. Elution was isocratic with the same buffer.

## III. Interaction of Lactoferrin with Immobilized IDA-Fe(III)

Since lactoferrin is a specific iron-binding protein, it is not surprsing to find it bound to immobilized IDA-Fe(III); only a change in pH could disrupt such interaction (Figure 2). However, the binding was not as strong as some other milk proteins, such as casein, which required the stripping of metal ions from the IDA by EDTA to effect elution (8). Furthermore, the iron-saturated form of lactoferrin could also bind, although with slightly less affinity (Figure 2). These findings again indicate that the specific metal-binding sites of lactoferrin were not involved in the interaction with immobilized IDA-Fe(III). There was iron transfer from the immobilized IDA-Fe(III) to apo-lactoferrin, since the recovered protein from the IDA-Fe(III) column was found to have the characteristic absorption peak at 465 nm.

The above studies using simple nonbiological chemically-defined chelators to immobilize transition metal ions demonstrated the ability of lactoferrin to interact with sequestered metal ions. Next we extend the model system to include bioactive protein surface domains as the effective chelating ligand to study biospecific metal ion transfer.

## IV. Interaction of Lactoferrin with Immobilized (GHHPH)$_5$G-Zn(II).

Histidine-rich glycoprotein is a major metal ion-binding protein originally found in serum and plateletes (11, 12). It can bind up to 13 moles of Cu(II) or Zn(II) with dissociation constants in the low micromolar range (13). Recently, we have found this protein in both human and

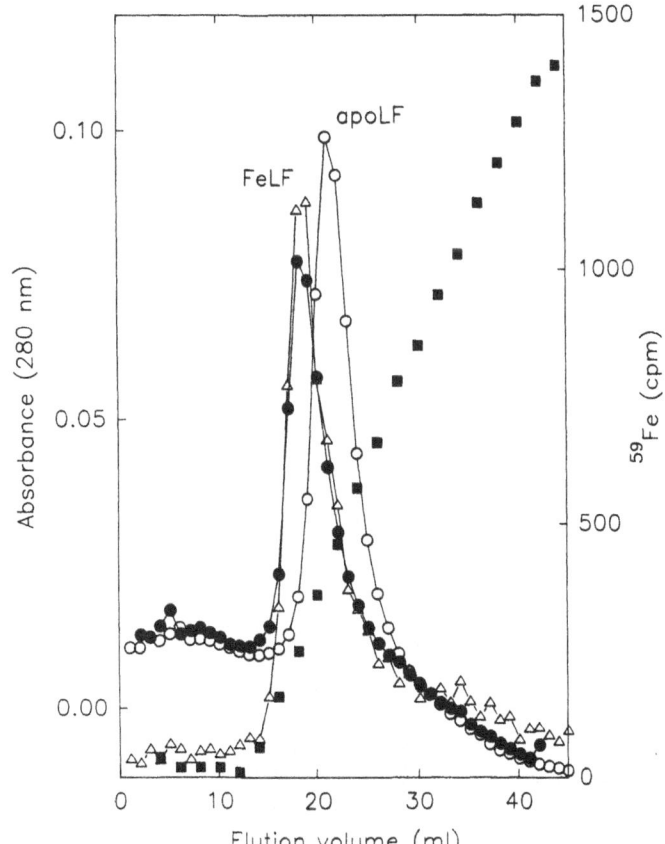

**Figure 2.** Interaction of apo- and Fe(III)-lactoferrin with immobilized IDA-Fe(III). Apo- or $^{59}$Fe(III)-lactoferrin was applied to a TSK Chelate 5PW-Fe(III) column (10 μm bead, 7.5 × 750 mm, Toso-Haas) equilibrated in 50 mM MES, 1 M NaCl, pH 6.0. Elution was by a pH gradient from 6.0 to 7.5 (50 mM PIPES, 1M NaCl, pH 7.5). Open triangle represents $^{59}$Fe in cpm, filled square represents pH.

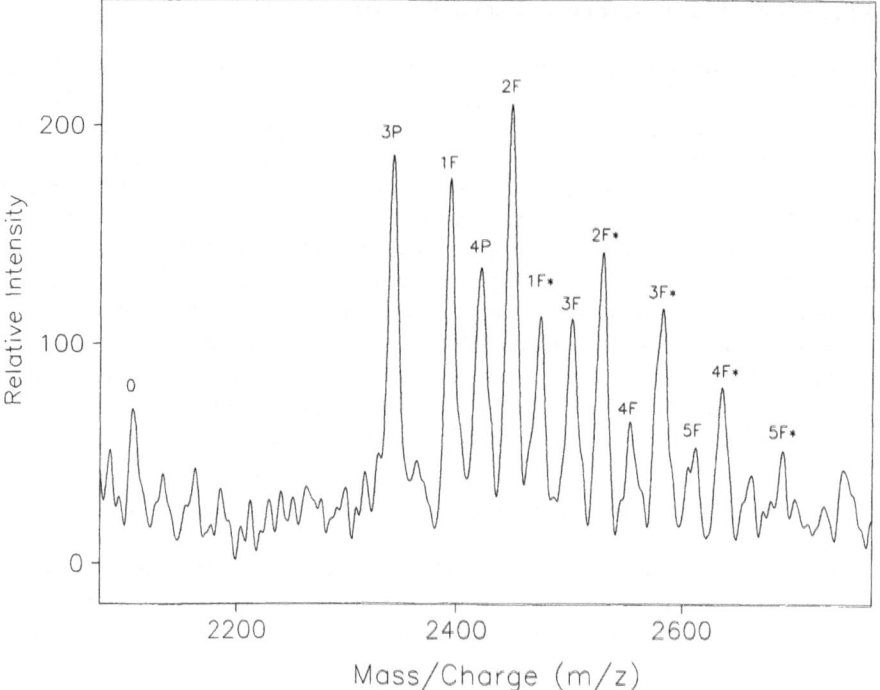

**Figure 3.** Matrix-assisted UV laser desorption time-of-flight mass spectrum of casein phosphopeptide after *in situ* iron binding. The synthetic casein phoshpopeptide (R1-K18) was purified by an ion-exchange column to contain mainly the tri- and tetra-phosphorylated forms (3P and 4P). Addition of Fe(II) showed the binding of multiple metal ions (1F-5F for the 3P peptide, 1F*-5F* for the 4P peptide).

bovine milk (14), and it should be expected to play a significant role in copper and zinc transport. By developing such enabling technology as matrix-assisted UV laser desorption/ionization time-of-flight mass spectrometry (MALDI TOF MS), we have characterized the metal ion-binding property of a synthetic peptide $(GHHPH)_5G$ representing the histidine-rich glycoprotein surface metal ion-binding domain (G365–H389 in the primary sequence) (2, 3, 5). The peptide immobilized on a surface has also been demonstrated to bind metal ions and interact with other proteins in a metal ion-dependent manner (1, 9).

When apo-lactoferrin was incubated with immobilized $(GHHPH)_5G$-$^{65}Zn(II)$ in pH 7.4 buffer at 37°C for 24 hr, 3.1% of the Zn(II) bound by the immobilized histidine-rich glycoprotein surface domain was specifically transferred to the lactoferrin. This seemingly small amount of metal ion transfer could become physiologically significant considering the relatively large quantity of lactoferrin existing in the milk. Indeed, the role of lactoferrin in Zn(II) ion transport has been suggested by others (15, 16).

### V. Interaction of Lactoferrin with Immobilized Human Casein-Fe(II) and Immobilized Casein Phosphopeptide-Fe(II)

Casein has long been known to play an important role in regulating iron bioavailability in the milk (17–20). We have also found that casein is the highest affinity binding milk protein to immobilized IDA-Fe(III) (8). By MALDI TOF MS, we have demonstrated that the phosphorylated form of the N-terminal R1-K18 peptide in the whole tryptic digest of human β casein was responsible for the iron-binding, the nonphosphorylated form of the same peptide did not bind any appreciable amount (1, 7, 8). The synthetic multiple phosphorylated form of R1-K18 casein peptide was found to show similar specific Fe(II)-binding properties. Figure 3 shows the mass spectrum of a R1-K18 synthetic casein peptide with no phosphate group (peak 0, molecular mass 2088.4 Da), with three phosphate groups (peak 3P, molecular mass 2088.4 + 3x79.98 Da), and with four phosphate groups (peak 4P, molecular mass 2088.4 + 4x79.98 Da), after *in situ* Fe(II)-binding study. The nonphosphorylated peptide had insignificant degree of metal-binding, whereas the triphosphorylated peptide (3P) could bind up to five Fe(II) (1F-5F), and the tetraphosphorylated peptide (4P) could bind at least five Fe(II) (1F*-5F*).

When apo-lactoferrin was incubated with immobilized [59]Fe(II)-casein proteins or immobilized [59]Fe(II)-casein R1-K18 phosphopeptides at pH 7.0 at 37°C for 18 hr, 13.3% of the casein protein-bound iron and 11.4% of the casein phosphopeptide-bound iron were transferred to the apo-lactoferrin. However, this represented less than 3% of iron-saturation for the amount of lactoferrin added. When [59]Fe(III)-saturated lactoferrin were incubated with the metal ion-free forms of immobilized casein proteins or immobilized casein R1-K18 phosphopeptides under the same conditions, 17.7% of the lactoferrin-bound iron was transferred to the immobilized casein protein and 23.9% to the immobilized casein phosphopeptide. This represented about 20% iron saturation for the casein protein or casein phosphopeptide immobilized. Therefore there seems to be much more ready iron transfer from lactoferrin to casein protein or casein phosphopeptide than the reverse. This biospecific transfer of iron from lactoferrin to casein is both surprising and significant, considering the avidity with which iron

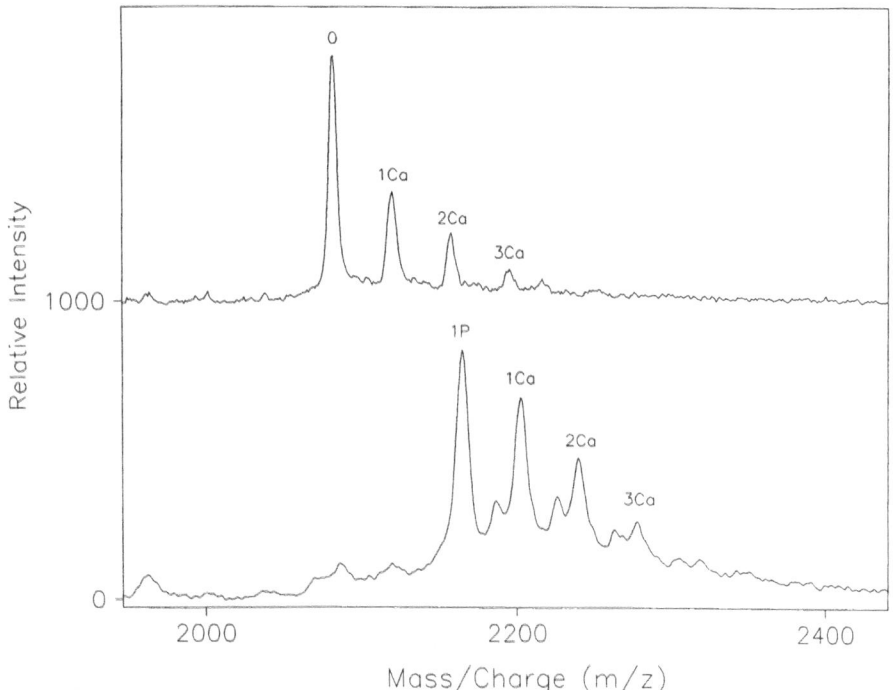

**Figure 4.** Matrix-assisted UV laser desorption time-of-flight mass spectra of casein peptides to show calcium binding. The top panel shows the R1-K18 casein peptide (0) with various numbers of bound Ca(II) (1–3 Ca). The lower panel shows the monophosphorylated R1-K18 casein peptide (1P) with various numbers of bound Ca(II) (1–3 Ca).

is bound to lactoferrin at neutral pH. This phenomenon is by no means an artefact since such redistribution of extrinsically added iron to lactoferrin in the milk has been observed to occur (19, 20).

Calcium has been known to interfere with iron absorption (21), the exact mechanism behind this is unknown. By MALDI TOF MS, we have found that the casein R1-K18 peptide could bind Ca(II) irrespective of phosphorylation (Figure 4), therefore the interactions between lactoferrrin and casein protein or casein phosphopeptide were also studied in the presence of Ca(II). The transfer of $^{59}$Fe(II) from immobilized casein protein to apo-lactoferrin was slightly decreased to 11.7% (cf. 13.3%), and transfer from immobilized casein phosphopeptide to apo-lactoferrin decreased to 9.6% (cf. 11.4%) in the presence of 5 mM CaCl$_2$. However, the reverse transfer of iron, i.e. from $^{59}$Fe(III)-lactoferrin to immobilized casein protein or immobilized casein phosphopeptide, was much more affected by the presence of Ca(II), it was reduced to 1.7% (cf. 17.7%) or 8.6% (cf. 23.9%) respectively. Ca(II) may affect the interaction between the basic lactoferrin protein and the acidic casein protein or casein phosphopeptide, or it may also affect the intrinsic iron-binding property of the casein phosphopeptide. This is currently under investigation.

This transfer of metal ions between lactoferrin and casein protein or casein phosphopeptide was studied in parallel entirely in solution without immobilization of any one component. As shown in Figure 5, when apo-lactoferrin was incubated with $^{59}$Fe(II)-casein phosphopeptide and then subsequently separated on a Superose 12 (Pharmacia) column, there was a time- and temperature-dependent transfer of $^{59}$Fe(II) from the casein phosphopeptide to the lactoferrin,

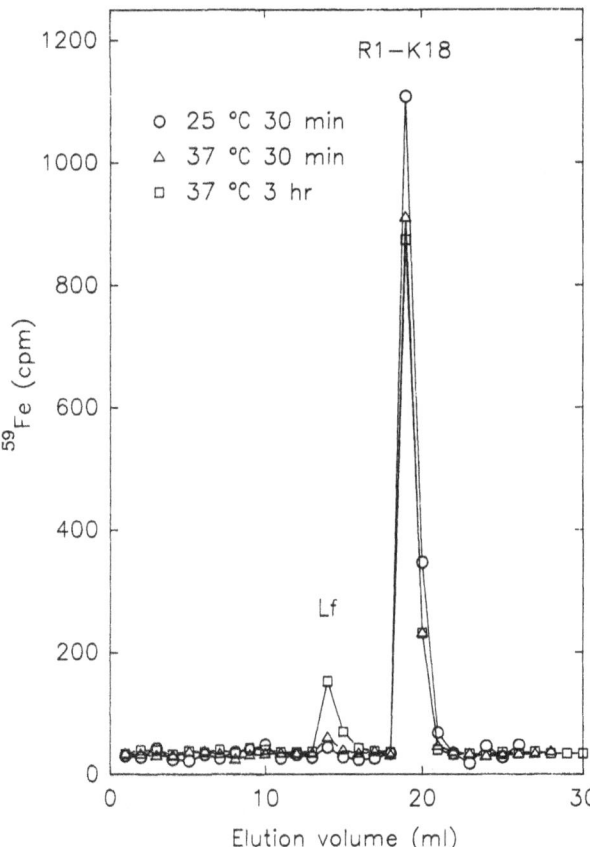

**Figure 5.** Iron transfer from $^{59}$Fe(II)-loaded casein phosphopeptide to apo-lactoferrin was incubated with the $^{59}$Fe(II)-loaded casein R1-K18 phosphopeptide at pH 7.0 for various periods of time either at room temperature or at 37°C. A Superose 12 high performance size exclusion column (10 μm bead size, 1 × 30 cm, Pharmacia) was used to separate the $^{59}$Fe(II) bound to lactoferrin from the $^{59}$Fe(II) bound to casein phosphopeptide.

the percentages of transfer closely agreed with those obtained with the immobilized system. This validates the model system of designing and building artificial biomimetic surfaces with bioactive protein surface domains to study protein-protein interaction, its modulation by metal ions, and biospecific metal ion transfer.

## CONCLUSION

A simple model system was used to investigate the interaction of lactoferrin with sequestered trnasition metal ions. Initially, simple nonbiological compounds like iminodiaetate was used to chelate the metal ions. It was shown that lactoferrin bound to the chelated metal ions mainly via the surface histidine residues rather than by the specific metal ion-binding sites. However, there was some transfer of the chelated metal ions to these specific sites when they were initially unoccupied. This transfer was further investigated with model protein surface metal ion-binding domains. Two examples of well known metal ion-binding proteins in the milk were selected: histidine-rich glycoprotein and casein. In each case, the protein surface exposed metal ion-binding domain was targeted by new analytical technologies, chemically synthesized and immobilized. After demonstration of metal ion-binding by such immobilized protein surface domains, biospecific metal ion transfers to and from lactoferrin were studied. By such initially simple model system, which serves as a foundation to build on, the interaction between lactoferrin and sequestered metal ion and the mechanism of biospecific metal ion transfer can be studied.

## REFERENCES

1. Hutchens TW, Yip TT. Model protein surface domains for the investigation of metal ion-dependent macromolecular interactions and biospecific metal ion transfer. Methods: A Companion to Methods in Enzymology, 1992, 4: 79–96.

2. Hutchens TW, Nelson RW, Allen MH, Li CM, Yip TT. Peptide-metal ion interactions in solution : Detection by laser desorption time-of-flight mass spectrometry and electrospray ionization mass spectrometry. Biol. Mass Spectrom. 1992, 21: 151–159.

3. Hutchens TW, Nelson RW, Yip TT. Recognition of transition metal ions by peptides: Identification of metal-binding peptides in proteolytic digest maps by UV laser desorption time-of-flight mass spectrometry. FEBS Letters 1992, 296: 99–102.

4. Hutchens TW, Allen MH, Li CM, Yip TT. Occupancy of a C2-C2 type "zinc-finger" domain by copper: Direct observation by electrospray ionization mass spectrometry. FEBS Letters 1992, 309: 170–174.

5. Hutchens TW, Nelson RW, Li CM, Yip TT. Synthetic metal-binding protein surface domains for metal ion-dependent interaction chromatography. I. Analysis of bound metal ions by matrix-assisted UV laser desorption time-of-flight mass spectrometry. J. Chromatogr. 1992, 604: 125–132.

6. Hutchens TW, Yip TT, Nelson RW. Identification of conserved protein surface metal-binding sites in related proteins by mass spectrometry. Techniques in Protein Chemistry IV, Angeletti RH, ed, Academic Press, 1993, pp. 33–40.

7. Yip TT, Hutchens TW. Mapping and sequence-specific identification of phosphopeptides in unfractionated protein digest mixtures by matrix-assisted UV laser desorption/ionization mass spectrometry. FEBS Letters 1992, 308: 149–153.

8. Yip TT, Hutchens TW. Protein phosphorylation: Sequence-specific identification of in vivo phosphorylation sites by MALDI-TOF mass spectrometry. Techniques in Protein Chemistry IV, Angeletti RH, ed, Academic Press, 1993, pp. 201–210.

9. Hutchens TW, Yip TT. Synthetic metal-binding protein surface domains for metal ion-dependent interaction chromatography. II. Immobilization of synthetic metal-binding peptides from metal ion transport proteins as a model bioactive protein surface domain. J. Chromatogr. 1992, 604: 133–141.

10. Hutchens TW, Yip TT. Metal ligand-induced alterations in the surface structures of lactoferrin and transferrin probed by interaction with immobilized copper(II) ions. J. Chromatogr. 1991, 536: 1–15.

11. Morgan WT. Human serum histidine-rich glycoprotein I. Interactions with heme, metal ions and organic ligands. Biochim. Biophys. Acta 1978, 533: 319–333.

12. Leung LLK, Harpel PC, Nachman RL, Rabellino EM. Histidine-rich glycoprotein is present in human platelets and is released following thrombin stimulation. Blood 1983, 62: 1016–1021.

13. Morgan WT. Interactions of the histidine-rich glycoprotein of serum with metals. Biochemistry 1981, 20: 1054–1061.

14. Hutchens TW, Yip TT. Identification of histidine-rich glycoprotein in human colostrum and milk. Pediatr. Res. 1992, 31: 239–246.

15. Blakeborough P, Salter DN, Gurr MJ. Zinc binding in cow's milk and human milk. Biochem. J. 1983, 209: 505–512.

16. Ainscough EW, Brodie AM, Plowman JE. Zinc transport by lactoferrin in human milk. Am. J. Clin. Nutr. 1980, 33: 1314–1315.

17. Lonnerdal B. Iron and breast milk. In: Stekel A, ed. Iron Nutrition in Infancy and Childhood, New York, Raven Press, 1984, pp. 95–117.

18. Hegenauer J, Saltman P, Ludwig D, Ripley L, Ley A. Iron-supplemented cow milk. Identification and spectral properties of iron bound to casein micelles. J. Agric. Food Chem. 1979, 27: 1294–1300.

19. Gislason J, Jones B, Lonnerdal B, Hambraeus L. Iron absorption differs in piglets fed extrinsically and intrinsically [59]Fe-labeled sow's milk. J. Nutr. 1992, 122:1287–1292.

20. Davidson LA, Litov RE, Lonnerdal B. Iron retention from lactoferrin-supplemented formulas in rhesus monkeys. Pediatr. Res. 1990, 27: 176–180.

21. Hallberg L, Rossander-Hulten L, Brune M, Gleerup A. Bioavailability in man of iron in human and cow's milk in relation to their calcium contents. Pediatr. Res. 1992, 31: 524–527.

# BACTERICIDAL ACTIVITY OF DIFFERENT FORMS OF LACTOFERRIN

D.T. Akin, M.Q. Lu, S.J. Lu, S. Kendall, J. Rundegren, and R.R. Arnold

Dental Research Centers
Emory University School of Postgraduate Dentistry and
University of North Carolina at Chapel Hill
School of Dentistry

Mucosal surfaces provide the portal of entry for most pathogenic microorganisms. In protection the mammalian host has concentrated a number of defense factors in the exocrine secretions that bathe mucosal surfaces. Understanding how these factors operate to prevent pathologic consequences of microbial challenge and what mechanisms a pathogen uses to evade this host defense is essential to the development of interventive approaches. Lactoferrin (LF) and secretory IgA (sIgA) are both secreted in significant quantities at most mucosal surfaces and in milk. Both of these factors are reported to have bacteriostatic activity (1–8) and interact with each other and other defense factors (including lysozyme, complement, and lactoperoxidase) to modify their effectiveness (9, 10). In addition purified iron-free LF has been shown to have microbicidal activity against a variety of bacteria and yeast via mechanisms that are not reversed by the addition of free iron (11–16). A factor present in many commercial LF-enriched preparations, which co-purifies with sIgA, has been reported to result in synergistic bactericidal activity and these impure LF preparations are able to kill bacterial strains that are resistant to purified LF (17).

Since LF and sIgA coexist at mucosal surfaces, their interaction may provide a distinct component of host defense. Previous attempts to investigate this potential have led to conflicting results with reports that affinity purified sIgA can either block or enhance the effects of LF on bacteria (5–8, 18). Several groups have suggested the presence of sIgA-LF complexes in human secretions (19, 20), but the biological significance of these complexes has not been addressed.

Preliminary observations in this laboratory employing heparin-affinity chromatography for the purification of intact lactoferrin (21) noted that the void volume of exhaustive passage of human colostrum over heparin invariably contained LF and the majority of the sIgA. Affinity purification of the sIgA using anti-α heavy chain polyclonal IgG resulted in a consistent LF contaminant as detected by anti-LF ELISA. LF-free sIgA could be obtained by anti-LF affinity. Interestingly, the eluate from the anti-LF column contained significant IgA. In addition the IgA depleted heparin void volume contained IgA-free LF with no affinity for heparin. The antimicrobial and physicochemical properties of these two LF containing molecular species in comparison to that of intact free LF are the subject of this study.

*Lactoferrin: Structure and Function*
Edited by T.W. Hutchens *et al.*, Plenum Press, New York, 1994

## A Large Molecular Weight IgA-Lf Complex Isolated from Human Milk

The IgA-LF complex isolated by its affinity for both anti-α heavy chain and anti-LF contained a large species of MW > 670 kD identified by molecular sieve chromatography on a Pharmacia Superose 6 FPLC column and by SDS-PAGE under non-reducing conditions (Fig 1 a, b). IgA-LF complex was isolated from human colostral single milk samples by successive affinity chromatography using anti-human α heavy chain (DAKO Corp. Copenhagen, Denmark, Lot No. 016) and rabbit affinity purified anti-human milk LF IgA (prepared in this laboratory) coupled to CNBr-activated Sepharose 4B (Pharmacia, Uppsala, Sweden). The complex was eluted with 0.1 M glycine-HCl, pH 2.5 and immediately neutralized to pH 7.2 with 1 M Tris. The sample was adjusted to 1 M NaCl and an 300 µl aliquot (30 µg) injected onto a Pharmacia Superose 6 10/50 column equilibrated in 0.01 M phosphate buffer, pH 7.4 containing 1 M NaCl. Elution was performed in the same buffer at 0.5 ml/min, psi–120. Elution positions of the following standard proteins were as indicated: thyroglobulin (670 k), gammaglobulin (158 k), ovalbumin (44 k), myoglobin (17 k), and vitamin B12 (1350). IgA-LF complex (6.7 µg, lane 4) isolated as described above was analyzed by SDS-PAGE under non-reducing conditions on an 7.5% polyacrylamide gel. Bands were visualized using silver stain as directed by BioRad. Secretory IgA (112 µg) (lanes 2 and 3) were enriched by affinity chromatography using rabbit anti-human heavy chain and depleted of contaminating LF and complex by adsorption using rabbit anti- human milk LF affinity chromatography. Human milk LF (167 µg, lane 1) was purified by heparin affinity as previously described (13).

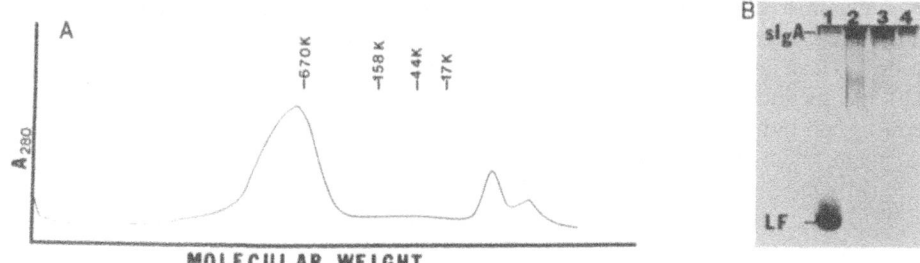

**Figure 1.** Molecular sieve chromatography and SDS-PAGA analyses of the anti-loctoferrin, anti-α heavy chain eluate from the exhaustive heparin void volume of colostrum (IgA-LF Complex).

## SDS-PAGE Analysis of IgA-LF Complex under Reducing Conditions and Identification of sIgA and LF Components by Transimmunoblot

SDS-PAGE under reducing conditions followed by transimmunoblotting confirmed the presence of heavy chains exhibiting mobilities consistent with both $\alpha_1$ and $\alpha_2$, secretory component, LF, and $\lambda$, and k light chains (Fig. 2). Aliquots of one µg each of sIgA (lane 1), iron-free IgA-LF complex (lane 2) and iron-free LF-enriched heparin eluate (> 0.6 M NaCl) (lane 3) were separated by SDS-PAGE on 10% polyacrylamide mini- gels after boiling for 10 min in presence of 2 % SDS and 5 % 2-mercaptoethanol. Gels were reacted with silver stain (A) or immunoblotted and reacted with the following antibodies: rabbit anti-human α heavy chain (B) and rabbit anti-human α-heavy chain and anti-human secretory component from DAKO Corp. (Copenhagen, Denmark Lot No. 014) (C), rabbit anti- human λ light chain from DAKO Corp. (D), rabbit anti-human k light chain from DAKO Corp. (E) and rabbit anti-human LF (F). Antigen-antibody complexes were visualized using a 1:1000 dilution of alkaline-phos-

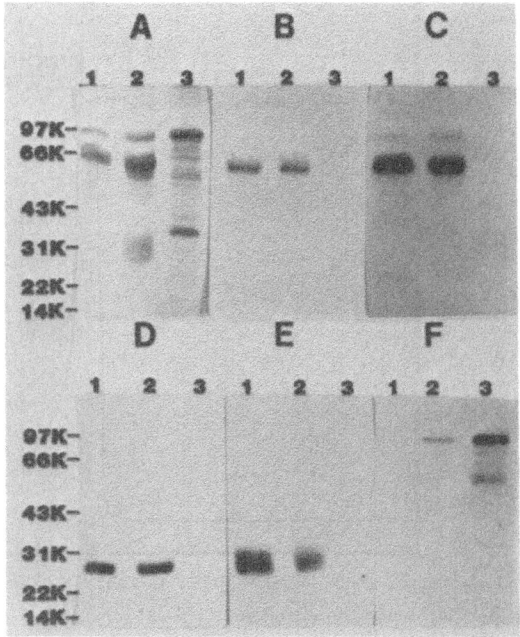

**Figure 2.** Comparison of IgA-LF complex (lanes 2) with sIgA (lanes 1) and heparin eluate (> 0.6 M NaCl) (lanes 3) by transimmunoblotting with anti-α (panel B), anti-α and anti-sectretory component (C), anti-λ light chain (D), anti-κ (E), and anti-LF according to the text. Panel A is silver stained, reducing SDS-PAGE>

phatase-conjugated goat anti-rabbit IgG (Sigma Chemical Co., St. Louis, MO Lot No. 27F-8950) and nitroblue tetrazolium and 5-bromo-4-chloro-3-indolyl (29).

The large molecular weight specie represents a covalent association of sIgA and LF, and its size is consistent with a tetrameric IgA molecule containing one or more each of secretory component and LF molecules. The isolated complex was stable in the presence of high salt concentrations (1 M NaCl), 5 M guanidine, pH 8.3, 2%SDS with boiling for 10 min, and at low pH (pH 2.5) for time periods of 1 hr or more. Consistent with the reducing SDS- PAGE data, elution profiles (Figure 3) on molecular sieve chromatography (FPLC, Pharmacia Superose 6) treatment of the complex with 20 mM DTT resulted in dissociation of IgA and LF and the collected fractions contained free LF (MW = 78,000) and IgA subunits (MW = 55,000 for α heavy, MW~ 25,000 for λ and k light chains) as determined by SDS-PAGE and transimmunoblots. Although ionic interactions may play a role in the initial attractions of LF and IgA *in vivo* as others have suggested (20), it is clear that specific mechanisms are needed to generate the covalent bonding which stabilizes the complex.

## Bactericidal Activity of IgA-LF Complex

Since LF has been shown to possess antibacterial activity in its iron-free form, the complex was dialyzed against 40 mM EDTA, 0.2 M phosphate, 0.076 M acetate, pH 4.0 as reported for the production of iron-free LF (17). The undialyzed and iron-free IgA- LF complexes were then tested for antibacterial activity against two organisms: *Streptococcus mutans* NCTC 10449 serotype c which is sensitive to bactericidal effects of purified iron-free LF (12) and E. coli 08 which is resistant to these effects (17) (Figs. 4a and b, respectively). Bactericidal effects were inconsistently observed with undialyzed IgA- LF complex. Iron-free complex, however, consistently exhibited marked bactericidal activity against both *S. mutans* and *E. coli* when compared to the same concentrations of similarly treated sIgA or LF (Fig. 4). In addition,

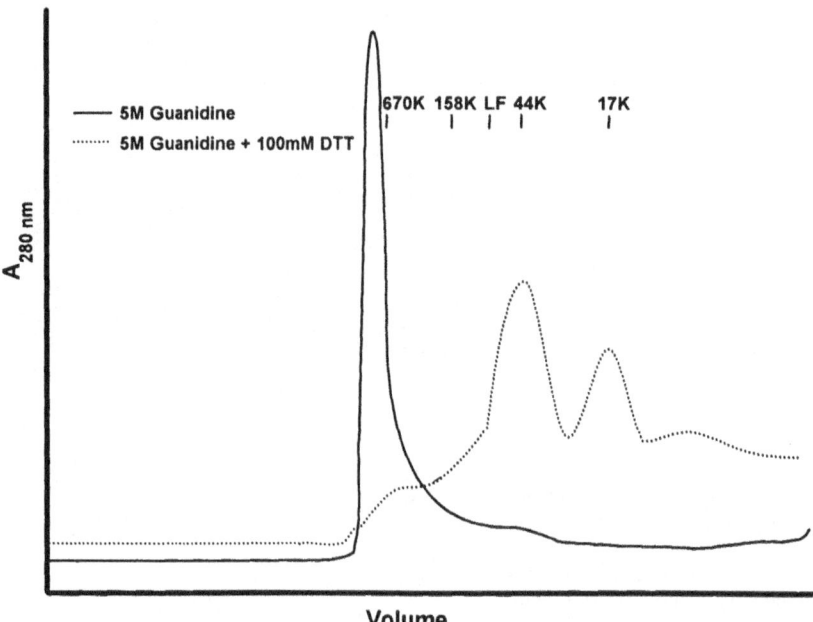

**Figure 3.** Molecular sieve profiles of IgA-LF complex under reducing and non-reducing conditions in 6 M guanidine on a Superose-6 column.

cooperativity between IgA and LF in the complex was not duplicated when uncomplexed, iron-free LF and sIgA were combined *in vitro*, indicating that the natural, covalent complexing of sIgA and LF is crucially important for its ability to function as a bactericidal molecule. Bactericidal effects were observed at concentrations of 10–20 µg/ml with IgA-LF complex preparations from ten different donor sources. Iron saturated IgA-LF complexes were inactive against either bacterium suggesting that the LF iron-binding capabilities are important to the antimicrobial properties of the complex (Fig. 4 c, d).

### Dose Dependence of Bactericidal Activity of IgA-LF Complex

Bactericidal assays using *S. mutans* NCTC 10449 (A) or *E. coli* 08 (B) were performed as described above. (A) Reaction mixtures contained 5 µg/ml (×), 10 µg/ml (◆), 20 µg/ml (■) iron-free IgA-LF complex or equivalent volumes distilled water (●). (B) Reaction mixtures contained 10 µg/ml (×), 20 µg/ml (◆), 26 µg/ml (■) iron-free IgA-LF complex or equivalent volumes distilled water(●).

The activity of iron- free IgA-LF complex was found to be concentration- dependent (Fig. 5). A decreasing lag period was apparent as concentration was increased in the bactericidal assays with *S. mutans* (Fig. 5a). The maximum bactericidal concentration of iron-free complex was capable of killing *S. mutans* with an undetected lag period (15 min). In comparison, bactericidal activity attributed to iron-free LF was reported to required a 15 min lag period even at maximal concentrations (12).

Bactericidal assays of IgA-LF complex activity using the Gram negative *E. coli* 08 exhibited different characteristics of killing, suggesting that the rate of killing was decreased with decreasing concentration (Fig. 5b). No evidence was observed to suggest a significant lag period before the loss in viability occurred with this organism. In addition, *E. coli* 08 was found

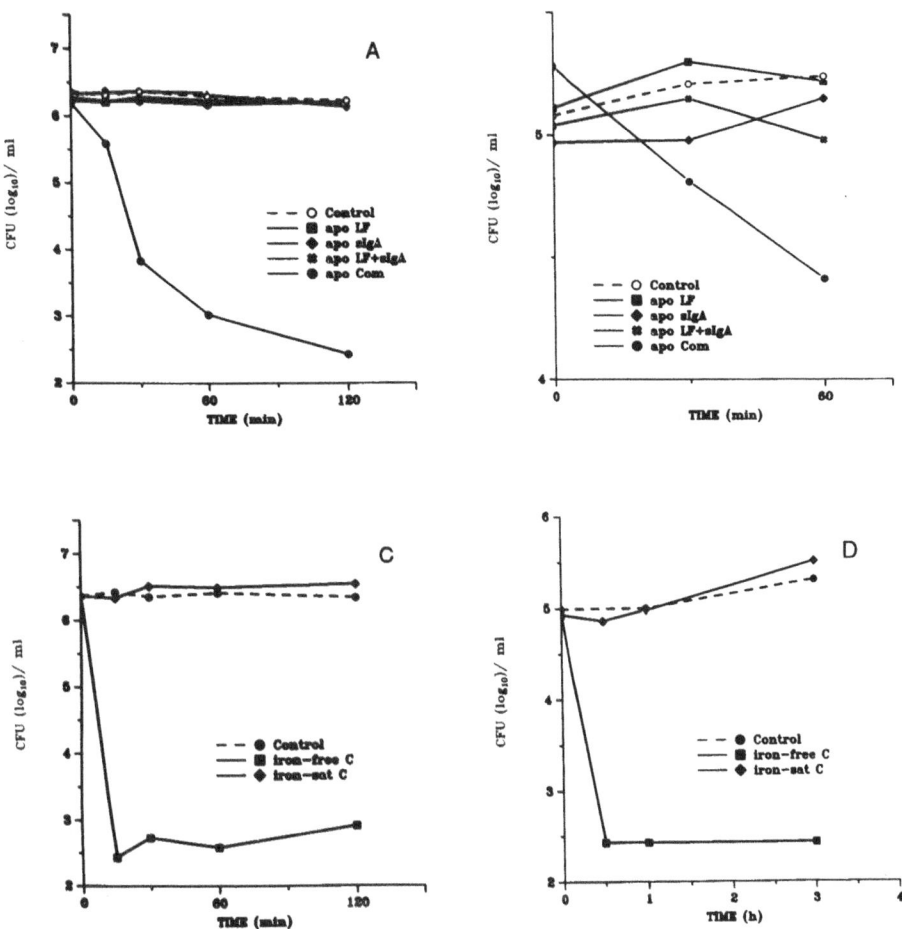

**Figure 4.** Bactericidal activity of IgA-LF complex. (A) *S. mutans* NCTC 10449 serotype c were grown in Todd-Hewitt broth to early-mid exponential phase, washed in saline and diluted in 1 ml saline reaction mixtures to approximately $1 \times 10^6$ colony forming units (CFU)/ml. The reaction mixtures contained 10 µg/ml iron-free LF (■), 10 µg/ml similarly dialyzed "iron-free" sIgA (♦) , 8 µg/ml "iron-free" sIgA + 2 µg/ml iron-free LF (an 1:1 molar ratio) (×), 10 µg/ml iron-free IgA-LF complex (●), or the equivalent volume of distilled water (○), 100 µl aliquots were removed at the indicated time periods, diluted 1:10 in saline and plated on trypticase soy agar containing 1% sucrose using the spiral plater system (Spiral Plater Systems, Inc. Cincinnati, Ohio). After a 48 h incubation at 37°, CFU/ml were counted using the spiral system laser counter (Spiral Plater Systems). (B) *E. coli* 08 were grown in Todd-Hewitt broth to early-mid exponential phase, washed in saline and diluted in 1 ml saline reaction mixtures to approximately $1 \times 10^5$ CFU/ml. The reaction mixtures contained 20 µg/ml iron-free LF (■), 20 µg/ml "iron-free" sIgA (♦), 16 µg/ml "iron-free" sIgA + 4 µg/ml iron-free LF (×), 20 µg/ml iron-free sIgA complex (●), or equivalent volumes of distilled water (○). Aliquots were removed at the indicated time periods and evaluated as described in 4 (A) after incubation on trypticase soy agar plates at 37° for 24 h. (C) *S. mutans* NCTC 10449 were grown and diluted into reaction mixtures as described in 4 (A). Reaction mixtures contained 20 µg/ml iron-free IgA-LF complex (■), 20 µg/ml iron-saturated IgA-LF complex (♦), or equivalent volumes of distilled water (●). (D) E. coli 08 were grown and diluted into reaction mixtures as described in 4 (B). Reaction mixtures contained 20 µg/ml iron-free IgA-LF complex (■), 20 µg/ml iron-saturated IgA-LF complex (♦), or equivalent volumes of distilled water (○).

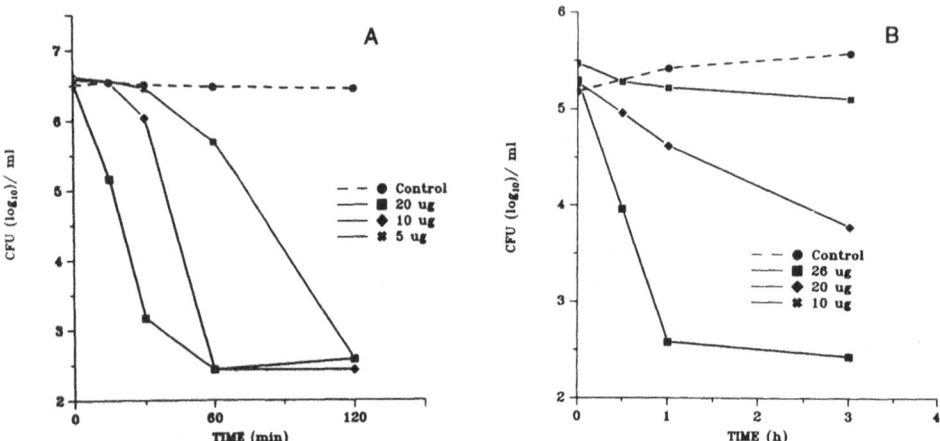

**Figure 5.** Dose dependence of bactericidal activity of IgA-LF complex. Bactericidal assays using *S. mutans* NCTC 10449 (A) or *E. coli* 08 (B) were performed as described above. (A) Reaction mictures contained 5 µg/ml (×), 10 µg/ml (♦), 20 µg/ml (■) iron-free LgA-LF complex or equivalent volumes distilled water (●), (B) Reaction mixtures contained 10 µg/ml (×), 20 µg/ml (♦), 26 µg/ml (■) iron-free IgA-LF complex or equivalent volumes distilled water (●).

to be resistant to purified iron-free LF bactericidal activity at appropriate concentrations (100–200 µg/ml) as previously reported (17).

### Effect of sIgA on Bactericidal Activity of IgA-LF Complex

The effects of purified preparations of both sIgA and LF on the bactericidal activity of iron-free complex were also investigated (Fig. 6). Addition of iron-free LF had little effect at equivalent concentrations (10–20 µg/ml) in assays with either bacterium and an apparent additive effect at higher concentrations (50-200 ug/ml) in the assay using *S. mutans* (data not shown); however, bactericidal activity was very sensitive to inhibition by addition of purified sIgA (Fig. 6a,b). Inhibition was demonstrated at very low levels of sIgA (1–10 µg/ml), and this inhibition was more efficient with homologous IgA (obtained from the same milk sample as the IgA-LF complex being assayed) than with sIgA obtained from a different milk sample. The bactericidal activity of purified iron-free LF against *S. mutans* is similarly inhibited by LF-free sIgA under these assay conditions (18).

### Characteristics of Non-complex, Heparin Void Volume LF

The residual void volume after exhaust heparin and anti-α-affinity was further subjected to anti-LF affinity chromatography and the eluate examined for bactericidal activity and SDS-PAGE (Fig. 7) and transimmunoblot analyses. SDS-PAGE revealed minor bands consistent with intact LF (78 kDa) and the 56 kDa fragment (44) and two major bands, one at 33 kDa (44) and the other at 42 kDa. Bands of 56 kDa and 33 kDa were consistently present as minor components of the 0.6 M NaCl heparin eluate of LF, but were only detected under reducing conditions. The 42 kDa specie was never detected in the heparin affinity purified, LF preparations. This LF fragment was > 150 times more bactericidal against *S. mutans*.

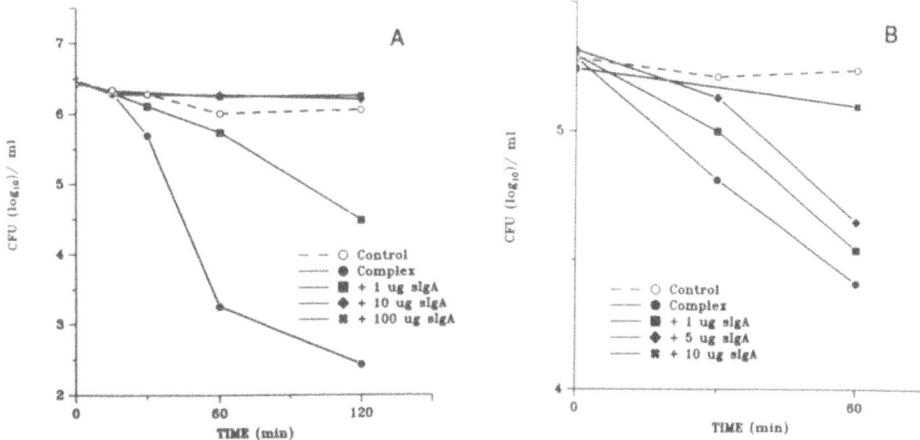

**Figure 6.** Bactericidal assays of *S. mutans* NCTC 10449 (A) or *E. coli* 08 (B) were performed in the presence of varying concentrations of purified sIgA. (A) Reaction mixtures contained 10 μg/ml iron-free complex (●), 10 μg/ml iron-free complex + 1 μg/ml sIgA (■), 10 μg/ml iron-free complex + 10 μg/ml sIgA (◆), 10 μg/ml iron-free complex + 100 μg/ml sIgA (×), or equivalent volumes of distilled water (O). (B) Reaction mixtures contained 20 μg/ml iron-free complex (●), 20 μg/ml iron-free complex + 1 μg/ml sIgA (■), 20 μg/ml iron-free complex + 5 μg/ml sIgA (◆), 20 μg/ml iron-free complex + 10 μg/ml sIgA (×), or equivalent volumes of distilled water (O).

## DISCUSSION

The ability of LF to associate with other factors in secretions has been previously demonstrated by the presence in human milk of LF exhibiting altered immunoelectrophoretic mobilities. These associations were suggested to be ionic and/or hydrophobic in nature and occurred spontaneously between LF and acidic proteins when mixed *in vitro* (22, 23). Similar results with human tears and seminal plasma have indicated the presence of LF in association with IgA and other proteins (19, 24). In addition, we have detected the presence by immunoelectrophoresis of an IgA-LF complex naturally found in relatively low levels in human milk which exhibited identical mobility to that of isolated IgA-LF complex. No evidence was found, however, for the presence of a similar species when purified sIgA and LF were combined *in vitro* (data not shown). Watanabe, et al. (20) have reported that LF was released from enriched

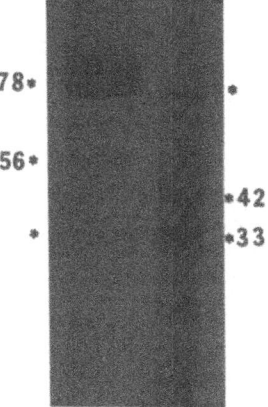

Figure 7.

human milk sIgA preparations by disulfide bond cleavage, suggesting that LF could be naturally bound to sIgA.

The bactericidal activity of the IgA-LF complex seems to be dependent on the availability of free iron-binding sites on the LF molecule. Although sufficient iron is present in human milk to fully saturate LF, LF exists predominantly in an iron-free form in human milk (25, 26) where ratios of citrate and bicarbonate discourage iron-binding to LF (72) and insure that LF (and presumably LF contained in IgA-LF complex) is present in its bactericidal, iron-free form.

sIgA is present in human colostrum and early milk at very high levels (10-30 mg/ml). Specific antibody levels in milk mirror the exposure of the mother to various organisms. sIgA from human milk has been credited with the ability to inhibit growth of enteric bacteria in the presence of LF when serotype-specific antibody is present (5–8). Commensal stains of bacteria, however, were found to be resistant to this effect despite the presence of anti-O specific antibody (28), a fact which may explain the successful colonization of the infant gut by commensal strains. sIgA alone does not exhibit bactericidal effects even at approximately 100 times the molar concentration at which the IgA-LF complex is effective; however, our results indicate that the complexing of sIgA with iron-free LF generates a potent bactericidal factor. It will be of interest to examine the IgA component of IgA-LF complex more fully to determine the role of antigen specificity and the effects on both commensal and pathogenic strains. The inhibitory effects of sIgA on the bactericidal activity of IgA-LF complex suggest that it may be capable of modulating this function *in vivo*. Interestingly, both bacteria used for this initial screening were effectively protected by milk sIgA. *S. mutans* is an inhabitant of the oral cavity which is reported to play a major role in caries production (30). *E. coli* 08 was isolated from a neonatal intestinal infection. Previous exposure of the donating mothers to these organisms or cross-reacting antigens was confirmed by the presence of agglutinating activity and fluorescent IgA antibody titers against each organism in the sIgA fraction.

Our findings confirm the presence in human milk of an IgA-LF complex which is stabilized by covalent bonds. Its protective role in milk is strongly suggested by potent bactericidal activity against Gram negative and Gram positive bacteria. Oral immunization of mothers has previously been shown to result in the appearance of specific IgA antibody in milk and other secretions (31–34) and may also be expected to result in specific IgA-LF complex production. Protection provided to the infant may extend beyond the intestinal tract based on the observations that small amounts of colostral poliovirus-specific sIgA have been detected in infant serum (35). Oral immunization of mothers may, therefore, be an excellent means of conveying protection to the infant both via the agglutinating and toxin-neutralizing functions of sIgA antibody and the bactericidal activity of specific IgA complexed to LF. Such immunization schemes have been shown to be effective against many enteric bacteria that pose significant public health threats (especially in underdeveloped countries) and against bacterial and viral pathogens of the respiratory tract (31–34, 36–41).

The presence of an IgA-LF complex has also been suggested in human tears (19) raising the possibility that IgA-LF complexes may exist and provide host protection at all mucosal surfaces. IgA at mucosal sites is provided via a common mucosal immune system whereby mucosal stimulation by antigen results in specific IgA production at many mucosal sites (42, 43). IgA-LF complex, then may be able to provide effective protection (either naturally or via oral or local immunization) against a myriad of pathogens that enter and/or colonize the host at mucosal surfaces.


## ACKNOWLEDGEMENTS

We wish to thank Ms. Ramona Kellam for processing this manuscript and Todd Arnold for his input on graphics. This work was supported by NIH grants RO1-DE06869 and F32-DE06869.

## REFERENCES

1. A.B. Otnaess and I. Orstavik, Infect. Immun. 33, 459, (1981).

2. E. Weinberg, Science 184, 952 (1974).

3. J.J. Bullen, H.J. Rogers and E. Griffiths, Curr. Top. Microbiol. Immun., 80, 1, (1978).

4. J.M. Dolby and P. Hoover, J. Hyg. (Camb). 83, 255, (1979).

5. H.J. Rogers, Immunology, 30, 425, (1976).

6. J.M. Dolby and S. Stephens, Acta Paediatr. Scand., 72, 577, (1983).

7. J. Bullen, H. Rogers, L. Leigh, Brit. Med. J., 1, 69, (1972).

8. G. Spik, A. Cheron, J. Montreuil and J. Dolby, Immunology 35, 663, (1978).

9. C.P. Thompson, B.M. McCarty, N.D. Chasteen, Biochim. Biophys. Acta 22, 530, (1986).

10. E. Ainscough, A. Brodie, J. Plowman, S. Bloor, J. Loehr and T. Loehr, Biochemistry, 19, 4072, (1980).

11. R. Arnold, J. Russell, W. Champion and J. Gauthier, Infect. Immun., 32, 655, (1981).

12. R.R. Arnold, M.F. Cole and J.R. McGhee, Science, 197, 263, (1977).

13. R.R. Arnold, J.E. Russell, W.J. Champion, M. Brewer and J.L. Gauthier, Infect. Immun., 35, 792, (1982).

14. J.R. Kalmar and R.R. Arnold. Infect. Immun. 56, 2552, (1982).

15. C.A. Bortner, R.D. Miller and R.R. Arnold. Infect. Immun. 51, 373, (1986).

16. C.A. Bortner, R.R. Arnold and R.D. Miller. Can. J. Microbiol. 35, 1048, (1989).

17. M.A. Motley and R.R. Arnold, in Recent Advances in Mucosal Immunology, J. Mestecky, J.R. McGhee, J. Bienenstock and P.L. Ogra, Eds. (Plenum Publishing Corp., 1987), Park A, pp. 591–599.

18. R.R. Arnold, J.E. Russell, S.M. Devine, M. Adamson and K.M. Pruitt, in Cariolog Today. Int. Congr., Zurich, B. Guggenhein, Eds., (Karger, Basel, 1984), 75.

19. R.M. Broekhuyse, Invest. Opthalmol. 13, 550, (1974).

20. T. Watanabe, H. Nagura, K. Watanabe and W.R. Brown, FEBS Lett. 168, 203 (1984).

21. L. Bläckberg and O. Hernell, FEBS Lett., 109, 180, (1980).

22. A. Hekman, Biochim. Biophys. Acta., 251, 380, (1971).

23. P.L. Masson, La Lactoferrine. Proteine des Secretions Externes et des Leucocytes Neutrophiles, Arsica, Bruxelles, 1970.

24. T.K. Roberts and B. Boettcher, J. Reprod. Fertil., 18, 347, (1969).

25. P.L. Masson and J.F. Heremans, Eur. J. Biochem., 6, 579, (1968).

26. B. Reiter, J.H. Brock and E.D. Steel, Immunology, 28, 83, (1975).

27. E. Griffiths and J. Humphreys, Infect. Immun., 15, 396, (1977).

28. S. Stephens, J.M. Dolby, J. Montreuil and G. Spik, Immunology, 41, 597 (1980).

29. M.S. Blake, K.H. Johnston, G.J. Russell-Jones and E.C. Gotschlich, Anal. Biochem, 136, 1715, (1984).

30. S. Hamada and H.D. Slade, Microbiol. Rev., 44, 331, (1980).

31. J. Mestecky, J.R. MeGbee, R.R. Arnold, S.M. Michalek, S.J. Prince and J.L. Babb, J. Clin. Invest., 61, 731, (1978).

32. S. Ahlstedt, B. Carlsson, S.P. Fällström, L.A. Hanson, J. Holmgren, G.Lidin-Janson, B.S. Lindblad, U. Jopdal, B. Kaijser, A. Sohl-Åkerlund and C. Wadsworth, Ciba Found. Synp. 46, Excerpta Medica, Amsterdam, (1977).

33. R.M. Goldblum, S. Ahlstedt, BN. Carlsson, L.A. Hanson, J. Jodal, G. Lidin-Janson and A. Sohl, Nature, 257, 797, (1975).

34. R.C. Montgomery, B.R. Rosner and J. Cohn, Immun. Comm., 3, 143, (1974).

35. S.S. Ogra, D. Weintraub and P.L. Ogra, Adv. Exp. Med. Biol., 107, 95, (1978).

36. H.L. Amos and E. Taylor, J. Exp. Med., 25, 507, (1917).

37. C.G. Bull and C.M. McKee, Am. J. Hyg., 9, 490, (1929).

38. R.D. Rossen, J.A. Kasel and R.B. Couch, Prog. Med. Virol., 13, 194, (1971).

39. J. Holmgren, A.-M. Svennerholm, O. Ouchterlony, A., Anderson, G. Wallerström and U. Westerberg-Berndtsson, Infect. Immun., 12, 1331–1340, (1975).

40. E.S. Fubara and R. Freter, J. Immunol., 111, 395, (1973).

41. L.A. Hanson, S. Ahlstedt, B. Carlsson, B. Kaijser, P. Larsson, I. Mattsby-Baltzer, A. Sohl-Åkerlund, C. Svanborg-Eden and A.-M. Svennerholm, Adv. Exp. Med. Biol., 107, 165, (1978).

42. J. Mestecky, J.R. McGhee, S.M. Michalek, R.R. Arnold, S.S. Crago and J.L. Babb, Adv. Exp. Med. Biol., 107, 185, (1978).

43. M.J. Parmley and A.E. Beer, J. Diary Sci., 60, 655, (1977).

44. T.W. Hutchens, J.F. Henry, and T.-T. Yip. Proc. Natl. Acad. Sci. U.S.A. 88, 2994, (1991).

# THE EFFECTS OF LACTOFERRIN ON GRAM-NEGATIVE BACTERIA

Richard T. Ellison III

Division of Infectious Diseases
Department of Medicine
University of Massachusetts School of Medicine
Worcester, Massachusetts 01655

## ABSTRACT

Lactoferrin is an iron-binding protein found in human mucosal secretions as well as the specific granules of polymorphonuclear leukocytes. A variety of functions have been ascribed to the protein, and it appears to contribute to antimicrobial host defense. In particular, it has been shown to have direct effects on pathogenic microorganisms including bacteriostasis and the induction of microbial iron uptake systems. Still its overall physiologic role remains to be defined. It has appeared logical that antimicrobial activity of the protein arises from sequestration of environmental iron thereby causing nutritional deprivation in susceptible organisms. This argument is buttressed by the finding that selected highly virulent pathogens have evolved techniques to subvert this effect and use the protein as an iron source. However, recent observations indicate that the protein has additional properties that contribute to host defense. Work by several groups has shown that the protein synergistically interacts with immunoglobins, complement, and neutrophil cationic proteins against Gram-negative bacteria. Further, both the whole protein and a cationic N-terminus peptide fragment directly damage the outer membrane of Gram-negative bacteria suggesting a mechanism for the supplemental effects. This review will summarize these diverse observations with a consideration of how the in vitro work relates to the physiological role of the protein.

## INTRODUCTION

Human lactoferrin and transferrin are monomeric metal-binding glycoproteins with an approximate molecular weight of 75,000 to 80,000[1-4]. Lactoferrin was first identified as a major component of breast milk and colostrum, and was subsequently identified within other mucosal fluids[1,5-7]. The recognition that these mucosal fluids inhibit microbial growth and the identification of lactoferrin within the specific granules of polymorphonuclear leukocytes has suggested that this protein has antimicrobial activity and contributes to antibacterial host defense. Early studies with purified human lactoferrin supported this hypothesis showing that the protein was bacteriostatic for *Staphylococcus* and *Bacillus* strains[8,9]. Bullen and coworkers extended these observations, showing that in vitro bacteriostasis of human milk for *Escherichia coli* could be replicated by purified lactoferrin and sIgA[10]. Additionally, they observed that the inhibitory effect of milk and lactoferrin could be eliminated by saturating the materials with iron. These observations of a bacteriostatic effect of lactoferrin that is reversed by iron saturation was subsequently duplicated by other investigators for a variety of other Gram-negative bacteria[11-17]. In this manuscript the present work on lactoferrin's activities against Gram-negative bacteria will be reviewed with a focus on possible mechanisms of injury and on the role these activities play in overall antimicrobial host defense.

## IN VIVO OBSERVATIONS WITH LACTOFERRIN

Several in vivo studies have been performed to assess the antimicrobial activities of lactoferrin. Bullen found that intestinal colonization of suckling guinea pigs by orally administered *E. coli* is inhibited by normal guinea pig milk containing both lactoferrin and transferrin[10]. The protective effect of the milk is lost if the pigs are concurrently fed hematin hydrochloride, suggesting that the milk contains a host defense system that is blocked by high levels of iron. Subsequently, in a small study Sawatzki and coworkers noted that the administration of intravenous purified mouse lactoferrin decreased the intrasplenic growth of *S. typhimurium* in mice[18]. In this study C3D2F1 mice received 4 mg lactoferrin intravenously every 12 hours beginning 8 hours before an intraperitoneal challenge with *S. typhimurium*. They were found to have a rise in plasma lactoferrin from 1.8 to 32 g/ml and a 1.5 log decrease in splenic bacteria at 24 hours as compared to untreated controls[18]. In a comparable model, Zagulski and coworkers found that the intravenous administration of purified human or bovine lactoferrin effectively protected CFW mice from a subsequent lethal intravenous injection of $2 \times 10^8$ *E. coli*[19]. However, in this study, experiments were not performed to determine the ability of lactoferrin to influence either an established infection or concurrent bacterial challenge. In contrast, Czirok et al. reported that neither intraperitoneal, intramuscular, and oral lactoferrin could protect mice from a lethal intraperitoneal injection of *E. coli* when the proteins was given concurrently or after initiation of experimental infection (although in this study neither the mouse strain, the bacterial strain, or the bacterial inoculum size were characterized)[20].

LaForce has shown that there is a marked increase in intrapulmonary lactoferrin concentrations following experimental pulmonary bacterial challenge with *E. coli*[21]. In more recent unpublished studies, this line of investigation has been extended to address the host defense function of endogenous lactoferrin (personal communication F. Marc LaForce). In a murine model of *E. coli* pneumonia, the concurrent intratracheal installation of specific anti-transferrin or anti-lactoferrin antisera impairs intrapulmonary clearance of the Gram-negative organism. In these studies, high-titer rabbit anti-murine transferrin and anti-murine lactoferrin immunoglobulin significantly decreased the 4 hour clearance of a $10^6$ cfu inoculum of *E. coli* in HA ICR mice as compared to normal immunoglobulin controls (91% bacteria killed in control mice versus 45 and 50% killed in mice given anti-transferrin and anti-lactoferrin antisera, p < 0.001). The rabbit anti-transferrin and anti-lactoferrin immunoglobulin had no measurable effect on the number of alveolar macrophages within lung lavage fluid or on the influx of polymorphonuclear leukocytes into the lungs. This work suggests that the addition of anti-transferrin and anti-lactoferrin antisera interfered with normal extracellular host defenses in these mice.

Thus, while there are conflicting observations between studies, as well as some methodologic limitations, the available in vivo data tends to support the hypothesis that lactoferrin contributes to antimicrobial host defense.

## EFFECT OF LACTOFERRIN ON BACTERIAL IRON METABOLISM

As noted above, the major hypothesis advanced for a mechanism of antimicrobial activity is that lactoferrin induces an iron deficient environment that limits growth of bacteria. As lactoferrin is secreted by polymorphonuclear leukocytes, it binds free iron and thereby limits the amount of iron available for bacterial uptake. The work supporting this fundamental role of lactoferrin has been the subject of several extensive reviews in the late 1970's and early 1980's and will not be reviewed in detail here[1,7,22]. However, a critical issue that has only more recently been identified is the recognition that selected bacterial pathogens have the capacity to use lactoferrin and transferrin as iron sources[23-33]. The human pathogens *Neisseria meningitidis, Neisseria gonorrhoeae, Branhamella catarrhalis*, and *Hemophilus influenzae* have each been found to have specific iron-regulated cell surface receptors for human lactoferrin and transferrin. In these species, the receptors for lactoferrin and transferrin are separate bacterial proteins, and each demonstrates greater specificity for the human protein than the related bovine protein. Overall, this ability to acquire iron from host iron-binding proteins appears to be a major virulence determinant for these pathogens giving them a selective advantage in causing invasive disease. In studies with *Neisseria* species, organisms that are pathogenic have been shown to have a greater ability to acquire iron from lactoferrin and transferrin than do related commensal species[34]. In work with *N. meningitidis*, the addition of exogenous iron-saturated transferrin and lactoferrin allows enhanced growth of the organism and the development of more severe infection[26].

Viewed in isolation, these studies appear to contradict a role for lactoferrin and transferrin in antibacterial host defense. What contribution can the iron-binding proteins make to host defense if bacteria can use the proteins as growth factors? As a response, it is important to distinguish an overall effect the two proteins have on organisms that are occasional pathogens or commensal flora from an effect against highly virulent human pathogens. It must be recognized that the human host is normally exposed to $10^{11}$ to $10^{12}$ commensal bacteria on a continual basis at the sites where there is contact with the environment; specifically the skin, mucous membranes, and respiratory and gastrointestinal epithelium. Under normal physiologic conditions, these bacteria reside on the surface without invading tissue because of the effects of general nonspecific defense mechanisms. Thus, when the total microbial environment to which that the human host is exposed is considered, the nonspecific defense mechanisms are in fact exceedingly effective even though they are occasionally breached by "pathogenic" microbes.

A useful definition of "commensal" microflora are those microbes whose growth is controlled by general nonspecific host defense mechanisms without the need for the implementation of the more potent host defenses used in the control of an "infection". A "pathogen" is an organism that can breach the nonspecific defenses to establish an infection that may or may not then be controlled by a more specific inflammatory response involving immunoglobulins, complement, macrophages, polymorphonuclear leukocytes, and lymphocytes. A relevant example of this distinction between a "commensal" and a "pathogenic" microorganism is *Pasteurella hemolytica*[27]. This Gram-negative bacteria is found in the upper respiratory tract of cattle and sheep and is "pathogenic" for these species causing epidemic pneumonia in sheep and cattle, as well as sepsis in lambs. In contrast it has only rarely been identified as a cause of human disease, and when isolated in cultures from humans is normally considered a "commensal". Recent work by Ogunnarino and Schryver has shown that this organism has a transferrin receptor that has high affinity for bovine transferrin but low affinity for the human, horse, pig or avian transferrins[27]. Thus this organism is a "pathogen" for a species where it can obtain iron from transferrin and overcome this host defense system. The organism is a "commensal" in species where it cannot utilize this iron source.

## ROLE OF LACTOFERRIN WITHIN PHAGOCYTIC CELLS

While the activity of lactoferrin in sequestering iron is certainly a major component of its contribution to host defense against Gram-negative bacteria, there is now evidence of additional effects of the protein. The high concentration of lactoferrin within the specific granules of neutrophils, the exocytosis of a portion of the specific granule pool during the neutrophil response to inflammation, and the incorporation of specific granules into the neutrophil phagolysosome all suggest that lactoferrin contributes significantly to the function of this key inflammatory cell[35,36]. In support of this hypothesis, there is a large body of evidence that indicates that lactoferrin directly modulates both the production and function of neutrophils and monocytes (summarized elsewhere in this symposium). Secondly, a major component of the antimicrobial activity of phagocytic cells is the production of superoxide and hydroxyl radicals. In its ability to chelate iron, lactoferrin appears to play an important role within both neutrophils and macrophages in modulating the Haber–Weiss reaction that generates hydroxyl radicals (also summarized in detail elsewhere in the symposium).

A third aspect of lactoferrin's relationship to phagocytic cell function is the apparent ability of the protein to contribute to oxygen-independent antimicrobial properties of these cells. Although the present evidence is all indirect, three lines of investigation support a role for lactoferrin in the antimicrobial function of polymorphonuclear leukocytes and macrophages. First, lactoferrin has been found to augment the in vitro killing of E. coli by bactenecin Bac7, a cationic antimicrobial peptide present within bovine neutrophils[37]. This would suggest that these two neutrophil associated proteins could act in synergy within the neutrophil phagolysosome against Gram-negative organisms. Second, in studies with human blood monocytes, lactoferrin has been found to enhance both the uptake and intracellular killing of the parasite Trypanosoma cruzi [38,39]. The mechanism of enhanced antimicrobial killing activity is not fully defined, although the killing is dependent on the presence of iron and superoxide. Third, and more relevant to the consideration of activity of lactoferrin against Gram-negative bacteria, work by Byrd and Horowitz shows that apo-lactoferrin enhances intracellular killing of Legionella pneumophila by human blood monocytes[40]. This effect is not seen with transferrin[41], and is not seen with iron-saturated lactoferrin. It is relevant that human monocytes and macrophages do not produce lactoferrin. Instead these cells acquire lactoferrin from the local environment through surface receptors[40,42]. Thus, the ability of exogenous lactoferrin to enhance monocyte killing of Legionella suggests that the active exocytosis of lactoferrin by the neutrophils during inflammation may serve as an indirect mechanism by which the neutrophil enhances the activity of monocytes and macrophages.

## OTHER ANTIBACTERIAL EFFECTS OF LACTOFERRIN

Beyond the interaction of lactoferrin with phagocytes, several lines of evidence have indicated that lactoferrin has still further effects on Gram-negative bacteria that appear independent of an effect on bacterial iron metabolism. First, the antimicrobial activity of IgG, IgM or sIgA against E. coli can be enhanced by concurrent bacterial exposure to lactoferrin[12,17,43,44]. This has been shown under a variety of in vitro conditions, including work in whole boiled milk (where the pH, osmolarity and ionic composition are relatively physiologic)[12]. It has been suggested that this synergistic interaction is due to antibodies directed against bacterial siderophores, thereby further limiting bacterial iron acquisition. However, this possibility has not been confirmed, and there is evidence that the enhancing antibodies are directed at LPS O antigens[17,44,45].

Second, Arnold and his colleagues have reported that lactoferrin has direct bactericidal activity against selected Gram-negative bacteria at lactoferrin concentrations below 50 $\mu$M[46–52].

Similar to the bacteriostasis of lactoferrin, this reported bactericidal activity is inhibited by iron saturation, but is not reproduced by exposure of bacteria to an iron-poor environment alone[48]. Activity against *Legionella pneumophila* is inhibited by the addition of magnesium or calcium, is greatest at acidic pH, and demonstrable on log-phase but not plate grown bacteria[49,51]. Still, the physiological relevance of these observations against Gram-negative organisms remains unclear as: 1) the bactericidal effects have been noted in non-physiologic saline or distilled water[46,49,51,53]; 2) the activity against Gram-negative bacilli appears to vary markedly with differing lactoferrin preparations and is influenced by apparent contaminants of lactoferrin preparations[52].

Third, there is now a large body of evidence that lactoferrin can directly bind to bacterial cells, and that binding relates to antimicrobial activity. Using an indirect adsorption assay to measure lactoferrin binding, Dalmastri and Visca and their coworkers found a correlation between bacteriostasis due to lactoferrin and bacterial cell adsorption of the protein[54,55]. These studies of bacterial binding of lactoferrin have been further extended by Naidu and colleagues who have used a [125]I-labelled protein binding assay to show that lactoferrin binds to strains of *E. coli*, *Aeromonas hydrophila*, and *Shigella flexneri*[56-59]. Lactoferrin binding was found to vary between bacterial species and between different strains within species. The binding appeared to be specific and saturable with both *A. hydrophila* and *S. flexneri*, and in western blot studies with horse radish peroxidase labeled lactoferrin there is apparent binding to outer membrane porin proteins.

Fourth, there is now evidence from my laboratory that lactoferrin damages the outer membrane of Gram-negative bacteria. In studies with strains of *E. coli*, *S. typhimurium*, and *V. cholerae* we have found that lactoferrin causes the release of structural LPS molecules from the Gram-negative outer membrane and an increase in bacterial susceptibility to hydrophobic antibiotics and the host antimicrobial enzyme lysozyme[60-62].

## STRUCTURE OF THE GRAM-NEGATIVE BACTERIAL CELL

As a framework in which to consider these apparently diverse antibacterial effects of lactoferrin, it is relevant to briefly review current knowledge on the structure of Gram-negative microorganisms. Although many of these bacteria have other structural surface elements, all share a characteristic trilaminar cell wall (Figure 1). This cell wall is composed of: 1) an inner cytoplasmic phospholipid bilayer membrane with integrated membrane transport proteins; 2) a periplasmic space containing a rigid peptidoglycan matrix [murein sacculus] essential to maintenance of the 3 dimensional shape of the bacterial cell; and 3) an outer lipid bilayer membrane that includes specific transport proteins, and transmembrane pore proteins [porins]. The outer membrane is distinct within biological systems in being asymmetric. The membrane's inner leaflet is composed of neutral phospholipids characteristic to other membranes, and the outer leaflet of unique lipopolysaccharide molecules (that are also known as endotoxin). These lipopolysaccharide (LPS) molecules also have 3 domains; a lipid A region that inserts into the membrane bilayer, a region of charged core sugar molecules with free phosphate groups present at the membrane surface, and a polysaccharide side chain extending beyond the membrane surface composed of a variable number of repeating sugar subunits. This sidechain serves to provide a hydrophilic surface coat for the bacterial cell. To counterbalance the strongly anionic core oligosaccharides, the outer leaflet contains stabilizing divalent cations (predominantly calcium and magnesium) that appear to be integrated into the membrane[63].

Over the last 30 years there has been extensive study of factors which affect the Gram-negative membrane as a mechanism to injure or kill these bacteria. Early studies with synthetic chelators found that they had significant activity against enteric Gram-negative bacilli[64,65]. EDTA releases up to 50% of the LPS from the bacterial outer membrane, increases membrane

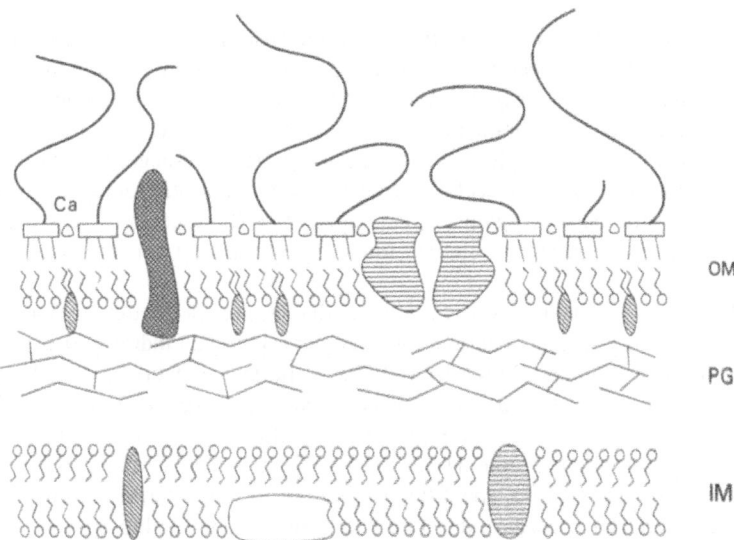

**Figure 1.** General Structure of Gram-negative bacterial cell wall structure. OM denotes the outer membrane composed of an outer leaflet with lipopolysaccharide molecules and stabilizing calcium or magnesium ions, an inner membrane with neutral phospholipids and lipoproteins, and membrane spanning transport and porin proteins. PG denotes the peptidoglycan framework, or murein sacculus, that gives a rigid shape-defining structure to the Gram-negative organism. Lipoproteins appear to link the outer membrane to this framework. IM denotes the inner or cytoplasmic membrane which is composed of a neutral phospholipid bilayer with integral membrane transport proteins.

permeability to hydrophobic molecules, and sensitizes the bacteria to the effects of host defense systems including serum complement and lysozyme[66,67]. In chelating the membrane associated cations, EDTA appears to destabilize the outer leaflet of the outer membrane causing the release of the charged LPS molecules. When these released LPS molecules are replaced by nonpolar phospholipids from the inner leaflet, the protective hydrophilic O-polysaccharide coat is lost and the outer membrane becomes more permeable to hydrophobic agents (such as rifampin) that are normally excluded[64,65]. These changes in membrane permeability, as well as the removal of steric hindrances due to the LPS O-polysaccharide sidechains, also appear to explain EDTA-induced bacterial sensitization to complement and lysozyme[66,67]. The membrane attack complex of complement creates a membrane spanning pore of polymeric C9 molecules, and the elimination of the LPS polysaccharide chains with EDTA treatment allows for more efficient positioning of the polymeric C9 complex into the outer membrane. Similarly, lysozyme acts to enzymatically degrade the bacterial peptidoglycan matrix. However, under normal conditions this 14,000 molecular weight enzyme cannot penetrate through the bacterial outer membrane to reach the peptidoglycan layer. When the outer membrane permeability is increased, the enzyme penetrates to lyse peptidogylcan and structurally damage the bacterial cell.

Compounds which are polycationic in nature also alter the Gram-negative outer membrane and in selected instances are bactericidal. Well defined examples of such agents are poly-l-lysine and the antimicrobial agent polymyxin B. More relevant to the activity of lactoferrin, numerous proteins isolated from neutrophils including bactericidal/permeability-increasing protein (BPI), cathepsin G, azurocidin, the defensins and bactenecins are also highly cationic and damage the Gram-negative membrane[68-70]. These agents have differing properties, but in general they also increase the permeability of the outer membrane to hydrophobic agents, lysozyme and complement[65,71]. Polymyxin B causes both permeability change and LPS release, while BPI and the defensins enhance membrane permeability without significant release of LPS. The activity of these agents appears to be due to an ability to bind to anionic core elements of LPS molecules thereby altering the position and function of LPS within the membrane. Differences in the binding and membrane insertion characteristics of the agents may then contribute to the observed variability in outer membrane effects[72]. These cationic neutrophil-associated proteins have bactericidal activity that appears to be related to effects on the Gram-negative inner membrane, although the precise mechanisms of activity of remains poorly characterized[69,70,73–75].

## OUTER MEMBRANE EFFECTS OF LACTOFERRIN

Lactoferrin is a high affinity chelator that binds a wide variety of metal ions[1], and thus it could be hypothesized that it might act similar to EDTA to alter the Gram-negative outer membrane. Additionally, noting its high pI (~8.5), lactoferrin could also have an effect similar to the membrane-active polycationic compounds.

We performed a series of experiments to address these hypotheses, and found that lactoferrin does damage the Gram-negative outer membrane[60]. In studies with defined *E. coli* and *Salmonella* strains (*E. coli* CL99-2, *S. montevideo* SL5222, *S. typhimurium* SL696 and SH4247) in which LPS molecules can be intrinsically radiolabeled[76–78], it was found that human lactoferrin can release LPS from the bacterial cell[60]. This lactoferrin-mediated LPS release is demonstrable in a pH range from 6 to 7.5, is dose dependent and most apparent at levels of lactoferrin above 1 mg/ml. This effect on the outer membrane is not associated with bacterial cell death. The effect is blocked by iron saturation of the protein, and is also influenced by the amount of free $Ca^{2+}$ and $Mg^{2+}$ present in the environment. LPS release occurs at calcium concentrations below 160 μM, but is absent at a level of 1.3 mM, a concentration approximating that of serum[61].

To determine if the release of bacterial LPS by lactoferrin alters the permeability of the outer membrane, we tested lactoferrin's effect on bacterial susceptibility to rifampin, an antibiotic normally excluded by the Gram-negative outer membrane[60,71]. We found that lactoferrin synergistically increases bacterial killing due to a subinhibitory concentration of rifampin. Moreover, we found that human lactoferrin at concentrations of 1 mg/ml ($\approx 10^{-5}$ M) and above also enhances the ability of lysozyme to kill strains of *V. cholerae*, *E. coli* and *S. typhimurium*[62]. Similar to its effect on lactoferrin-mediated LPS release, iron saturation of the protein blocks this ability to enhance the activity of either rifampin or lysozyme. Similarly, calcium and magnesium also inhibit these effects in a dose dependent fashion.

While these studies indicate that lactoferrin modifies the Gram-negative bacterial outer membrane, they do not define the mechanism of action. Subsequent work was undertaken to determine whether lactoferrin damages the outer membrane through its chelating activity or through a polycationic effect. As the ability of lactoferrin to chelate calcium had not been defined, we undertook formal studies of the ability of lactoferrin to chelate calcium. In equilibrium dialysis studies with $^{45}Ca^{2+}$ under conditions similar to those used in the LPS release and lysozyme synergy work, neither lactoferrin nor transferrin has any ability to chelate the cation[62]. As calcium appears to be one of major cations present in the bacterial outer membrane[63], this observation argues strongly against the hypothesis that lactoferrin's membrane effect is due to its ability to act as an analogue of EDTA.

To then determine whether lactoferrin influences Gram-negative bacteria through a direct interaction with the bacterial membrane in a manner similar to the polycationic agents, we repeated studies of lactoferrin's interaction with lysozyme against *E. coli* in a dialysis chamber[62]. In these experiments contact between the bacterial cell and the proteins was controlled by the presence of a 6,000 molecular weight exclusion membrane. In this system, we found that as expected, the activity of lysozyme requires direct bacterial cell contact. Similarly, as expected, we found that a chelating resin (Chelex 100) can increase bacterial sensitivity to lysozyme independent of contact with the bacterial cell. Finally, we found that lactoferrin's effect on the outer membrane requires direct physical contact with the bacterial cells; it can not enhance the effect of lysozyme when it is separated from the cell by a dialysis membrane but does so when in direct contact.

Overall, this body of work indicates that lactoferrin modifies the Gram-negative bacterial outer membrane. The protein causes LPS release from multiple bacterial strains at concentrations and pH conditions that can be expected physiologically, and simultaneously alters the permeability characteristics of the membrane. The experimental observations are consistent with the prior studies showing that the antimicrobial properties of lactoferrin are inhibited by saturating the protein with iron. The inability of lactoferrin to bind calcium and the findings in the dialysis chamber studies suggest that these effects are not related to the protein's chelating ability but instead to a direct membrane interaction.

As lactoferrin and lysozyme are found in highest concentration at the same sites in vivo, the synergy between the proteins noted above may not be solely of mechanistic interest. The concurrent damage of both the outer membrane and the peptidoglycan layer by the two proteins should enhance bacterial susceptibility to other host defense systems such as complement, immunoglobulins and neutrophils. Further, this may represent a model to explain other reported interactions of lactoferrin. Such a synergistic effect may contribute to the previously reported observations that lactoferrin enhances the activity both of sIgA[17,43,44,79,80] and of bactenecin Bac7[37].

## LACTOFERRIN LPS BINDING

Lactoferrin could interact with several components of the Gram-negative bacterial outer membrane in mediating its damage. As noted above, selected Gram-negative bacteria appear to have protein receptors that specifically bind lactoferrin. However, such an interaction would not readily explain the ability of lactoferrin to cause LPS release or alter membrane permeability. In contrast, the apparent site of action of the polycationic membrane-active agents is the LPS molecule[81-84]. These agents directly bind to LPS, thereby altering the conformation and permeability characteristics of the outer membrane. Thus, if lactoferrin affects the outer membrane by a mechanism similar to the polycationic agents, it would be anticipated that lactoferrin should bind LPS.

During the last 2 years, 3 different groups have now demonstrated that lactoferrin does have the ability to bind LPS[62,85,86]. Binding has been shown by the use of an iodinated, photoactivable LPS derivative[85], through studies with intrinsically radiolabelled LPS molecules and immobilized lactoferrin[62], and through bioassays[85,86]. While general binding curves have been generated, an exact calculation of the number of LPS binding sites and the $K_d$ for the interactions between lactoferrin and LPS has not been possible because LPS exists as a mixed population of molecules of varying size and because free LPS molecules aggregate in solution[87].

That lactoferrin binds LPS provides support for a model that the protein can damage the Gram-negative outer membrane by acting as a polycationic agent (Figure 2). In this model lactoferrin would come in contact with the bacterial outer membrane and insert into the membrane to interact with the anionic core region of the LPS molecule. This interaction would lead to a disruption of the normal LPS-cation interaction and release of LPS from the bacterial cell. Replacement of the lost LPS molecules with neutral phospholipid molecules from the inner membrane would retain overall membrane integrity but would result in a change in membrane permeability.

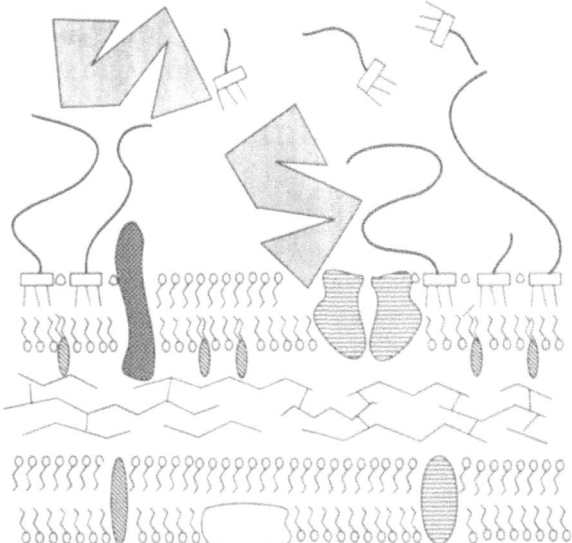

**Figure 2.** Proposed Model of Lactoferrin-mediated Outer Membrane damage. Lactoferrin (indicated by large shaded angular structure with two clefts) directly binds to the outer membrane and causes displacement of LPS molecules. The integrity of the membrane may be maintained by replacement of the lost LPS molecules with neutral phospholipids, but this changes alters the general membrane permeability to hydrophobic agents.

Any model of the interaction of lactoferrin with the outer membrane must also be consistent with recent observations that lactoferrin appears to bind to outer membrane porin proteins of Gram-negative organisms[58]. As noted above, in studies with iodinated lactoferrin Naidu and coworkers have shown reversible and specific binding of lactoferrin to *E. coli, A. hydrophila,* and *S. flexneri* whole bacterial cells[57,59]. After disruption and separation of bacterial outer membranes, immunoblot probing indicates that bound lactoferrin is associated with the outer membrane porin proteins.

These findings raise a question as to where lactoferrin interacts with the Gram-negative outer membrane. However, the work supports a model of a direct interaction between lactoferrin and the bacterial cell beyond that related to specific lactoferrin receptors. Further, these observations are actually not inconsistent with lactoferrin disrupting the LPS structure of the outer membrane. As integral outer membrane components, the bacterial porin proteins have an intimate interaction with LPS molecules. If lactoferrin binds to a porin protein it may disrupt the integrity of the adjacent membrane. Additionally, studies of porin proteins have shown that they have a high affinity for LPS, and LPS molecules remain associated with the porin proteins even after boiling and SDS-PAGE separation[88]. Thus, it is possible that the association of lactoferrin to porin proteins may be related to both of these proteins binding to an LPS intermediary. Further work will be necessary to elucidate the precise nature of the interaction between lactoferrin and the Gram-negative outer membrane.

It should be noted that the ability of lactoferrin to bind LPS may be an important property of the protein independent of the relevance of this interaction to its direct antimicrobial effects. Lactoferrin interferes with the ability of LPS to "prime" neutrophils for superoxide formation[86], and binding to LPS allows for increased uptake of lactoferrin by HL-60 cells[85]. It is of interest that another cationic neutrophil protein, BPI, can bind to LPS and neutralize its ability to activate neutrophils[84]. It is possible that the ability of lactoferrin to bind to and alter the effects of LPS may be another mechanism that contributes to the ability of lactoferrin to protect mice from a lethal challenge with E. coli[19]. As before, additional work is needed to define the importance of this property of lactoferrin.

## ANTIMICROBIAL ACTIVITY OF LACTOFERRICIN

If the outer membrane effect of lactoferrin is due to a mechanism of action similar to the other neutrophil derived cationic proteins such as BPI, the defensins, and the bactenecins; then cationic domains of the molecule must be important to the activity. The X-ray crystallographic analysis of human lactoferrin by Anderson et al indicates that one surfaced exposed region of the N-terminus of the protein has nine arginine and lysine residues in close proximity[89]. This is the principal cationic domain of the protein, and it can be hypothesized that it would be important to the outer membrane activity of lactoferrin.

This hypothesis has been confirmed by recent work with peptide fragments of lactoferrin. In studies with bovine lactoferrin, Tomita et al found that the bacteriostatic properties of the protein increase after enzymatic digestion with porcine pepsin[90]. The bacteriostatic properties are associated with a specific low-molecular weight peptide fragment that was subsequently purified by HPLC. Sequence analysis found that this peptide fragment is 25 amino acids long and has exact homology with an amino-terminal segment of the whole bovine lactoferrin sequence as derived from cDNA analysis (Figure 3)[91–93]. This segment corresponds precisely with the surface-exposed cationic domain identified in human lactoferrin, and has been designated as lactoferricin.

In collaborative studies with the Morinaga Milk Company, my laboratory characterized the antimicrobial properties of bovine lactoferricin[94]. To determine the effects of lactoferricin on the Gram-negative outer membrane we tested its ability to release LPS from each of the previously studied bacterial strains. Lactoferricin causes dramatically greater release of LPS

```
                      5           10          15          20          25
HUMAN LF[1]    G R R R R S V Q W C A V S Q P E A T K C F Q W Q R N
BOVINE LF[2]     A P R K N V R W C T I S Q P D S F K C P R W Q W R
MURINE LF[3]   A - K A T T V R W C A V S N S E E K C L R W Q N E

HUMAN TF[1]      V P D K T V R W C A V S E H E A T K C Q S F R C H


                    30          35          40          45
HUMAN LF      M R K V R - - - G P P V S C I K R D S P I Q C I Q A
BOVINE LF     M K K L G - - - A P S I T C V R R A F A L E C I R A
MURINE LF     M R K V G - - - G P P L S C V K K S S T R Q C I Q A

HUMAN TF      M K S V I P S D G P S V A C V K K A S Y L D C I R A


                50          55          60          65          70          75
HUMAN LF      I A E N R A D A V T L D G G F I Y E A G L A P Y K L
BOVINE LF     I A E K K A D A V T L D G G M V F E A G R D P Y K L
MURINE LF     I V T N R A D A M T L D G G T M F D A G K P P Y K L

HUMAN TF      I A A N E A D A V T L D A G L V Y D A Y L A P N N L
```

**Figure 3.** N-terminus sequences of human, bovine, and murine lactoferrin (LF), as well as human transferrin (TF). The boxed sequence indicates regions of complete homology. The antimicrobial bovine lactoferricin peptide is delineated by the double underline. Amino acid residues indicated in bold are positively charged. (References: (1) Eur. J. Biochem. 1984, 145: 659–676; (2) Nucleic Acids Res. 1990, 18: 7167; (3) J. Biol. Chem. 1987, 262: 10134–10139).

**Table 1.** Release of LPS from Three Bacterial Strains by Bovine Lactoferrin and Lactoferricin[a]

| Bacteria and growth medium | 30 minute % H Release (Mean ± SEM) | | |
| --- | --- | --- | --- |
| | HBSS-CM buffer[b] | Bovine lactoferrin (2 mg/ml) | Lactoferricin (100 µg/ml) |
| E. coli CL99 1-2, WMS-III | 1.1 ± 1.1 (n = 3) | 0.8 ± 0.8 (n = 3) | 26.6 ± 3.4[c] (n = 3) |
| S. montevideo SL5222, Luria + 2 mM calcium | 3.4 ± 1.9 (n = 3) | 24.6 ± 2.0[d] (n = 3) | 39.8 ± 10.6[d] (n = 3) |
| S. typhimurium SL696, Luria + 2 mM calcium | 4.6 ± 1.0 (n = 7) | 22.8 ± 6.8[e] (n = 5) | 45.5 ± 1.8[f] (n = 7) |

[a]from[94]
[b]Hanks balanced salt solution lacking calcium and magnesium
[c]p < 0.05 vs HBSS-CM and bovine lactoferrin
[d]p < 0.05 vs HBSS-CM
[e]p < 0.005 vs HBSS-CM
[f]p < 0.05 vs bovine lactoferrin p < 0.0001 vs HBSS-CM
[g]p < 0.0001 vs bovine lactoferrin and HBSS-CM

than does the whole protein (Table 1). Similar to whole lactoferrin, lactoferricin increases bacterial susceptibility to lysozyme[94]. Additionally, the peptide fragment appears to have a similar ability to bind intrinsically radiolabelled LPS (Figure 4).

Although in these properties lactoferricin is similar to lactoferrin, it differs in having marked bactericidal activity[94]. When studied in 1% Bacto Peptone media, lactoferricin is bactericidal for a variety of bacteria and yeast at concentrations between 2 and 20 µg/ml. Time-kill studies against an E. coli strain show an inoculum effect and possibility two phases of killing. The majority of bacterial cells are killed within 1 hr but killing continues through 24 hrs (Figure 5). This bactericidal activity of lactoferricin is inhibited by the same levels of calcium and magnesium that inhibit the outer membrane effects of lactoferrin. In contrast, while iron saturation completely blocks the bacteriostatic effects of lactoferrin, ferric chloride only partially inhibits the bactericidal effect of lactoferricin[94]). Finally, ultrastructural analysis

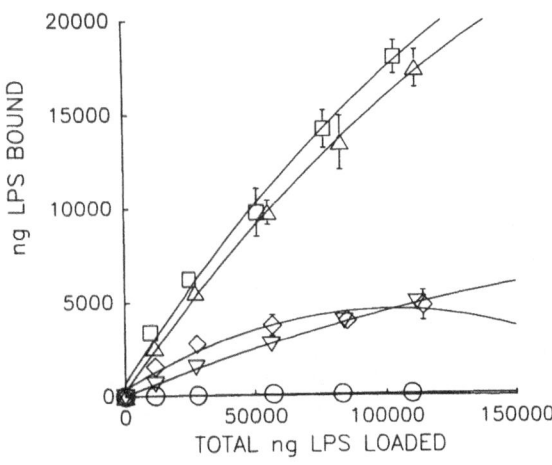

**Figure 4.** Binding of E. coli CL99 1–2 [³H]-LPS to bovine lactoferrin (□), lactoferricin (△), poly-1-lysine (◊), and bovine serum albumin (▽) that had each been bound to Sepharose beads, and to Tris-blocked Sepharose beads (○) (mean ± SEM, 3 experiments). The binding curves are plotted using second-order linear regression. Adapted from [94].

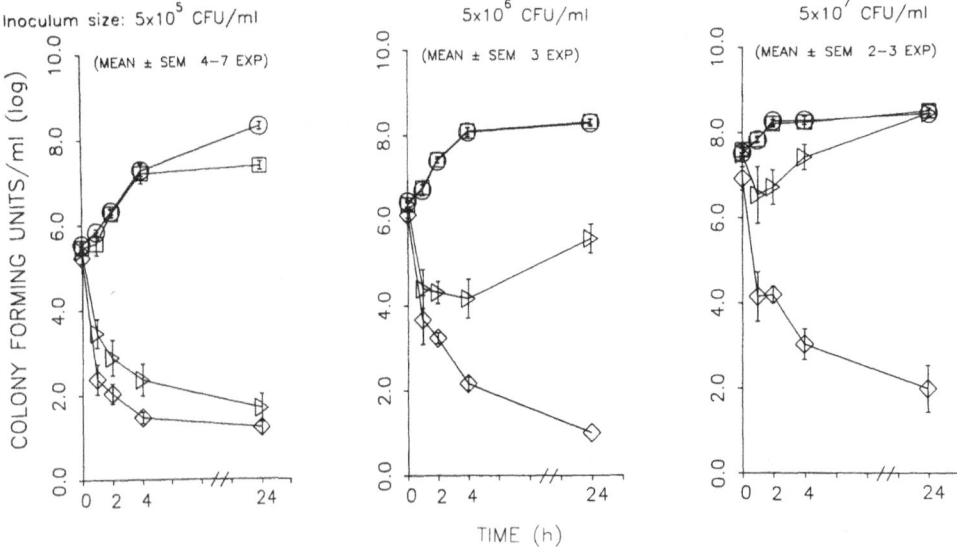

**Figure 5.** Killing of *E. coli* CL99 1–2 by varying concentrations of lactoferricin in 1% Bacto Peptone. Each panel shows the activity against a different initial bacterial inoculum with $5 \times 10^5$ cfu on the left, $5 \times 10^6$ cfu in the center, and $5 \times 10^7$ cfu on the right. ○, Bacto Peptone; ▢, lactoferricin (1.0 µg/ml); Δ, lactoferricin (10 µg/ml); ◊, lactoferricin (100 µg/ml); from[94.]

by transmission electron microscopy of *E. coli* cells exposed to lactoferricin shows marked effects. Lactoferricin causes the immediate development of electron dense membrane "blisters", and there is a subsequent "coagulation" of cytoplasmic elements (Figure 6).

These studies with bovine lactoferricin have several implications. First, they suggest that the amino terminal domain of lactoferrin that comprises lactoferricin makes an important contribution to the outer membrane effects of lactoferrin; and may be the sole domain contributing to this activity. This N-terminus domain is distinct from the metal binding clefts of lactoferrin[89] indicating that the protein must then be at least bifunctional in its antimicrobial activity. As there is striking homology between the N-termini of bovine, human, and murine lactoferrin and human transferrin, it is probable that similar active peptides can be generated from all these proteins (Figure 3). Bellamy has reported the isolation of a human lactoferricin peptide with similar bactericidal activity that corresponds to the first 47 residues of the N-terminus of lactoferrin[91].

Second, the broad bactericidal activity of lactoferricin and its effects on bacterial morphology are quite distinct from the effects of whole lactoferrin and more similar to the activities of the defensins[70]. This would suggest a slightly different model for the antimicrobial activity of lactoferricin (Figure 7). It is probable that the peptide lactoferricin also directly binds to the outer membrane of the cell to cause LPS release and altered outer membrane permeability. However, while the large lactoferrin molecule is sterically restricted, the small size of lactoferricin (3,126 molecular weight) should allow it to penetrate through the membrane and interact with the inner membrane. It is likely that an effect on the inner membrane may then mediate the bactericidal effect as occurs with defensins[70], but this awaits experimental confirmation.

Third, that lactoferricin is released from lactoferrin after pepsin treatment suggests that it could be released in vivo within the gut lumen. Alternatively, as the peptide can be released by the aspartic protease of *Penicillium duponti*[91], it is possible that other host proteases, such

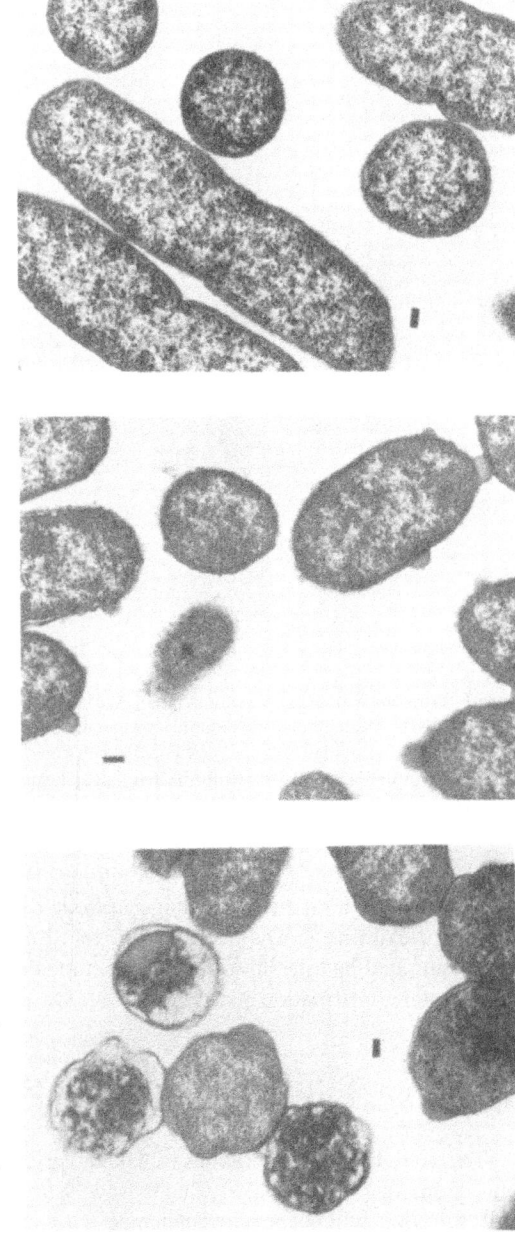

**Figure 6.** Transmission electron microscopy of *E. coli* CL99 1–2 cells (*Upper*) incubated for 2 h in 1% Bacto Peptone, (*Middle*) incubated for 0 h in 1% Bacto Peptone with 100 µg/ml lactoferricin, and (*Lower*) incubated for 0.5 h in 1% Bacto Peptone with 100 µg/ml lactoferricin (markers indicate 100 nm).

as those within a phagocytic cell could release the peptide fragment. Thus, the overall antimicrobial activity of lactoferrin may represent a summation of the antimicrobial activities of both the whole protein and the free peptide fragment.

Fourth, while iron does partially inhibit the activity of lactoferricin, the inhibition occurs at high iron levels suggesting a nonspecific interaction similar to that of calcium and magnesium (discussed in greater detail below). In contradistinction, iron completely blocks multiple outer membrane effects of whole lactoferrin at low concentrations[60,62]. As the N-terminus domain is not associated with the chelating activity of lactoferrin, there is

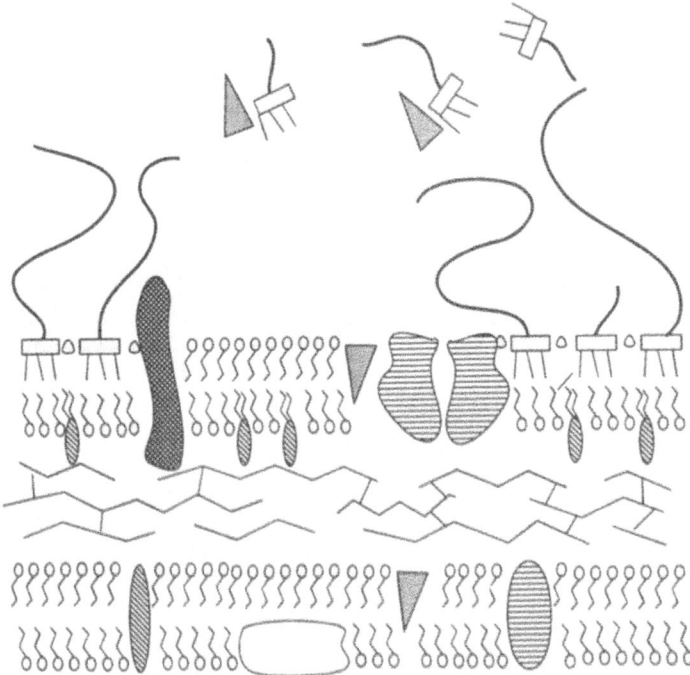

**Figure 7.** Proposed Model of the Antimicrobial Activity of Lactoferricin. Similar to whole lactoferrin, lactoferricin (indicated by shaded triangles) binds to the outer membrane causing LPS release and altered outer membrane permeability. With its small size, lactoferricin molecules are likely to penetrate through the outer membrane and murein sacculus to reach and damage the bacterial cytoplasmic membrane leading to bacterial cell death.

no direct explanation for these observations. However, when lactoferrin becomes iron saturated there is a significant change in three-dimensional structure and a diminution in molecular flexibility[95]. The decrease in overall molecular flexibility may limit the ability of iron-saturated lactoferrin to interact with the bacterial cell. This hypothesis will require experimental confirmation.

## IMPORTANCE OF CALCIUM AND MAGNESIUM

There are high concentrations of calcium and magnesium in most physiologic environments. Consequently, it is likely that these in vitro observations on the activity of lactoferrin and lactoferricin will not be relevant in many sites in vivo. However, work by Pollack indicates that the $Ca^{2+}$ level within the macrophage phagolysosome is less than 100 μM, and is an environment where lactoferrin should alter the Gram-negative outer membrane and lactoferricin should be bactericidal[96]. Additionally, it is notable that both $Ca^{2+}$ and $Mg^{2+}$ (at concentrations between 1 and 10 mM) have been shown to block the antimicrobial activity of other characterized cationic neutrophilic antimicrobial proteins including BPI, azurocidin, cathepsin G, and defensins[69,97–100]. Thus, the influence of free $Ca^{2+}$ and $Mg^{2+}$ on lactoferrin's activity are similar to their effects on a family of host defense proteins, and is probably related to an interference with protein binding to the bacterial cell[69]. The divalent cation concentrations within the neutrophil phagolysosome have not been characterized, but it is reasonable to

**Table 2.** Antimicrobial Activity of Bovine Lactoferricin against Selected Bacteria and Yeast in Bacto-peptone*

| Bacterial strains | Minimal inhibitory concentration (µg/ml) | Minimal bactericidal concentration (µg/ml) |
|---|---|---|
| *E. coli* CL99 1-2 | | |
| prep 1 | 4 | 8 |
| prep 2 | 13 | 17 |
| *S. typhimurium* SL696 | | |
| prep 1 | 5 | 8 |
| prep 2 | 21 | 21 |
| *S. montevideo* SL5222 | | |
| prep 1 | 3 | 9 |
| prep 2 | 13 | 13 |
| *S. typhimurium* 6749 | 1.6 | 3.3 |
| *S. typhimurium* SH7641 | 1.6 | 1.6 |
| *E. coli* K12 UB1005 | 1.6 | 1.6 |
| *E. coli* K12 UB1005 DC-2 | 1.6 | 1.6 |
| *E. coli* ATCC 25922 | 3.3 | 3.3 |
| *P. aeruginosa* ATCC 2783 | 3.3 | > 125 |
| *P. aeruginosa* PAO-1 | 3.3 | > 125 |
| *S. aureus* ATCC 29213 | 6.6 | 13.2 |
| *L. monocytogenes* EGD | 1.6 | 3.3 |
| *L. monocytogenes* 4b, maritime | 6.6 | 13.2 |
| *Candida albicans* 6372 | 0.8 | 0.8 |
| *Candida albicans* 6434 | 0.8 | 0.8 |

[a]Mean 2–6 experiments. *E. coli* CL99 1-2, *S. typhimurium* SL696, and *S. montevideo* SL5222 were tested against two separate preparations of lactoferricin purified by 2 different HPLC chromatographic schema. All other isolates were tested against a single lactoferricin preparation. *E. coli* strains UB1005 and UB1005 DC-2 are a laboratory parent and polymyxin B hypersusceptible mutant[81]. *P. aeruginosa* PAO-1 and *L. monocytogenes* EGD are defined laboratory isolates. *L. monocytogenes* 4b, maritime is a clinical epidemic strain; and the *C. albicans* strains are blood culture isolates from the University of Colorado Health Sciences Center clinical microbiology laboratory. Adapted from [94].

hypothesize that for these neutrophil proteins to function the calcium levels are probably comparable to that of the macrophage. Thus, this is another site where lactoferrin could cause outer membrane damage. Even more recently, defensin-like peptides have been isolated both from bovine trachea (tracheal antimicrobial peptide) and from murine small intestine paneth cells (cryptdins)[101,102]. While the effect of calcium on the activity of these new peptides has not been investigated, given their structural homology with the defensin peptides it is likely that they are also calcium inhibitable. Thus, their production within the trachea and the small intestine suggests that microenvironments in these locations are also likely to have low calcium concentrations where these peptides and other neutrophil-associated cationic proteins can be active.

## SUMMARY

In vivo studies suggest that lactoferrin can contribute to antimicrobial host defense. This argument is supported by in vitro data that indicates that the protein can limit the availability of iron to bacteria, enhance the function of phagocytic cells, and directly damage the Gram-negative bacterial outer membrane. Mechanistic studies suggest that the outer membrane activity of the protein relates to a direct interaction with the bacterial cell, potentially on the basis of the highly cationic nature of the protein. This hypothesis is supported by evidence that a cationic N-terminal domain of the protein that is distinct from the chelating region of the molecule can be cleaved to release a peptide fragment with bactericidal activity. The peptide itself appears to share a mechanism of action common to a large family of other cationic neutrophil-associated peptides.

Thus, lactoferrin can be considered at least bifunctional in its contribution to antimicrobial host defense. It is not clear whether one or the other of the two antimicrobial mechanisms of action is predominant. It may well be that they are independent functions active against different microorganisms at different physiologic sites. Selected bacteria may be affected by both functions of the protein, and others resistant to both. Further, given the influence of calcium and magnesium on the direct membrane effects of lactoferrin, it is probable that this mechanism is relevant only in selected physiologic environments. It is obvious that much more information is needed before we fully understand the host defense function of this multipotent protein.

## REFERENCES

1. Bullen JJ, Rogers HJ, Griffiths E. (1978) Role of iron in bacterial infection. *Curr Top Microbiol Immunol* **80**:1–35.

2. Moguilevsky N, Retegui LA, Masson PL. (1985) Comparison of human lactoferrins from milk and neutrophilic leucocytes. Relative molecular mass, isoelectric point, iron-binding properties and uptake by the liver. *Biochem J* **229**:353–359.

3. Baker EN, Rumball SV, Anderson BF. (1987) Transferrins: insights into structure and function from studies on lactoferrin. *Trends Biochem Sci* **12**:350–353.

4. Aisen P, Leibman A. (1972) Lactoferrin and transferrin: a comparative study. *Biochim Biophys Acta* **257**:314–323.

5. Groves ML. (1960) The isolation of a red protein from milk. *J Am Chem Soc* **82**:3345–3360.

6. Johansson B. (1960) Isolation of an iron-containing red protein from human milk. *Acta Chem Scand* **14**:510–512.

7. Bullen JJ. (1981) The significance of iron in infection. *Rev Infect Dis* **3**:1127–1138.

8. Masson PL, Heremans JF, Prignot JJ, Wauters G. (1966) Immunohistochemical localization and bacteriostatic properties of an iron-binding protein from bronchial mucus. *Thorax* **21**:538–544.

9. Oram JD, Reiter B. (1968) Inhibition of bacteria by lactoferrin and other iron-chelating agents. *Biochimica et Biophysica Acta* **170**:351–365.

10. Bullen JJ, Rogers HJ, Leigh L. (1972) Iron-binding proteins in milk and resistance to *Escherichia coli* infection in infants. *Br Med J* **1**:69–75.

11. Reiter B, Brock JH, Steel ED. (1975) Inhibition of *Escherichia coli* by bovine colostrum and post-colostral milk. II. The bacteriostatic effect of lactoferrin on a serum-susceptible and serum-resistant strain of *E. coli*. *Immunology* **28**:83–95.

12. Spik G, Cheron A, Montreuil J, Dolby JM. (1978) Bacteriostasis of a milk-sensitive strain of *Escherichia coli* by immunoglobulins and iron-binding proteins in association. *Immunology* **35**:663–671.

13. Bishop JG, Schanbacher FL, Ferguson LC, Smith KL. (1976) In vitro growth inhibition of mastitis-causing coliform bacteria by bovine apo-lactoferrin and reversal of inhibition by citrate and high concentrations of apo-lactoferrin. *Infect Immun* **14**:911–918.

14. Nonnecke BJ, Smith KL. (1984) Inhibition of mastitic bacteria by bovine milk apo-lactoferrin evaluated by in vitro microassay of bacterial growth. *J Dairy Sci* **67**:606–613.

15. Griffiths E, Humphreys J. (1977) Bacteriostatic effect of human milk and bovine colostrum on *Escherichia coli*: importance of bicarbonate. *Infect Immun* **15**:396–401.

16. Stuart J, Norrell S, Harrington JP. (1984) Kinetic effect of human lactoferrin on the growth of *Escherichia coli* O111. *Int J Biochem* **16**:1043–1047.

17. Stephens S, Dolby JM, Montreuil J, Spik G. (1980) Differences in inhibition of the growth of commensal and enteropathogenic strains of *Escherichia coli* by lactotransferrin and secretory immunoglobulin A isolated from human milk. *Immunology* **41**:597–603.

18. Sawatzki G, Hoffman F, Kubanek B. The role of iron binding proteins, lactoferrin and transferrin, in *Salmonella typhimurium* infections in mice. In: *Structure and function of iron storage and transport proteins*, edited by Urushizaki, I. and et al, Elsevier Science Publishers B.V., 1983, p. 435–439.

19. Zagulski T, Lipinski P, Zagulska A, Broniek S, Jarzabek Z. (1989) Lactoferrin can protect mice against a lethal dose of *Escherichia coli* in experimental infections *in vivo*. *Br J Exp Path* **70**:697–704.

20. Czirok E, Milch H, Nemeth K, Gado I. (1990) In vitro and in vivo (LD$_{50}$) effects of human lactoferrin on bacteria. *Acta Microbiologica Hungarica* **37**:55–71.

21. LaForce FM, Boose DS, Ellison RTIII. (1986) Effect of aerosolized *Escherichia coli* and *Staphylococcus aureus* on iron and iron-binding proteins in lung lavage fluid. *J Infect Dis* **154**:959–965.

22. Finkelstein RA, Sciortino CV, McIntosh MA. (1983) The role of iron in microbe-host interactions. *Rev Infect Dis* **5**:S759–S777.

23. Schryvers AB. (1989) Identification of the transferrin- and lactoferrin-binding proteins in *Haemophilus influenzae*. *J Med Microbiol* **29**:121–130.

24. Lee BC, Bryan LE. (1989) Identification and comparative analysis of the lactoferrin and transferrin receptors among clinical isolates of gonococci. *J Med Microbiol* **28**:199–204.

25. Schryvers AB, Lee BC. (1988) Comparative analysis of the transferrin and lactoferrin binding proteins in the family *Neisseriaceae*. *Can J Microbiol* **35**:409–415.

26. Schryvers AB, Gonzalez GC. (1989) Comparison of the abilities of different protein sources of iron to enhance *Neisseria meningitidis* infection in mice. *Infect Immun* **57**:2425–2429.

27. Ogunnariwo JA, Schryvers AB. (1990) Iron acquisition in *Pasteurella haemolytica*: expression and identification of a bovine-specific transferrin receptor. *Infect Immun* **58**:2091–2097.

28. Schryvers AB, Morris LJ. (1988) Identification and characterization of the human lactoferrin-binding protein from *Neisseria meningitidis*. *Infect Immun* **56**:1144–1149.

29. Blanton KJ, Biswas GD, Tsai J, Adams J, Dyer DW, Davis SM, Koch GG, Sen PK, Sparling PF. (1990) Genetic evidence that *Neisseria gonorrhoeae* produces specific receptors for transferrin and lactoferrin. *J.Bacteriol.* **172**:5225–5235.

30. Stevenson P, Williams P, Griffiths E. (1992) Common antigenic domains in transferrin-binding protein 2 of *Neisseria meningitidis*, *Neisseria gonorrhoeae*, and *Haemophilus influenzae* type b. *Infect Immun* **60**:2391–2396.

31. Banerjee-Bhatnagar N, Frasch CE. (1990) Expression of *Neisseria meningitidis* iron-regulated outer membrane proteins, including a 70-kilodalton transferrin receptor, and their potential for use as vaccines. *Infect Immun* **58**:2875–2881.

32. Padda JS, Schryvers AB. (1990) N-linked oligosaccharides of human transferrin are not required for binding to bacterial transferrin receptors. *Infect Immun* **58**:2972–2976.

33. Dolby JM, Stephens S. (1983) Antibodies to *Escherichia coli* O antigens and the in vitro bacteriostatic properties of human milk and its IgA. *Acta Paediatr Scand* **72**:577–582.

34. Pollock JJ, Shoda J, McNamara TF, Cho M-I, Campbell A, Iacono VJ. (1984) In vitro and in vivo studies of cellular lysis of oral bacteria by a lysozyme-protease-inorganic monovalent anion antibacterial system. *Infect Immun* **45**:610–617.

35. Wright DG, Gallin JI. (1979) Secretory responses of human neutrophils: exocytosis of specific (secondary) granules by human neutrophils during adherence in vitro and during exudation in vivo. *J Immunol* **123**:285–294.

36. Joiner KA, Ganz T, Albert J, Rotrosen D. (1989) The opsonizing ligand on *Salmonella typhimurium* influences incorporation of specific, but not azurophil, granule constitutents into neutrophil phagosomes. *J Leukocyte Biol* **109**:2771–2782.

37. Skerlavaj B, Romeo D, Gennaro R. (1990) Rapid membrane permeabilization and inhibition of vital functions of Gram-negative bacteria by bactenecins. *Infect Immun* **58**:3724–3730.

38. Lima MF, Kierszenbaum F. (1985) Lactoferrin effects on phagocytic cell function. I. Increased uptake and killing of an intracellular parasite by murine macrophages and human monocytes. *J Immunol* **134**:4176–4183.

39. Lima MF, Kierszenbaum F. (1987) Lactoferrin effects on phagocytic cell function. II. The presence of iron is required for the lactoferrin molecule to stimulate intracellular killing by macrophages but not to enhance the uptake of particles and microorganisms. *J Immunol* **139**:1647–1651.

40. Byrd TF, Horwitz MA. (1991) Lactoferrin inhibits or promotes *Legionella pneumophila* intracellular multiplication in nonactivated and interferon gamma-activated human monocytes depending upon its degree of iron saturation. Iron-lactoferrin and nonphysiologic iron chelates reverse monocyte activation against *Legionella pneumophila*. *J Clin Invest* **88**:1103–1112.

41. Byrd TF, Horwitz MA. (1989) Interferon gamma-activated human monocytes downregulate transferrin receptors and inhibit the intracellular multiplication of *Legionella pneumophila* by limiting availability of iron. *J Clin Invest* **83**:1457–1465.

42. Britigan BE, Serody JS, Hayek MB, Charniga LM, Cohen MS. (1991) Uptake of lactoferrin by mononuclear phagocytes inhibits their ability to form hydroxyl radical and protects them from membrane autoperoxidation. *J Immunol.* **147**:4271–4277.

43. Rainard P. (1986) Bacteriostasis of *Escherichia coli* by bovine lactoferrin, transferrin and immunoglobulins (IgG1, IgG2, IgM) acting alone or in combination. *Vet Microbiol* **11**:103–115.

44. Rogers HJ, Synge C. (1978) Bacteriostatic effect of human milk on *Escherichia coli*: the role of IgA. *Immunology* **34**:19–28.

45. Fitzgerald SP, Rogers HJ. (1980) Bacteriostatic effect of serum: role of antibody to lipopolysaccharide. *Infect Immun* **27**:302–308.

46. Arnold RR, Cole MF, McGhee JR. (1977) A bactericidal effect for human lactoferrin. *Science* **197**:263–265.

47. Arnold RR, Brewer M, Gauthier JJ. (1980) Bactericidal activity of human lactoferrin: sensitivity of a variety of microorganisms. *Infect Immun* **28**:893–898.

48. Arnold RR, Russell JE, Champion WJ, Brewer M, Gauthier JJ. (1982) Bactericidal activity of human lactoferrin: differentiation from the stasis of iron deprivation. *Infect Immun* **35**:792–799.

49. Bortner CA, Miller RD, Arnold RR. (1986) Bactericidal effect of lactoferrin on *Legionella pneumophila*. *Infect Immun* **51**:373–377.

50. Kalmar JR, Arnold RR. (1988) Killing of *Actinobacillus actinomycetemcomitans* by human lactoferrin. *Infect Immun* **56**:2552–2557.

51. Bortner CA, Arnold RR, Miller RD. (1989) Bactericidal effect of lactoferrin on *Legionella pneumophila*: effect of the physiological state of the organism. *Can J Microbiol* **35**:1048–1051.

52. Motley MA, Arnold RR. (1987) Cofactor requirements for expression of lactoferrin bactericidal activity on enteric bacteria. *Adv Exp Med Biol* **216A**:591–599.

53. Hammill R, Brown D, Watson D, Musher D, Baughn R. (1987) The observed bactericidal effect of lactoferrin is an artifact of pH. *27th Interscience Conference on Antimicrobial Agents and Chemotherapy* **Abstract** 560.

54. Dalmastri C, Valenti P, Visca P, Vittorioso P, Orsi N. (1988) Enhanced antimicrobial activity of lactoferrin by binding to the bacterial surface. *Microbiologica* **11**:225–230.

55. Visca P, Dalmastri C, Verzili D, Antonini G, Chiancone E, Valenti P. (1990) Interaction of lactoferrin with *Escherichia coli* cells and correlation with antibacterial activity. *Med Microbiol Immunol* **179**:323–333.

56. Naidu SS, Erdei J, Czirok E, Kalfas S, Gado I, Thoren A, Forsgren A, Naidu AS. (1991) Specific binding of lactoferrin to *Escherichia coli* isolated from human intestinal infections. *APMIS* **99**:1142–1150.

57. Kishore AR, Erdei J, Naidu SS, Falsen E, Forsgren A, Naidu AS. (1991) Specific binding of lactoferrin to *Aeromonas hydrophila*. *FEMS Microbiol.Lett.* **83**:115–119.

58. Tigyi Z, Kishore AR, Maeland JA, Forsgren A, Naidu AS. (1992) Lactoferrin-binding proteins in *Shigella flexneri*. *Infect Immun* **60**:2619–2626.

59. Gado I, Erdei J, Laszlo VG, Paszti J, Czirok E, Kontrohr T, Toth I, Forsgren A, Naidu AS. (1991) Correlation between human lactoferrin binding and colicin susceptibility in *Escherichia coli*. *Antimicrob.Agents Chemother*. **35**:2538–2543.

60. Ellison RT III, Giehl TJ, LaForce FM. (1988) Damage of the outer membrane of enteric Gram-negative bacteria by lactoferrin and transferrin. *Infect Immun* **56**:2774–2781.

61. Ellison RT III, LaForce FM, Giehl TJ, Boose DS, Dunn BE. (1990) Lactoferrin and transferrin damage of the Gram-negative outer membrane is modulated by $Ca^{2+}$ and $Mg^{2+}$. *J Gen Microbiol* **136**:1437–1446.

62. Ellison RT III, Giehl TJ. (1991) Killing of Gram-negative bacteria by lactoferrin and lysozyme. *J Clin Invest* **88**:1080–1091.

63. Coughlin RT, Tonsager S, McGroarty EJ. (1983) Quantitation of metal cations bound to membranes and extracted lipopolysaccharide of *E. coli*. *Biochemistry* **22**:2002–2007.

64. Leive L. (1974) The barrier function of the Gram-negative envelope. *Ann NY Acad Sci* **235**:10–127.

65. Nikaido H, Vaara M. (1985) Molecular basis of bacterial outer membrane permeability. *Microbiol Rev* **49**:1–32.

66. Joiner KA. (1985) Studies on the mechanism of bacterial resistance to complement-mediated killing and on the mechanism of bactericidal antibody. *Curr Top Microbiol Immunol* **121**:99–133.

67. Bryan CS. (1974) Sensitization of *E. coli* to the serum bactericidal system and to lysozyme by ethyleneglycoltetraacetic acid. *Proc Soc Exp Biol Med* **1451**:1431–1433.

68. Vaara M, Vaara T. (1983) Polycations as outer membrane-disorganizing agents. *Antimicrob Agents Chemother* **24**:114–122.

69. Mannion BA, Weiss J, Elsbach P. (1990) Separation of sublethal and lethal effects of the bactericidal/permeability increasing protein on *Escherichia coli*. *J Clin Invest* **85**:853–860.

70. Lehrer RI, Barton A, Daher K, Harwig SSL, Ganz T, Selsted ME. (1989) Interaction of human defensins with *Escherichia coli*. Mechanisms of bactericidal activity. *J Clin Invest* **84**:553–561.

71. Hancock REW, Wong PGW. (1984) Compounds which increase the permeability of the *Pseudomonas aeruginosa* outer membrane. *Antimicrob Agents Chemother* **26**:48–52.

72. Vaara M, Viljanen P. (1985) Binding of polymyxin B nonapeptide to Gram-negative bacteria. *Antimicrob Agents Chemother* **27**:548–554.

73. Boman HG. (1991) Antibacterial peptides: key components needed in immunity. *Cell* **65**:205–207.

74. Lehrer RI, Ganz T. (1990) Antimicrobial polypeptides of human neutrophils. *Blood* **76**:2169–2181.

75. Lehrer RI, Ganz T, Selsted ME. (1991) Defensins: endogenous antibiotic peptides of animal cells. *Cell* **64**:229–230.

76. Goldman RC, Leive L. (1980) Heterogeneity of antigenic-side-chain length in lipopolysaccharide from *Escherichia coli* O111 and *Salmonella typhimurium* LT2. *Eur J Biochem* **107**:145–153.

77. Hukari RI, Helander M, Vaara M. (1986) Chain length heterogeneity of lipopolysaccharide released from *Salmonella typhimurium* by ethylenediaminetetraacetic acid or polycations. *Eur J Biochem* **154**:673–676.

78. Joiner KA, Grossman N, Schmetz M, Leive L. (1986) C3 binds preferentially to long-chain lipopolysaccharide during alternative pathway activation by *Salmonella montevideo*. *J Immunol* **136**:710–715.

79. Samson RR, Mirtle C, McClelland DBL. (1979) Secretory IgA does not enhance the bacteriostatic effects of iron-binding or vitamin B12-binding proteins in human colostrum. *Immunology* **38**:367–373.

80. Boesman-Finkelstein M, Finkelstein RA. (1985) Antimicrobial effects of human milk: inhibitory activity on enteric pathogens. *FEMS Microbiol Lett* **27**:167–174.

81. Rocque WJ, Fesik SW, Haug A, McGroarty EJ. (1988) Polycation binding to isolated lipopolysaccharide from antibiotic-hypersusceptible mutant strains of *Escherichia coli*. *Antimicrob Agents Chemother* **32**:308–313.

82. Peterson AA, Fesik SW, McGroarty EJ. (1987) Decreased binding of antibiotics to lipopolysaccharides from polymyxin-resistant strains of *Escherichia coli* and *Salmonella typhimurium*. *Antimicrob Agents Chemother* **31**:230–237.

83. Danner RL, Joiner KA, Rubin M, Patterson WH, Johnson N, Ayers KM, Parrillo JE. (1989) Purification, toxicity, and antiendotoxin activity of polymyxin B nonapeptide. *Antimicrob Agents Chemother* **33**:1428–1434.

84. Marra MN, Wilde CG, Griffith JE, Snable JL, Scott RW. (1990) Bactericidal/permeability-increasing protein has endotoxin-neutralizing activity. *J.Immunol.* **144**:662–666.

85. Miyazawa K, Mantel C, Lu L, Morrison DC, Broxmeyer HE. (1991) Lactoferrin-lipopolysaccharide interactions. Effect on lactoferrin binding to monocyte/macrophage-differentiated HL-60 cells. *J Immunol* **146**:723–729.

86. Cohen MS, Mao J, Rasmussen GT, Serody JS, Britigan BE. (1992) Interaction of lactoferrin and lipopolysaccharide (LPS): effects on the antioxidant property of lactoferrin and the ability of LPS to prime human neutrophils for enhanced superoxide formation. *J Infect Dis* **166**:1375–1378.

87. Brade L, Brandenburg K, Kuhn H-M, Kusumoto S, Macher I, Rietschel ET, Brade H. (1987) The immunogenicity and antigenicity of lipid A are influenced by its physicochemical state and environment. *Infect Immun* **55**:2636–2644.

88. Rocque WJ, Coughlin RT, McGroarty EJ. (1987) Lipopolysaccharide tightly bound to porin monomers and trimers from *Escherichia coli* K-12. *J Bacteriol* **169**:4003–4010.

89. Anderson BF, Baker HM, Norris GE, Rice DW, Baker EN. (1989) Structure of human lactoferrin: crystallographic structure analysis and refinement at 2.8 angstrom resolution. *J Mol Biol* **209**:711–734.

90. Tomita M, Bellamy W, Takase M, Yamauchi K, Wakabayashi H, Kawase K. (1991) Potent antibacterial peptides generated by pepsin digestion of lactoferrin. *J Dairy Sci* **74**:4137–4142.

91. Bellamy W, Takase M, Yamauchi K, Wakabayashi H, Kawase K, Tomita M. (1992) Identification of the bactericidal domain of lactoferrin. *Biochim Biophys Acta* **1121**:130–136.

92. Pierce A, Colavizza D, Benaissa M, Maes P, Tartar A, Montreuil J, Spik G. (1991) Molecular cloning and sequence analysis of bovine lactotransferrin. *Eur J Biochem* **196**:177–184.

93. Goodman RE, Schanbacher FL. (1991) Bovine lactoferrin mRNA: sequence, analysis, and expression in the mammary gland. *Biochem Biophys Res Comm* **180**:75–84.

94. Yamauchi K, Tomita M, Giehl TJ, Ellison RTIII. (1993) Antibacterial activity of Lactoferrin and a pepsin-derived lactoferrin peptide fragment. *Infect Immun* **61**:719–728.

95. Baker EN, Anderson BF, Baker HM, Haridas M, Jameson GB, Norris GE, Rumball SV, Smith CA. (1991) Structure, function and flexibility of human lactoferrin. *Int J Biol Macromol* **13**:122–129.

96. Pollack C, Straley SC, Klempner MS. (1986) Probing the phagolysosomal environment of human macrophages with a $Ca^{2+}$-responsive operon fusion in *Yersinia pestis. Nature (Lond.)* **322**:834-836.

97. Weiss J, Muello K, Victor M, Elsbach P. (1984) The role of lipopolysaccharides in the action of the bactericidal/permeability increasing neutrophil protein on the bacterial envelope. *J Immunol* **132**:3109–3115.

98. Campanelli D, Detmers PA, Nathan CF, Gabay JE. (1990) Azurocidin and a homologous serine protease from neutrophils. Differential antimicrobial and proteolytic properties. *J Clin Invest* **85**:90–-915.

99. Odeberg H, Olsson I. (1976) Mechanisms for the microbicidal activity of cationic proteins of human granulocytes. *Infect Immun* **14**:1269–1275.

100. Lehrer RI, Ganz T, Szklarek D, Selsted ME. (1988) Modulation of the in vitro candidacidal activity of human neutrophil defensins by target cell metabolism and divalent cations. *J Clin Invest* **81**:1829–1835.

101. Diamond G, Zasloff M, Eck H, Brasseur M, Maloy WL, Bevins CL. (1991) Tracheal antimicrobial peptide, a cysteine-rich peptide from mammalian tracheal mucosa: peptide isolation and cloning of a cDNA. *Proc Natl Acad Sci USA* **88**:3952–3956.

102. Eisenhauer PB, Harwig SSSL, Lehrer RI. (1992) Cryptdins: Antimicrobial defensins of the murine small intestine. *Infect Immun* **60**:3556–3565.

# FAECAL FLORA IN THE NEWBORN

## Effect of Lactoferrin and Related Nutrients

B.A. Wharton,[†*] S.E. Balmer,[‡] and P.H. Scott[§]

[†] Department of Human Nutrition
University of Glasgow, Glasgow, Scotland, United Kingdom

[‡] Sorrento Maternity Hospital
Milk Bank
Birmingham, United Kingdom

[§] Selly Oak Hospital
Biochemistry Department
Birmingham, United Kingdom

## SUMMARY

Bifidobacteria, lactobacilli and staphylococci are the predominant organisms in the faeces of breast fed babies whereas in formula fed babies coliforms, enterococci and bacteroides predominate.

In vitro studies suggest that the mechanisms responsible are probably related to the acid base properties of the formula and 'immunological' proteins such as lactoferrin and sIgA. In human babies however the addition of bovine lactoferrin to an infant formula has little effect on the faecal flora and does not move it in the direction of the breast fed baby.

There are various possible explanations of this lack of effect, e.g., inactivation of the lactoferrin when it is added to a formula, and immunological responses in the intestine to a foreign protein. We consider the most likely explanation is that other factors necessary for the optimum activity of lactoferrin were not present or in inappropriate concentration, e.g. sIgA, lysozyme, citrate, bicarbonate. If human lactoferrin is added to an infant formula it may be these other factors will require attention if the lactoferrin is to have a significant effect. An iconoclastic interpretation which cannot be completely excluded is that the hypothesis of lactoferrin bacteriostatic activity is based on in vitro studies and is not a reflection of what happens in babies.

The faecal flora of a breast fed baby is very different from that of a baby receiving either cow's milk or a modern infant formula (Figure 1). Among breast fed babies bifidobacteria

---

* Current address: Old Rectory, Belbroughton, Worcs., DY9 9TF, U.K.

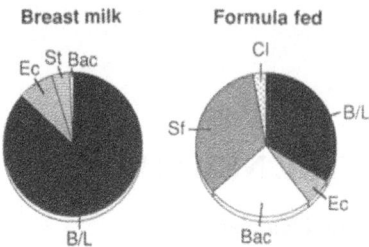

**Figure 1.** Faecal flora (mean proportion of total bacterial counts) at 14 days of age in breast fed babies and those receiving a modern infant formula. B/L: Bifidobacteria and Lactobacilli; Ec: *E. coli*; St: Staphylococci; Bac: Bacteroides; Sf: Enterococci including *Strep faecalis*; Cl: Clostridia. (from data in reference 1.)

lactobacilli and staphylococci are the predominant organisms, whereas in formula fed babies the predominant organisms are enterococci, coliforms, and bacteroides (l).

## MECHANISMS

Why is this? Some mechanisms are related to the physico-chemical properties of the two foods, particularly their acid base properties, and others to the properties of the food proteins, but these mechanisms are related.

### Acid Base Properties

The preponderance of lactobacilli in the stools of breast fed babies has been explained broadly along the following lines. Not all of the lactose and the small amount of oligosaccharide in breast milk are absorbed, indeed normal breast fed babies excrete reducing sugars in their stools. The large bowel, therefore, contains small amounts of lactose and oligosaccharide which favour the growth of lactobacilli and as acid is generated from the bacterial metabolism of the carbohydrate, the resulting low pH also favours the growth of lactobacilli rather than *E. coli*. The low phosphate and casein contents of breast milk limit its buffering capacity so maintaining the lower pH. In addition various substances such as bicarbonate and citrate form buffer systems which apart from their buffering properties may affect the intestinal flora via interactions with iron and lactoferrin.

There is not much information about the acid base properties of infant formulas. We have made a number of determinations on representative formulas. Generally the values are between those of breast milk and cow's milk, and the demineralised whey formula had the lowest buffering capacity, presumably reflecting the lower casein and very low phosphate content (Table 1). Care is necessary in extrapolating from these *in vitro* observations to the metabolic effects on the babies *in vivo*. In most milks and formulas the measurable titratable acid is mainly citric acid which after absorption is rapidly metabolised to carbon dioxide and so does not contribute to the net load of metabolic acid. The major sources of acid in the intermediary metabolism of the baby are from the endogenous production of organic acids, of sulphuric acid from sulphur amino acids. and from hydrogen ions released during bone deposition (2–4). These strictures do not apply when considering the effect of a diet **in the gut** although its early intraluminal digestion may well alter its acid-base characteristics and hence its microbiological effect.

The number of studies relating acid base properties of a diet to faecal flora is small. When sodium bicarbonate was added to cow's milk to bring its pH to between 7.2 and 7.4 the

**Table 1.** Acid Base Properties of Breast Milk, Cow's Milk and Infant Formulas

| Formula | pH[a] | Titratable acidity (mmol of bicarbonate per litre)[b] | Buffering capacity (mmol HCl per litre)[c] | Bicarbonate (mmol/litre) | Citrate (mmol/litre)[d] |
|---|---|---|---|---|---|
| **Breast milk** | 7.0 | 34 | 16 | 8 | 5 |
| **Infant Formulas** | | | | | |
| Added lactose[e] | 6.6 | 66 | 28 | 6 | 6 |
| Added maltodextrin[f] | 6.8 | 54 | 30 | 10 | 5 |
| Substituted fat[g] | 6.7 | 69 | 32 | 8 | 7 |
| Demineralised whey[h] | 6.8 | 50 | 22 | 3 | 6 |
| **Cow's milk** | 6.7 | 86 | 50 | 12 | 9 |

[a] Radiometer electrode
[b] Amount of bicarbonate required to raise pH to 7.45
[c] Amount of acid required to lower pH of milk to 5.0
[d] Enzymatic methods
[e] Cow's milk plus lactose
[f] Cow's milk plus maltodextrin
[g] Skimmed milk plus vegetable fat blend plus lactose
[h] Demineralised whey plus skimmed milk plus vegetable fat blend and lactose.

increased bacteriostatic effect *in vitro* of the milk was associated with a change in stool flora so that lactobacilli predominated (5). In some ways these changes are difficult to interpret in terms of the mechanisms discussed above because the addition of alkali would increase the buffering capacity, at least as measured by titration with hydrochloric acid as indeed we found when sodium and potassium citrate were added to a formula (3). Similar results were obtained when the pH was raised with trometamole (5), although this has not been confirmed in other studies (6–7).

## Food Proteins

**Immunoglobulin.** Cow's milk contains very little IgA—about 3% of that in human milk. Attempts have been made, however, to increase its immunoglobulin concentration by immunisation. Pregnant cows were immunised with a polyvalent *E. coli* vaccine and the whey proteins separated from colostrum collected during the first ten days of lactation. The whey proteins had anti *E. coli* activity as measured by passive haemagglutination and a mouse protection test.

There were three groups of clinical observations. Infants receiving 2 g per kg per day of the whey proteins had stools containing both intact and fragmented immunoglobulin which conserved their protective activity in mice. In further studies in Barcelona and Lille 156 children with *E. coli* gastroenteritis received whey proteins. *E. coli* disappeared from the stools of over half of the treated children compared to about a third of the controls (8).

The approach has not been taken much further since then and perhaps more modern methods (e.g., transgenic animals) will be used to lead to the secretion in cow's milk of antibodies to common enteric pathogens.

**Lactoferrin.** The possible role of lactoferrin is well known and is the subject of this book. Reference should be made to the other relevant chapters. A simplified explanation of the concept is given in Table 2. Our own work on lactoferrin is described below.

**Table 2.** The Lactoferrin–Faecal Flora Concept

| | |
|---|---|
| 1. | Lactoferrin, a whey protein, binds iron. |
| 2. | Bacteria cannot get the iron and therefore the iron dependent ones cannot multiply. |
| 3. | Non-iron dependent bacteria multiply preferentially in this ecosystem. |
| 4. | Some bacteria have enterochelins which take the iron from lactoferrin. The bacteria can then multiply. |
| 5. | If the bacteria are coated with sIgA the enterochelins are locked in and cannot take the iron from lactoferrin. Multiplication again ceases. |

## EFFECTS OF LACTOFERRIN AND RELATED NUTRIENTS ON FAECAL FLORA

Lactoferrin is one of the whey proteins. Its mechanism of action is thought to be via its iron-binding properties. It is therefore relevant to consider the effects of different combinations of lactoferrin, whey proteins not including lactoferrin, and iron. This has been done in the studies summarised here and published in detail elsewhere (1, 9–11 ).

The babies studied were born at Sorrento Maternity Hospital, Birmingham and received the "test" feed from birth throughout the observation period. The test feeds were a normal infant formula (whey predominant with iron, 7–9 mg per 100 ml) a whey formula without added iron, casein predominant formulas with and without added iron, a whey formula plus bovine lactoferrin (280 mg per 100 ml) and whey formula plus bovine lactoferrin plus iron. The results were compared with those from breast fed babies studied at the same time. Faeces were collected in nappy liners, weighed and emulsified in a transport medium (BHI broth containing 10% glycerol and 0.03% sodium formaldehyde sulphoxalate). This was immediately frozen and stored at −20°C. All specimens were analysed within one month of collection. The microbiological methods are described in Balmer and Wharton (1).

### Establishment of the Faecal Flora (Figure 2)

When comparing the effects of breast milk and a modern infant formula at 4 days of age the flora of the two groups were similar. Marked differences were apparent at 14 days and the pattern mostly persisted thereafter. By 14 days the bottle fed babies have a flora very similar to that of the adult with a predominance of coliforms, enterococci (e.g., *St. faecalis*) and bacteroides, whereas the breast fed baby flora is dominated by lactobacilli and bifidobacteria.

In the subsequent results, presented for ease of reference, only the faecal flora at 14 days is shown (see Figure 3).

### Whey Proteins (Excluding Lactoferrin)
(i.e., whey compared with casein groups in Figure 3)

Bifidobacteria and lactobacilli were present in greater numbers in the whey group than the casein group. They were also more common in the whey + F group than the casein + F group. On further analysis of the results more whey fed babies were colonised with bifidobacteria compared with casein fed group, more casein fed babies were colonised with bacteroides. It seems therefore that the whey predominant formula induced a faecal flora a little closer to that of breast fed babies than did a casein formula.

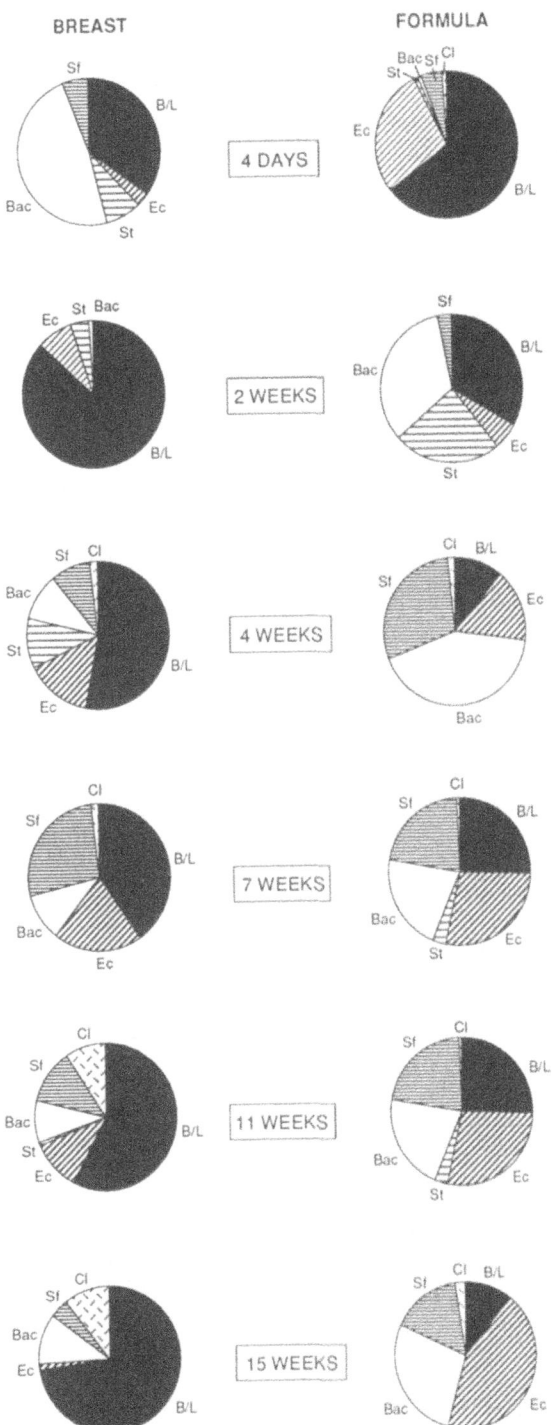

**Figure 2.** Faecal flora (mean proportion of total bacterial counts) at various ages in breast fed babies and those receiving a modern infant formula. (Whey predominant including iron). Bacteria: see legend to Figure 1. For each method of feeding there are 6 groups of babies, those studied at 4, 14 and 28 days and those studied at week 7, 11 and 15. (From data in references 1, 9, 10, and 11.)

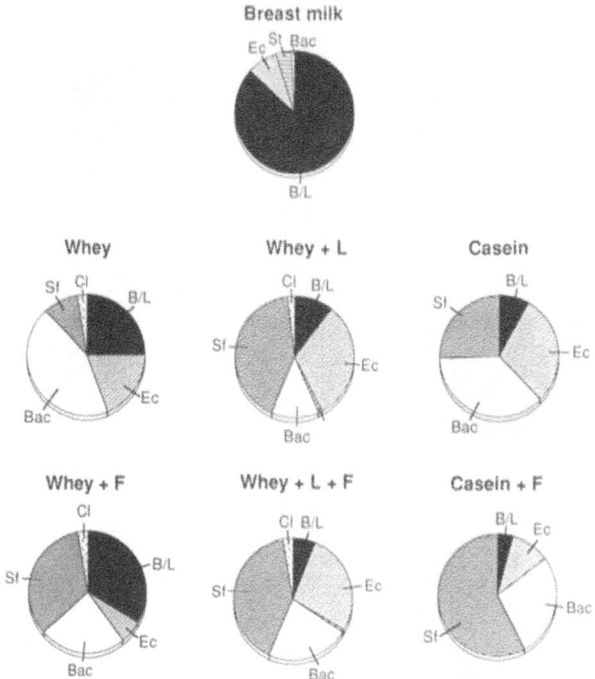

**Figure 3.** Faecal flora (mean proportion of total bacterial counts) at 14 days of age in breast fed babies and those receiving various routine and experimental infant formulas. Bacteria: see legend to Figure 1. Diet: Whey: Whey predominant infant formula, 60:40. Casein: Casein predominant formula, 80:20. +F: added iron, 7–9 mg per 100 mi. +L: added bovine lactoferrin 280 mg per 100 mi. (From data in references 1, 9, 10, and 11.)

### Bovine Lactoferrin
(i.e., whey compared with whey + L group in Figure 3)

The addition of lactoferrin (+ 280 mg per 100 ml) did not enhance the growth of bifidobacteria and lactobacilli. Greater numbers of *St. faecalis* and *E. coli* were observed. The addition of lactoferrin did not move the pattern of the faecal flora in the direction of the breast fed baby.

### Iron
(i.e., whey compared with whey + F group, casein compared with casein + F group, and whey + L compared with whey + L + F group)

In all three comparisons the addition of iron had little effect on the numbers of bifidobacteria and lactobacilli. In two of the comparisons the addition of iron favoured the growth of Strep faecalis at the expense of bacteroides and E. coli. In a further analysis in later weeks iron appeared to favour the growth of clostridia as well.

## WHY DID BOVINE LACTOFERRIN HAVE LITTLE EFFECT ON FAECAL FLORA

Clearly the addition of bovine lactoferrin did not move the faecal flora closer to that of breast milk. Indeed if anything it moved the flora further away. A recent report has confirmed our findings (12). After the addition of lactoferrin at 10 mg per 100 ml to the formula diet (a whey based one with iron) a flora similar to that of the standard formula was seen. This concentration of lactoferrin is well below that seen in breast milk. Lactoferrin at 100 mg per 100 ml was able to establish a bifidus flora in one half of the babies but only at the age of 3 months.

There are at least 4 possible explanations for the ineffectiveness (see Table 3).

**Table 3.** Possible Explanations Why Bovine Lactoferrin Had Little Effect on Faecal Flora

| | |
|---|---|
| 1. | The incorporation of lactoferrin into an infant formula reduces its activity in some way. |
| 2. | Bovine lactoferrin may have attracted foreign protein responses negating its effect. |
| 3. | Other features of the system prevented the lactoferrin achieving biological activity in the gut. |
| 4. | The hypothesis of lactoferrin bacteriostatic activity is based on in vitro studies and is not a reflection of what happens in babies. |

1. Was the lactoferrin inactivated? Two independent studies using different sources of bovine lactoferrin and different infant formulas have found lactoferrin to have little effect (10, 12). These repeated observations do not necessarily exclude (1) as a scientific explanation but it does seem empirically that the simple addition of bovine lactoferrin to a formula cannot be used to manipulate the faecal flora of newborn human babies. Perhaps the presently used food technology is inappropriate.

2. Foreign protein responses are possible but human and cow lactoferrin are said to be very similar in structure. Nevertheless they are handled differently by the monkey intestine and so this possibility cannot be completely excluded (13).

3. In our opinion this is the most likely explanation. *In vitro* lactoferrin does have activity alone but acts synergistically with sIgA to achieve greater bacteriostasis. This may be because specific sIgA locks in the enterochelins within bacteria so that they cannot obtain the small amounts of iron attached to lactoferrin (see Table 2).

   *In vitro* studies have also shown that bovine lactoferrin requires bovine antibodies (mainly IgG) and the complement cascade, lysozyme, and the presence of bicarbonate for it to be biologically active (14–17). None of these factors were present in the formula. These 'immunological' considerations are also related to simple physico-chemical constituents. Bovine colostrum contains more lactoferrin than mature human milk but it is not inhibitory *in vitro* at its natural pH because of the high concentration of citrate which competes with lactoferrin for iron and makes it available for bacterial growth (16). Indeed the bacteriostatic action of human milk due to the combined action of lactoferrin and antibody depends on the addition of bicarbonate to counteract the iron mobilising effect of citrate (14). The contents of bicarbonate and citrate in milk and their absorption and secretion in the intestine may, therefore, be very important in the immunological qualities of milk, but they have received little attention.

   The concentration of these substances in most formulas is not stated, but there are quite marked variation in the various characteristics (see Table 1). In view of the Harrison and Peat demonstration that the addition of bicarbonate to an infant formula affected faecal flora (5) and the sensitivity of bicarbonate concentrations to the *in vitro* demonstration of lactoferrin activity (14, 16) it could be that each of these variations in the acid base qualities of the formulas has microbiological effects *in vivo*. It would be a daunting task, logistically, however to determine their effect in a series of human babies.

If human lactoferrin, produced by whatever methods is to be effective in the bottle fed baby, then it may be that attention to other characteristics of the formula will be necessary, such as antibody, lysozyme, acid/base, bicarbonate, citrate, etc.

4. This iconoclastic explanation should not be too easily rejected. Other papers in this book suggest many other biological roles for lactoferrin Is it just conceivable that all these fascinating experiments with bacteria in test tubes are just epiphenomena with little relationship to the true role of lactoferrin in the human body?

## REFERENCES

1. Balmer SE, Wharton BA. (1989) Diet and faecal flora in the newborn: breast milk and infant formula. Arch Dis Child 64: 1672–77.

2. Kildeberg P, Winters RW. (1978) Diet and whole body base balance. Advances in Pediatrics 25:349-81.

3. Berger HM, Scott PH, Kenward C, Scott P and Wharton BA. (1978) Milk pH, acid base status and growth in babies. Arch Dis Child 53: 926–930.

4. Shaw JWL. (1987) The effect of dietary composition on whole body net base balance. In: Wharton BA. ed. Nutrition and feeding of preterm infants. Oxford: Blackwell, 173–190.

5. Harrison VC, Peat G. (1972) Significance of milk pH in newborn infants. Br Med J 4:515–518.

6. Bullen CL, Tearle PV, Willis AT. (1976) Bifidobacteria in the intestinal tract of infants: an in vivo study. J Med Microbiol 9:325–333

7. Bullen CL, Tearle PV. (1976) Bifidobacteria in the intestinal tract of infants: an in vitro study. J Med Microbiol 9:335–344.

8. Hilpert H, Gerber H, Amster H, Pahud JJ, Ballabriga A, Avcalis L, et al. (1977) Food and Immunology, Almqvist and Wiksell International, Stockholm, 182–196.

9. Balmer SE, Scott PH, Wharton BA. (1989) Diet and faecal flora in the newborn: casein and whey proteins. Arch Dis Chim 64: 1678-1684.

10. Balmer SE, Scott PH, Wharton BA. (1989) Diet and faecal flora in the newborn: lactoferrin. Arch Dis Child 64: 1685–1690.

11. Balmer SE. Wharton BA. (1981) Diet and faecal flora in the newborn. Iron. Arch Dis Child 68:1390–4.

12. Roberts AK, Chierici R, Sawatzki G, Hill MJ, Volpato S, Vigi V (1992). Supplementation of an adapted formula with bovine lactoferrin 1. Effect on the infant faecal flora. Acta Paediat Scand 8:119–24.

13. Davidson, LA, Litov, RE, Lönnerdal, B. (1990) Iron retention from lactoferrin-supplemented formulas in infant Rhesus monkeys. Pediatric Research 27:176–180.

14. Griffiths E, Humphreys J. (1977) Bacteriostatic effect of human milk and bovine colostrum on *E. coli*: importance of bicarbonate. Infect Immun 15: 396–401.

15. Spik G, Jorieux S, Mazurier J, Navarro J, Romond C, Montreuil J. (1984) Characterisation and biological role of human lactotransferrin complexes. In: Williams AF, Baum JD, eds. Human milk banking. Vevey: Nestle Nutrition and New York: Raven Press, 133–43.

16. Reiter B, Brock JH, Steel ED. (1975) Inhibitions of *E. coli* by bovine colostrum and post colostral milk. II. The bacteriostatic effect of lactoferrin on a serum susceptible and serum resistant strain. Immunology 28:83–95

# THE MONOCYTIC RECEPTOR FOR LACTOFERRIN AND ITS INVOLVEMENT IN LACTOFERRIN-MEDIATED IRON TRANSPORT

Henrik S. Birgens[1]

Department of Hematology L
Herlev Hospital
University of Copenhagen
DK-2730 Herlev, Denmark

## SUMMARY

Several studies suggest biological functions of the iron-binding neutrophilic glycoprotein lactoferrin that imply an initial interaction with cells from the monocyte/macrophage family. Among these, an important role of lactoferrin as responsible for the inflammatory-induced blood hyposideremia and accumulation of iron in the monocyte/macrophage system has been suggested mainly based on experiments in rodents.

In a series of experiments we have examined the binding of human lactoferrin to human monocytes. We have demonstrated the presence of a receptor binding including a high-affinity component and a low-affinity component. The affinity of the binding is compatible with a biological significance of this receptor ($K_D$ is about $10^{-8}$ M, and the number of receptors about $10^6$ per cell). More than 90 % of the lactoferrin will dissociate from the cell. The binding is not truly reversible since lactoferrin will lose its receptor-binding property after dissociation from the cell. The only observed change in the molecule is a small decrease in isoelectric point from 8.9 to 8.8.

Lactoferrin is able to translocate at least 50 % of its bound iron to intracellular ferritin in monocytes. These findings are compatible with the idea that lactoferrin might be involved in the pathogenesis of the disturbances in iron metabolism observed during inflammation.

## INTRODUCTION

The glycoprotein lactoferrin, first described in 1939 at the Carlsberg Laboratory in Copenhagen by Margrethe and SPL Sørensen (1), belongs to a family of iron-binding proteins

---

[1] Correspondence to: Henrik S. Birgens, MD, Department of Hematology L, Herlev Hospital, DK-2730 Herlev, Denmark. Fax - 44 53 2518.

also including serum transferrin and ovotransferrin (2). Although the chemical and the molecular structure (3, 4, 5), the nucleotide sequence of human lactoferrin cDNA (6), and the locations of lactoferrin in the human body are known (7), its biological function is still a subject for continuing discussion and research. However, a variety of biological effects have been suggested, mainly based on in vitro experimental data. A common feature to some of these seems to depend on or require an initial interaction of lactoferrin with cells from the mono-cyte/macrophage system (Table 1). In the years 1974 to 1977 Van Snick et al (8, 9, 10) found evidence in a series of experiments, that plasma lactoferrin might contribute to the hyposidere-mia observed during inflammatory conditions by a receptor-mediated translocation of iron from lactoferrin to intracellular ferritin in macrophages. Later, other functions involving an interaction of lactoferrin with a membrane receptor in macrophages as a prerequisite for an effect have been described (Table 1).

This paper describes a series of experiments dealing with the characterization of the human monocyte lactoferrin receptor and its significance in the light of the putative role of lactoferrin in iron metabolism.

## MATERIALS AND METHODS

Only the main principles of the methods employed are described below, whereas more detailed descriptions of methods are referred to. Lactoferrin was isolated from human milk and tested for purity as described (16). Radioactive labelling with $^{125}I$ was performed by a solid-phase lactoperoxidase/glucose oxidase method (17), and labelling with $^{59}Fe$ as $^{59}FeSO_4$ was performed as previously described (18). The target cell in these experiments was the monocyte isolated from peripheral blood by a combination of Lymphoprep$^R$ centrifugation and Percoll$^R$ gradient centrifugation (19). The monocyte content of this cell preparation was 50–60% as judged from a non-specific esterase staining. The remaining cells were virtually all lymphocytes, and therefore, control experiments were often nessesary with pure lymphocyte preparations prepared as described (19). All incubations of cells with lactoferrin were per-formed in either Hank's balanced salt solution, pH 7.4, supplemented with 2% human albumin or in RPMI buffer containing 2 mM L-glutamine, 15% fetal calf serum and supplemented with 1% (v/v) penicillin-streptomycin solution. Routinely, all incubations were performed at 37°C by gentle rotation of the tubes. The employed tubes were modified polyethylene tubes (Minisorb$^R$, Nunc, Denmark). This was important since most other plastic materials and glass bind lactoferrin in a receptor-like fashion, and the monocytes do not adhere to this material.

**Table 1.** Putative Biological Effects of Lactoferrin Involving an Interaction with the Monocyte/ Macrophage System*.

---

**a.** Increase of iron deposition as ferritin in the monocyte/macrophage system (10).

**b.** Inhibition of the monocyte/macrophage production of granulocyte-macrophage colony-stimulating activity (11).

**c.** Inhibition of the monocyte production of interleukin-1 (12).

**d.** Modulation of the primary antibody response (13).

**e.** Enhanced monocytic cytotoxicity (14).

**f.** Enhancement of phagocytosis and killing of intracellular parasites (15).

**g.** Inhibition of macrophage production of prostaglandin $E_2$ (23).

**h.** Inhibition of the monocytic production of hydroxyl radicals (20).

---

*Reference number in parentheses.

Separation of bound [125]I-lactoferrin from free [125]I-lactoferrin was done by a rapid high-speed centrifugation procedure followed by a washing with iced buffer (19).

## THE MONOCYTIC LACTOFERRIN RECEPTOR.

### Results and Discussion

In a mononuclear cell preparation incubated with [125]I-lactoferrin, the amount of cell-bound [125]I-lactoferrin is strongly dependent on the fractional monocyte content (Fig. 1) (19). The receptor-nature of this binding was demonstrated by an inhibition of [125]I-lactoferrin binding by the presence of excess of unlabelled lactoferrin (not shown on the figure) (19). Similar results have been obtained by other authors either in monocytes (20) or in other members of the human monocyte/macrophage family, i.e. adherent mononuclear cells (19, 21), alveolar macrophages (20, 22) and in breast milk macrophages (23). Additionally, these experiments agreed with previous animal studies (9, 24).

The binding of [125]I-lactoferrin to monocytes/macrophages is independent of the ambient temperature in the range 0°C to 37°C (22, 25) and to some extent dependent on the presence of $Ca^{2+}$ (21, 25).

In the analysis of the equilibrium binding data, cellular bound [125]I-lactoferrin was expressed as a ratio between bound and free [125]I-lactoferrin and plotted against either the log total concentration of lactoferrin in the incubation medium (26, 27) or against the amount of bound lactoferrin (28) (Fig. 2). The experiments were carried out over the concentration range of lactoferrin from $10^{-11}$ M to $10^{-5}$ M.

As demonstrated in Figure 2 A, in this concentration range of lactoferrin, the B/F ratio decreased to less than 5 % of its maximal value (19, 29). Thus, there was a large saturable component of binding and only a small non-saturable component, i.e. non-specific binding. The binding affinity of lactoferrin was not influenced by the labelling procedure, since the B/F ratio was identical whether it was obtained by addition of increasing concentrations of unlabelled lactoferrin to a tracer concentration of [125]I-lactoferrin or by adding [125]I-lactoferrin in increasing concentrations (19). It was also shown that the B/F ratio was almost identical whether the added lactoferrin was 100 % or 20% iron-saturated (19).

The B/F ratio of the saturable component of the binding was depressed to its half-maximal value at a lactoferrin concentration of about $10^{-8}$ M (Fig. 2 A). On the assumption of a reversible bimolecular reaction between lactoferrin and a homogenous class of receptors, this value represents the dissociation constant at equilibrium of the binding ($K_D$) (26). From this value, the number of receptors $R_o$ can be calculated according to the formula (26):

$$(B/F)_{max} = \lim [R_o]/K_D + [Lf] = [R_o]/K_{D'}, \text{ if } [Lf] \rightarrow 0$$

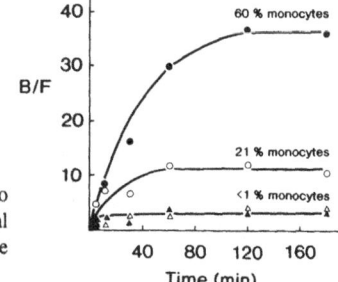

**Figure 1.** Time course of [125]I-lactoferrin binding ($1.3 \times 10^{-10}$ M) to mononuclear cells ($2 \times 10^5$ cells per ml) with varying fractional content of monocytes. The binding is expressed as a Bound over Free ratio of [125]I-lactoferrin as described (19).

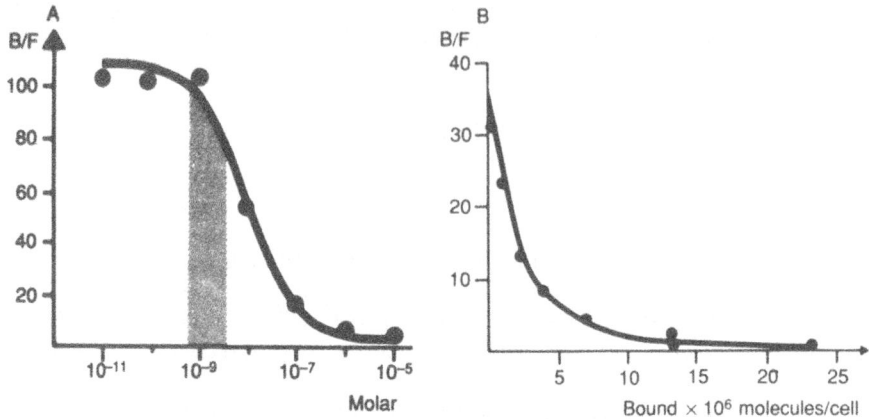

**Figure 2. A.** Equilibrium binding of $^{125}$I-lactoferrin to human monocytes. $5 \times 10^5$ cells were incubated with $^{125}$I-lactoferrin either alone ($10^{-11}$ to $10^{-9}$ M) or with $10^{-9}$ M $^{125}$I-lactoferrin and increasing amounts of unlabelled lactoferrin ($10^{-8}$ to $10^{-5}$ M). The abcissa represents the total extracellular lactoferrin concentration and the ordinate the Bound/Free ratio of $^{125}$I-activity (B/F) (19). The hatched area indicates the normal range of plasma lactoferrin concentration (16). **B.** Scatchard plot of binding data obtained by incubating monocytes with $^{125}$I-lactoferrin as mentioned above.

where [R$_o$] is the total number of receptors and [Lf] the concentration of lactoferrin. By using the estimated value of K$_D$, the number of receptors in monocytes was calculated in a series of experiments to vary from 1 to $5 \times 10^6$ per cell (19). However, one might question the assumptions for the calculations. First, in a Scatchard analysis of the binding data (Fig. 2 B), the curve had a curvilinear course suggesting the presence of at least two classes of receptors with different affinities, or the presence of negative cooperativity among the receptors (30). However, the latter was probably not the reason as demonstrated by Birgens et al (19) in dissociation studies, where the dissociation rate was unaffected by the occupancy of the receptors created by presence of a high concentration of lactoferrin in the dissociation medium. Furthermore, the binding was not truly reversible, since the lactoferrin molecule had decreased affinity to its monocyte receptor after its dissociation (31). Therefore, quantitative binding kinetics should be taken with some caution.

Several studies have stated the K$_D$ values and the number of lactoferrin receptors in different kinds of cells from the monocyte/macrophage family, mainly based on estimations from Scatchard analysis. As can be seen in Table 2, the obtained values of the binding data differ considerably among the different studies. The obtained affinty obtained by us (reciprocal K$_D$) in monocytes is about 100-fold higher and the number of receptors at least 20-fold lower than those measured in other members of the macrophage family. One reason for this divergence might be the different cell type investigated. Actually, when experiments employing identical cell types are compared, i.e. adherent mononuclear cells (Table 2), this difference in affinity is somewhat lower, but not eliminated. However, we also think that some of the difference is due to methodological causes. There is great variance in the lactoferrin concentrations employed. In the monocyte studies by Birgens et al (19, 29), the majority of the saturable binding was observed over the lactoferrin concentration range $10^{-10}$ to $10^{-7}$ M (Fig. 2 A). In the four other papers mentioned in table 2, the lactoferrin binding was not investigated (or mentioned) at lactoferrin concentrations below $10^{-8}$ M, and for three of them not below $10^{-7}$ M. Therefore, these studies might have missed a high-affinity component of the binding in the low-range of lactoferrin concentrations, and

**Table 2.** Quantitative Binding Data (Dissociation Constant $K_D$ and the Number of Binding Sites per Cell) and the Lactoferrin Concentration Ranges Employed in Studies Dealing with Lactoferrin Binding to Human Monocytes/Macrophages.

| Authors | Source of macrophages | KD (Molar) | N ($\times 10^6$) | Lactoferrin concentration range (Molar)* |
|---------|----------------------|------------|-------------------|------------------------------------------|
| Van Snick et al 1976, (9) | Peritoneal** | $1.5 \times 10^{-6}$ | 22 | $2 \times 10^{-7}$–$5 \times 10^{-6}$ |
| Bennett & Davis 1981, (21) | Adherent MNC*** | $2.7 \times 10^{-6}$ | 200 | $10^{-8}$–$2 \times 10^{-5}$ |
| Campbell 1982, (22) | Alveolar*** | $1.7 \times 10^{-6}$ | 54 | $10^{-7}$–$10^{-5}$ |
| Birgens et al 1983, (19) | Monocytes*** | $4.5 \times 10^{-9}$ | 1.6 | $10^{-11}$–$10^{-5}$ |
| Birgens et al 1983, (19) | Adherent MNC*** | $4.0 \times 10^{-7}$ | n.g. | $10^{-9}$–$10^{-4}$ |
| Britigan et al 1991, (20) | Monocytes*** | $3.6 \times 10^{-6}$ | 34 | $6 \times 10^{-7}$–$10^{-5}$ |

*If not given, then the approximate value as read from the figures.
**Mouse cells.
***Human cells.
MNC = mononuclear cells.
n.g. = not given.

might only have investigated a residual low-affinity component of binding with a high number of binding sites. The existence of such a low-affinity component with a high number of binding sites seems possible as judged from the curvilinear Scatchard plot described by Birgens et al (29). This possibility has been supported by studies in breast milk macrophages (23), where a curvilinear Scatchard plot was demonstrated by employing a lactoferrin concentration range of $10^{-10}$ to $10^{-6}$ M.

These aspects are important in the light of the physiological concentrations of lactoferrin found in various biological fluids and the concentrations observed during pathological conditions. The normal plasma concentration of lactoferrin has been measured by different authors (16, 32, 33) and has varied in these studies from $0.6 \times 10^{-9}$ to $4.5 \times 10^{-9}$ M. From the equilibrium binding curve (Fig. 2 A) the receptor occupancy can be deduced. Assuming a plasma lactoferrin concentration of $10^{-9}$ M, 5 to 10% of the receptors will be occupied by lactoferrin. However, during pathological conditions, considerably higher plasma concentrations can be measured. Thus, during bacterial infections lactoferrin concentrations might increase to more than 10 times the normal level (34, 35, 36). Similar high concentrations have been measured during larger burns (37) and during conditions with complement activation (37, 38). Thus, during such conditions the receptor occupancy might exceed 50% and in focal inflammatory areas with lactoferrin concentrations from $10^{-7}$ to $10^{-6}$ M, as measured in inflammatory joints (39) almost 100% of the receptors will be occupied.

Taken together, the observed affinity of the lactoferrin receptor with a $K_D$ value somewhere between $10^{-9}$ to $10^{-8}$ M (19), and the lactoferrin concentrations measured in biological fluids are compatible with a regulatory function of this receptor. In contrast, the relationship between $K_D$ values more than 50-fold higher as mentioned in other studies (see Table 2) and the concentrations of lactoferrin found in biological fluids are harder to understand, since even during pathological conditions only a few percentage of the receptors would be occupied.

The protein specificity of the monocyte receptor for lactoferrin has been examined by several authors. Although there are great molecular similarities between lactoferrin and serum transferrin (3, 40), the latter will not inhibit the binding of lactoferrin to the monocyte receptor (9, 19, 21). However, other proteins with quite diverging protein structure such as neutrophil elastase and cathepsin G shared the same binding site as lactoferrin in alveolar macrophages (22). This phenomenon indicates, that other properties than the protein structure are involved in this binding. Moguilevsky et al (41) suggested, that the charge of the molecule was important, since other cationic proteins such as elastase, cathepsin G, and dimeric lysozyme inhibited the binding of lactoferrin to macrophages. However, other binding forces must act as well, since lactoferrin released from monocytes will lose its receptor-binding properties although molecularly intact and still cationic (isoelectric point decrease from 8.9 to 8.8) (31).

As already mentioned, lactoferrin is dissociable from the monocytes (19) and after 16 hours of dissociation more than 90 % of the bound lactoferrin has dissociated fom the cell in a multiexponential fashion (42) (Fig. 3 lower curve). The binding is, however, as already mentioned, not truly reversible, since the monocyte binding will modify the lactoferrin molecule so it is unable to rebind to its receptor (31).

Finally, our present knowledge about the lactoferrin receptor gives rise to several lines of new questions. One line concerns the intracellular movements and handling of the lactoferrin molecule. Another line concerns the isolation of the lactoferrin receptor and subsequent generation of anti-receptor antibodies, which would be a valuable tool in further functional studies.

## THE ROLE OF THE MONOCYTE LACTOFERRIN RECEPTOR IN IRON TRANSPORT

### Results and Discussion

In a series of dissociation experiments including a homogenous system with human lactoferrin and human monocytes we were able to demonstrate that after incubation of the cells with $^{125}$I-$^{59}$Fe-lactoferrin only about 45 % of the $^{59}$Fe-activity from lactoferrin dissociated from the cells whereas almost 90% of the $^{125}$I-activity was released (Fig. 3) (42). We later showed that the $^{59}$Fe-activity was translocated to a cytosolic molecule with a molecular weight and antigenic properties as human ferritin (42). These findings confirmed previous animal studies from Van Snick and coworkers (10) and experiments with human monocytes (43). This iron transport was a monocytic phenomenon, since it was not observed in a pure lymphocyte cell preparation (42). Lactoferrin in plasma participating in this process is believed to originate from neutrophils (16, 44), but there is no evidence to suggest a difference in lactoferrin isolated from milk, as in these studies, or from neutrophils with respect to its interaction with the monocyte/macrophage system (45). Multiple forms of lactoferrin have been isolated from milk

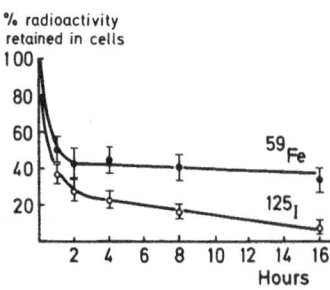

**Figure 3.** Time course of dissociation of $^{59}$Fe-activity and $^{125}$I-activity from human monocytes. The cells ($4 \times 10^6$ per ml) were incubated with $2 \times 10^{-7}$ M $^{59}$Fe-$^{125}$I-lactoferrin for 16 hours at 37°C and then exposed to lactoferrin-free medium. Each point is mean of four experiments ± SE.

and neutrophils (46), and only a subfraction apparently binds iron. It is still unknown, however, whether these molecular forms differ in their affinity to the monocyte receptor.

Subcellular fractionation of the incubated monocytes suggested internalization of lactoferrin into a myeloperoxidase-positive lysosomal fraction (42), compatible with previous studies demonstrating internalization of lactoferrin into human monocytes (20, 47). A transit of internalized lactoferrin through endosomes and/or lysosomes with low pH would offer an explanation of the release of iron from lactoferrin due to decrease in affinity of iron to lactoferrin in acidic environments (48).

Hyposideremia and deposition of iron bound into ferritin in the monocyte/macrophage system are cardinal features of the iron metabolic disturbances observed in anemia of chronic disorders such as cancer and inflammatory diseases (49). Additionally, iron kinetic studies have disclosed that impaired release of iron from these cells to the blood is the critical deficiency in inflammatory conditions (50, 51).

Much evidence suggests an important role of neutrophilic-derived lactoferrin in this context. First, apolactoferrin will induce hyposideremia after injection into rodents (8, 52). Second, neutrophils seem necessary for the development of these inflammatory-induced changes in iron metabolism whether the inflammatory condition was chemical induced with turpentine (53) or whether it was created by injection of interleukin 1 or endotoxin (52). Interesting, new experiments have shown, that chlorpromazine will prevent hyposideremia in mice, probably by inhibiting neutrophilic release of lactoferrin (54).

Based on these evidences and in the light of our own experimental data we have proposed the following hypothesis concerning the role of lactoferrin in inflammatory-induced changes in the iron metabolism (Fig. 4): During inflammation and conditions with increased neutrophil turnover a lot of released cytokines and/or inflammatory products such as interleukin 1 (55), endotoxin (56), TNF-alpha (57) and immune complexes (58) can induce release of lactoferrin from the neutrophils. Lactoferrin will bind to its membrane receptor in monocytes/macrophages. Captured in this position it might represent an iron-trapping mechanism preventing

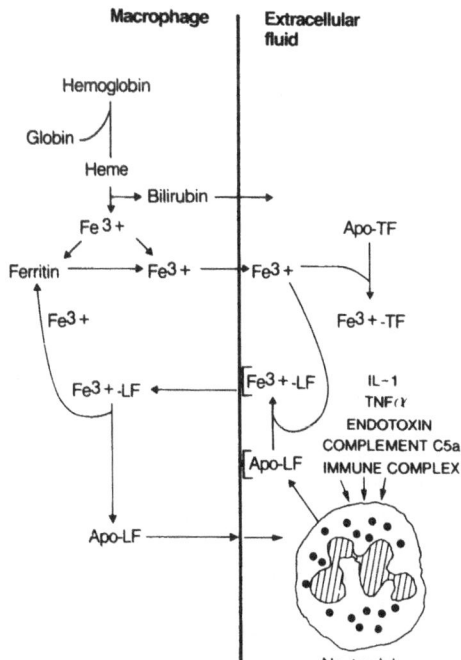

**Figure 4.** A proposed model for lactoferrin-mediated transport of iron from extracellular fluid to intracellular ferritin in monocytes during inflammatory conditions (see text).

the transfer of iron from the macrophage to serum transferrin. Iron released from senescent phagocytosed red cells might in this way be captured by membrane-bound lactoferrin and transferred back to intracellular ferritin (Fig. 4). This iron transport will not be a continuous ongoing process, since, as mentioned before, lactoferrin released from monocytes will lose its receptor-binding properties (31), and therefore lactoferrin will not undergo many cycles of iron transport.

However, several qustions remain to be answered. What is the significance of macrophage activation? The binding of lactoferrin to the macrophage receptor will increase considerably during activation (21). Which cell type is responsible for the clearing of lactoferrin from the blood? Lactoferrin injected into animals or man is rapidly removed from the circulation by the liver (59, 60). However, the main cell type responsible for this lactoferrin removal is still controversial. Prieels et al (59) reported that parenchymal cells are the main cell type reponsible for lactoferrin binding, and others have claimed that the sinusoidal cells are the most important (61). However, a defect iron processing during inflammatory condition is not confined to macrophages, reutilization of iron from hepatocytes is also being impaired (49). Lactoferrin receptor binding by hepatocytes has been described (62), and the handling by hepatocytes of lactoferrin-bound iron would be an intriguing object for new research.

## CONCLUSIONS

1. The iron-binding glycoprotein lactoferrin binds to human monocytes by a receptor-like binding.

2. The binding is complex, possibly including a high-affinity as well as a low-affinity component of binding.

3. After uptake and dissociation of lactoferrin from monocytes, the molecule loses its receptor-binding properties—a change accompanied by a small decrease in isoelectric point.

4. The quantitative binding data of the high-affinity binding agrees with a biological effect of lactoferrin in monocytes.

5. After binding and uptake of lactoferrin into monocytes about 50% of its iron is translocated to cytosolic ferritin.

## REFERENCES

1. Sørensen M, Sørensen SPL,(1939) The proteins in whey. *CR Trav Lab* Carlsberg **23**: 55–99.

2. Aisen P, Listowsky I, (1980) Iron transport and storage proteins. *Ann Rev Biochem* **49**: 357–393.

3. Metz-Boutigue MH, Jollés J, Mazurier J, Schoentgen F, Legrand D, Spik, G, Montreuil J, Jollés P, (1984) Human lactotransferrin: amino acid sequence and structural comparisons with other transferrins. *Eur J Biochem* **145**: 659-675.

4. Matzsumoto A, Yoshima H, Takasaki S, Kobata A, (1982) Structural study of the sugar chains of human lactoferrin: Finding of four novel complex-type asparagine-linked sugar chains. *J Biochem* **91**: 143–155.

5. Baker EN, Anderson BF, Baker HM, Haridas M, Jameson GB, Norris GE, Rumball SV, Smith CA (1991) Structure, function and flexibility of human lactoferrin. *Int J Biol Macromol* **13**: 122–129.

6. Powell MJ, Ogden JE (1990) Nucleotide sequence of human lactoferrin cDNA. *Nucleic Acids Research* **18**: 4013.

7. Masson PL, Heremans JF. (1966) Studies on lactoferrin, the iron-binding protein of secretions. *Prot Biol Fluids* **14**: 115–124.

8. Van Snick JL, Masson PL, Heremans JF. (1974) The involvement of lactoferrin in the hyposideremia of acute inflammation. *J Exp Med* **140**: 1068–1084.

9. Van Snick JL, Masson PL. (1976) The binding of lactoferrin to mouse peritoneal cells. *J Exp Med* **144**: 1568–1580.

10. Van Snick JL, Markowetz B, Masson PL. (1977) The ingestion and digestion of human lactoferrin by mouse peritoneal macrophages and transfer of its iron into ferritin. *J Exp Med* **146**: 817–827.

11. Broxmeyer HE, Smithyman A, Eger RR, Meyers PA, DeSousa M. (1978) Identification of lactoferrin as the granulocyte-derived inhibitor of colony-stimulating activity production. *J Exp Med* **148**: 1052–1067.

12. Zucalli JR, Broxmeyer HE, Morse C. (1989) Lactoferrin decreases monocyte-induced fibroblast production of myeloid-stimulating activity by suppressing monocyte release of interleukin-1. *Blood* **74**: 1531–1536.

13. Duncan RL, McArthur WP. (1981) Lactoferrin-mediated modulation of mononuclear cell activities: I. Suppression of the murine in vitro primary antibody response. *Cell Immunol* **63**: 308–320.

14. Nishiya K, Horwitz DA. (1982) Contrasting effects of lactoferrin on human lymphocyte and monocyte natural killer activity and antibody-dependent cell-mediated cytotoxicity. *J Immunol* **129**: 2519–2523.

15. Lima MF, Kierszenbaum F. (1985) Lactoferrin effects on phagocytic cell function. I. Increased uptake and killing of an intracellular parasite by murine macrophages and human monocytes. *J Immunol* **134**: 4176–4183.

16. Birgens HS. (1985) Lactoferrin measured by an ELISA technique: Evidence that plasma lactoferrin is an indicator of neutrophil turnover and bone marrow activity in acute leukaemia: *Scand J Haematol* **34**: 326–31.

17. Thorell JI, Johansson BG. (1971) Enzymatic iodination of polypeptides with $^{125}$I to high specific activity.*Biochim Biophys Acta* **251**: 363–369.

18. Masson PL, Heremans JF. (1968) Metal-combining properties of human lactoferrin (red milk protein). I The involvement of bicarbonate in the reaction. *Europ J Biochem* **6**: 579–584.

19. Birgens HS, Hansen NE, Karle H, Kristensen LØ. (1983). Receptor binding of lactoferrin by human monocytes. *Brit J Haematol* **54**: 383–391.

20. Britigan BE, Serody JS, Hayek MB, Charniga LM, Cohen MS. (1991) Uptake of human lactoferrin by mononuclear phagocytes inhibits their ability to form hydroxyl radical and protects them from membrane autoperoxidation. *J Immunol* **147**: 4271–4277.

21. Bennett RM, Davis J (1981) Lactoferrin binding to human peripheral blood cells: An interaction with B-enriched populations of lymphocytes and a subpopulation of adherent mononuclear cells. *J Immunol* **127**: 1211–1216.

22. Campbell EJ. (1982) Human leukocyte elastase, cathepsin G, and lactoferrin: Family of neutrophil granule glycoproteins that bind to an alveolar macrophage receptor. *Proc Natl Acad Sci* **79**: 6941–6945.

23. Bartal L, Padeh S, Passwell JH. ( 1987) Lactoferrin inhibits prostaglandin $E_2$ secretion by breast milk macrophages. *Pediatr Res* **21**: 54–57.

24. Markowetz B, Van Snick JL, Masson PL. (1979) Binding and ingestion of human lactoferrin by mouse alveolar macrophages. *Thorax* **34**: 209–212.

25. Birgens HS, Karle K, Hansen NE, Kristensen LØ. (1984) Lactoferrin binding to peripheral human leukocytes. *Prot Biol Fluids* **31**: 145–148.

26. Schaumburg BP, Bojesen E. (1968) Specificity and thermodynamic properties of the corticosteroid binding to a receptor of rat thymocytes in vitro. *Biochim Biophys Acta* **170**: 172–188.

27. Rodbard D, Munson PJ, Thakur AK. (1980) Quantitative characterization of hormone receptors. *Cancer* **46**: 2907- 2918.

28. Scatchard G. (1949) The attractions of proteins for small molecules and ions. *Ann NY Acad Sci* **51**: 660–672.

29. Birgens HS, Karle H, Hansen NE, Kristensen LØ. (1984) Lactoferrin receptors in normal and leukaemic human blood cells. *Scand J Haematol* **33**: 275–280.

30. DeMeyts P, Raffaele Bianco A, Roth J. (1976) Site-site interactions among insulin receptors. *J Biol Chem* **251**: 1877–1888.

31. Birgens HS, Kristensen LØ. (1990) Impaired receptor-binding and decrease in isoelectric point of lactoferrin after binding and uptake into human monocytes. *Europ J Haematol* **45**: 31–35.

32. Estevenon JP, Figarella C. (1983) A non-competitive enzyme immunoassay of human lactoferrin in biological fluids. *Clin Chim Acta* **129:** 311–318.

33. Adayemi EO, Hodgson, HJF. (1988) Augmented release of human leukocyte lactoferrin (and elastase) during coagulation. *J Clin Lab Immunol* **27:** 1–4.

34. Hansen NE, Karle H, Andersen V, Malmquist J, Hoff GE. (1976). Neutrophilic granulocytes in acute bacterial infection. Sequential studies on lysozyme, myeloperoxidase and lactoferrin. *Clin Exp Med* **26:** 463–468.

35. Baynes R, Bezwoda W, Bothwell T, Khan Q, Mansoor N. (1986) The non-immune inflammatory response: serial changes in plasma iron, iron-binding capacity, lactoferrin, ferritin and C-reactive protein. *Scand J Lab Invest.* **46:** 695–704.

36. Nuijens JH, Abbink JJ, Wachtfogel YT, Colman RW, Eerenberg AJM, Dors D, Kamp AJM, Strack van Schijndel RJM, Thijs LG, Hack CE. (1992) Plasma elastase, alfa$_1$-antitrypsin and lactoferrin in sepsis: Evidence for neutrophils as mediators in fatal sepsis. *J Lab Clin Med* **119:** 159–168.

37. Wolach B, Coates TD, Hugli TE, Baehner RL, Boxer LA. (1984) Plasma lactoferrin reflects granulocyte activation via complement in burn patients. *J Lab Clin Med* **103:** 284–293.

38. Lash JA, Coates TD, Lafuze J, Baehner RL, Boxer LA. (1983) Plasma lactoferrin reflects granulocyte activation in vivo. *Blood* **61:** 885–888.

39. Bennett RM, Skosey JL. (1977). Lactoferrin and lysozyme levels in synovial fluid. Differential indices of articular inflammation and degradation. *Arthrites and Rheumatism* **20:** 84–90.

40. Mazurier J, Metz-Boutigue MH, Jollés J, Spik G, Montreuil J, Jollés P. (1983) Human lactotransferrin: Molecular, functional and evolutionary comparisons with human transferrin and hen ovotransferrin. *Experientia* **39:** 135–141.

41. Moguilevsky N, Retegui LA, Courtoy PJ, Castracane CE, Masson PL. (1984) Uptake of lactoferrin by the liver. III. Critical role of the protein moiety. *Lab Invest* **50:** 335–340.

42. Birgens HS, Kristensen LØ, Borregaard N, Karle H, Hansen NE. (1988) Lactoferrin-mediated transfer of iron to intracellular ferritin in human monocytes. *Europ J Haematol* **41:** 52–57.

43. Moguilevsky N. Masson PL, Courtoy PJ. (1987) Lactoferrin uptake and iron processing into macrophages: a study in familial haemochromatosis. *Brit J Haematol* **66:** 129–136.

44. Hansen NE, Malmquist J, Thorell J. (1975) Plasma myeloperoxidase and lactoferrin measured by radioimmunoassay: Relations to neutrophil kinetics. *Acta Med Scand* **198:** 437–443.

45. Moguilevsky N, Retegui LA, Masson PL. (1985) Comparisons of human lactoferrins from milk and neutrophilic leukocytes. Relative molecular mass, isoelectric point, iron-binding properties and uptake by the liver. *Biochem J* **229:** 353–359.

46. Furmanski P, Li Z-P (1990) Multiple forms of lactoferrin in normal and leukemic human granulocytes. *Exp Hematol* **18:** 932–935.

47. Steinmann G, Broxmeyer HE, DeHarven E, Moore MAS. (1982) Immuno-electron microscopic tracing of lactoferrin, a regulator of myelopoiesis, into a subpopulation of human peripheral blood monocytes. *Brit J Haematol* **50:** 75–84.

48. Aisen P, Leibman A. (1972) Lactoferrin and transferrin: a comparative study. *Biochim Biophys Acta* **257:** 314–323.

49. Lee GR. (1983) The anemia of chronic disease. *Sem Hematol* **20:** 61–80.

50. Quastel MR, Ross JF. (1966) The effect of acute inflammation on the utilization and distribution of transferrin-bound and erythrocyte radioiron. *Blood* **28:** 738–757.

51. Fillet G, Cook JD, Finch CA. (1974) Storage iron kinetics VII. A Biological model for reticuloendothelial iron transport. *J Clin Invest* **53:** 1527–1533.

52. Goldblum SE, Cohen D, Jay M, McClain CJ. (1987) Interleukin 1-induced depression of iron and zinc: role of granulocytes and lactoferrin. *Am J Physiol* **252:** E27–32.

53. Kampschmidt RF, Upchurch H (1969) Lowering of plasma iron concentration in the rat with leukocytic extracts. *Am J Physiol* **216:** 1287–1291.

54. Bertini R, Wang JM, Mengozzi M, Willems J, Joniau M, Van Damme J, Ghezzi P (1991) Effects of chlorpromazine on PMN-mediated activities in-vivo and in-vitro. *Immunology* **72:** 138–143.

55. Klempner MS, Dinarello CA, Gallin JI. (1978) Human leukocytic pyrogen induces release of specific granule contents from human neutrophils. *J Clin Invest* **61:** 1330–1336.

56. Gutteberg TJ, Røkke O, Jørgensen T, Andersen O. (1988) Lactoferrin as an indicator of septicemia and endotoxemia in pigs. *Scand J Infect Dis* **20**: 659–666.

57. Koivuranta-Vaara P, Banda D, Goldstein IM. (1987) Bacterial lipopolysaccharide-induced release of lactoferrin from human polymorphonuclear leukocytes: Role of monocyte-derived tumor necrosis factor alpha. *Infect Immun* **55**: 2956–2961.

58. Leffell MS, Spitznagel JK. (1975) Fate of human lactoferrin and myeloperoxidase in phagocytizing human neutrophils: Effects of immunoglobulin G subclasses and immune complexes coated with latex beads. *Infect Immun* **12**: 813–820.

59. Prieels JP, Pizzo SV, Glasgow LR, Paulson JC, Hill RL. (1978) Hepatic receptor that specifically binds oligosaccharides containing fucosyl alpha13 N-acethylglucosamine linkages. *Proc Natl Acad Sci* **75**: 2215–2219.

60. Bennett RM, Kokocinski T, (1979) Lactoferrin turnover in man. *Clin Sci* **57**: 453–460.

61. Courtoy PJ, Moguilevsky N, Retegui LA, Castracane CE, Masson PL. (1984) Uptake of lactoferrin by the liver. II. Endocytosis by sinusoidal cells. *Lab Invest* **50**: 329–334.

62. McAbee DD, Esbensen K. (1991) Binding and endocytosis of apo- and holo-lactoferrin by isolated rat hepatocytes. *J Biol Chem* **266**: 23624–23631.

# STUDY ON THE BINDING OF LACTOTRANSFERRIN (LACTOFERRIN) TO HUMAN PHA-ACTIVATED LYMPHOCYTES AND NON-ACTIVATED PLATELETS

## Localisation and Description of the Receptor-Binding Site

Joël Mazurier, Dominique Legrand, Béatrice Leveugle, Elisabeth Rochard, Jean Montreuil, and Geneviève Spik

Université des Sciences et Technologies de Lille
Laboratoire de Chimie Biologique (Unité Mixte de Recherche n° 111 du CNRS)
59655 Villeneuve d'Ascq Cedex, France

## SUMMARY

Fluorescein isothiocyanate derivatization of human lactotransferrin on Lys-264 as well as covalent addition of sulfosuccinimidyl 2-(p-azidosalicylamido)ethyl-1,3'- dithiopropionate (SASD)[*] on Lys-74 inhibits the binding of the glycoprotein to both human PHA-activated lymphocytes and non-activated platelets. This suggests that the cell binding site of lactotransferrin is located in the vicinity of the lysine residues 74 & 264 and does not occur either through electrostatic or lectin interactions. In contrast, the derivatization of lactotransferrin using sulfosuccinimidyl 6-(4'-azido-2'-nitrophenyl-amino) hexanoate (sulfo-SANPAH), on Lys-281 does not modify the binding parameters of lactotransferrin to the cells. Molecular modeling showed the position of SASD, sulfo-SANPAH and fluorescein molecules at the surface of the protein and suggested that SASD and fluorescein could mask the two loop-containing regions of human lactotransferrin (residues 28–34 and 38–45). Elsewhere, a 6 kDa peptide covering the peptide chain from residues 4 to 52 was isolated and its inhibitory effect on the binding of lactotransferrin to both human PHA-activated lymphocytes and non-activated platelets was demonstrated. Inhibition of ADP-induced platelet aggregation by lactotransferrin (50% inhibition = 10 nM) was also found with the N-t fragment of lactotransferrin (residues 3–281; 50% inhibition = 2 μM) and with two synthetic peptides: KRDS tetrapeptide (50% inhibition = 350

---

[*] ABBREVIATIONS: FITC, fluorescein isothiocyanate; SASD, sulfosuccinimidyl 2-(p-azidosalicylamido)ethyl-1,3'-dithiopropionate; Sulfo-SANPAH, sulfosuccinimidyl 6-(4'-azido-2'nitrophenylamino) hexanoate; RP-HPLC, reverse phase high-performance liquid chromatography; PBS, phosphate-buffered saline; SDS-PAGE, sodium dodecyl sulfate polyacrylamide gel electrophoresis; Tris, tris(hydroxymethyl)aminomethane: PHA, phytohemagglutinin.

μM) and CFQWQRNMRKVRGPPVSC octodecapeptide (50% inhibition = 20 μM) corresponding to the lactotransferrin amino acid sequence 39–42 and 20–37, respectively.


## INTRODUCTION

Human lactotransferrin (also called lactoferrin) is a bilobed iron-binding glycoprotein present in milk (1, 2), in external secretions (3, 4) and in neutrophils (5). The primary structure of human lactotransferrin was first determined by Metz-Boutigue et al. (6) and then by cDNA cloning (7, 8). The 3D-structure of the protein is now well elucidated (9–12). The binding of lactotransferrin to numerous cells has been assayed and both high and low affinity binding sites have been described for enterocytes (13), liver cells (14) and blood cells except platelets (15). Low affinity ($K_d$ =$10^{-6}$ M) lactotransferrin binding sites were described on monocytes and macrophages, and it has been suggested that they occurred through electrostatic (16) or lectin interactions (17).

More recently, we have described a specific receptor for human lactotransferrin with an affinity constant close to $K_d = 10^{-8}$ M on both activated peripheral blood lymphocytes (18) and platelets (19). The membrane receptor is a glycoprotein of 105 kDa molecular mass consisting of a single polypeptide chain which has also been found on platelets (19) and interacts with a well defined domain of lactotransferrin (20–23). Furthermore, we have shown that lactotransferrin exhibits through the membrane receptor an inhibitory activity on ADP-induced platelet aggregation (19).

In the present paper, the molecular interactions between human lactotransferrin and the lactotransferrin receptor demonstrated at the surface of both the human phytohemagglutinin-stimulated peripheral blood lymphocytes (18) and the platelets were investigated. In order to obtain information about the receptor-binding site, masking reagents: FITC and two heterobifunctional reagents generally used for receptor-ligand crosslinking, were allowed to react with the lysine residues of lactotransferrin. The effects of lactotransferrin derivatization by SASD and sulfo-SANPAH on the receptor recognition were analyzed and the derivatized lysine residues were identified. Molecular modeling was used for setting FITC, SASD and sulfo-SANPAH molecules at the surface of the whole protein so that they could interact with the receptor-binding domain. The obtained results allowed the definition of the areas of human lactotransferrin that interact with the lactotransferrin receptor. Furthermore, proteolysis of the 30 kDa N-tryptic fragment of the protein, which is still able to bind to both activated peripheral-blood lymphocytes (20, 21) and platelets (19), was performed yielded a 6 kDa peptide covering residues 4 to 52.

Finally, using the two synthesized lactotransferrin loops [residues: 28–34 and 39–42], which are belonging to the receptor-binding site (24), we have demonstrated that they both possess, like human lactotransferrin, an inhibitory effect on platelet aggregation.


## MATERIALS AND METHODS

### Cell Preparation

Human peripheral blood mononuclear cells were obtained by Lymphoprep (Nycomed Pharma AS, Oslo, Norway) separation of heparinized venous blood samples drawn from healthy volunteers. The cells were prepared and stimulated with phytohemagglutinin (Industrie Biologique Française, Villeneuve la Garenne, France) as previously described (19). Platelets were isolated (25) from platelet-rich plasma. The platelets were washed three times by repeated centrifugations (2,000 g, 15 min, 20°C) in saline buffer pH 6.5 (36 mM citric acid, 5 mM glucose, 5 mM $CaCl_2$, 2 mM $MgCl_2$, 103 mM NaCl, 0.35% bovine serum albumin and 100 nM prostaglandin $E_1$) and the last pellet was resuspended in a minimum volume of the saline buffer.

## Isolation and Purification of the Iron-Binding Tryptic Fragments from Human Lactotransferrin

Lactotransferrin was prepared from pooled human milk and purified according to Spik et al. (25). Iron saturation of lactotransferrin was carried out according to Mazurier et al. (26). The iron-binding N-tryptic and C-tryptic and 6 kDa fragments were prepared from human lactotransferrin, as previously described by Legrand et al. (20–23).

## Derivatization of human lactotransferrin with FITC

Human lactotransferrin used for binding and competitive binding experiments was labeled with FITC from Sigma (Isomer I) as previously described (22).

## Derivatization of Human Lactotransferrin with Iodinated SASD and Sulfo-SANPAH

Human lactotransferrin was dissolved in 0.1 M sodium carbonate, pH 9.2 at the concentration of 12.5 nM and was incubated for 2 h at room temperature with hetero-bifunctional reagents (Pierce Chemical Co.), at concentrations ranging from 12.5 to 250 µM. The derivatized lactotransferrin solution was photolysed at 4°C for 10 min using the Minuvis Desaga UV lamp at a distance of 10 cm for inactivating the azide group and then stored at 4°C before use.

## Binding Assays and Competitive Binding Assays

The binding assays of native and derivatized lactotransferrin samples were performed in siliconized polypropylene tubes as already reported (18, 20, 23). Non-specific binding was measured in the presence of a 100-fold molar excess of unlabeled lactotransferrin. In order to prevent the non-specific binding of human lactotransferrin to cells or to plastic, all binding experiments were performed in presence of 1% (v:v) human serotransferrin (Behring, Marburg, FRG), since serotransferrin has no inhibitory effect on the binding of lactotransferrin (18).

## Proteolysis of FITC, SASD- and Sulfo-SANPAH-Derivatized Human Lactotransferrin and Peptide Purification

One µ mol of derivatized lactotransferrin (equimolar ratio SASD/protein) was passed through a Sephadex G25-fine column (2 cm × 20 cm) equilibrated with either 0.1 M glycine/0.1 M HCl, pH 2.4 or 0.1 M Tris/HCl, pH 8.2. Proteolytic enzymes (pepsin or trypsin) were added to the solution and incubation was performed for 16 h at 37°C under magnetic stirring. The fractions containing labeled peptides were concentrated under vacuum and filtered prior to RP-HPLC (Spectra-Physics model 8700 liquid chromatograph equipped with a Spectra-Physics model 8450 variable wavelength UV/visible detector connected to a model 4270 computing integrator). Labeled peptides were purified by RP-HPLC on a 10-µm Chromatem C14 column (Touzart & Matignon, Vitry sur Seine, France).

## Molecular Modeling Studies

The crystallographic data of diferric human lactotransferrin were kindly provided by Dr. Baker (Massey University, New Zealand). All calculations were carried out on an Evans & Sutherland PS 350 graphic station and a Vax 6320 host computer using the Sybyl 5.3 molecular modeling package (27). The structures of FITC, SASD and sulfo-SANPAH were built from fragments extracted from the Cambridge crystallographic databank (Cambridge, U.K.) and their geometry was minimized by using the Sybyl Search program (Mayer et al., 1987). The crystallographic data of rabbit serotransferrin used for comparing the three-dimensional structures of human lactotransferrin and rabbit serotransferrin were kindly provided by Dr. Lindley (London University, Great-Britain).

## Platelet Aggregation

Platelet aggregation was measured in an aggregometer Biodata. Typically, 250 µL of washed platelets (450,000/µL) in a Tyrode pH 7.4 buffer containing 0.35% of bovine serum albumin were incubated for 3 min at 37°C with 50 µl of a solution of lactotransferrin, N-t tryptic and C-t tryptic fragments or synthetic peptides. The final lactotransferrin or related peptides concentration was ranging from 0 to $1 \times 10^{-6}$ M. Then aggregation was induced by addition of 25 µl of fibrinogen solution (15 g/l) followed by the addition, 0.5 min later, of 30 µl of $2 \times 10^{-4}$ M ADP.

## RESULTS

### Binding Assays and Competitive Binding Assays of Non-derivatized and Derivatized Lactotransferrins

The effect of iodinated-SASD and sulfo-SANPAH conjugation to [125]I-labeled human lactotransferrin (ratio 1:1) on the binding of the protein to phytohemagglutinin-stimulated human lymphocytes is shown in Fig. l. As previously reported (18), the radiolabeled non-derivatized lactotransferrin exhibited a concentration-dependent and saturable binding curve to mitogen-activated lymphocytes with a Kd of 43 nM and a number of binding sites of 300,000. A very similar binding curve was obtained with the sulfo-SANPAH-derivatized lactotransferrin, while no more than 5–10% binding was observed for both FITC-, used as control, and iodinated-SASD- derivatized lactotransferrins. The results we obtained are in total agreement with the competitive binding experiments (data not shown) which have demonstrated that the binding of radiolabeled human lactotransferrin was 55% inhibited either by non-derivatized lactotransferrin or sulfo-SANPAH-derivatized lactotransferrin whereas no binding inhibition was observed for either FITC- or iodinated-SASD-derivatized lactotransferrins.

### Isolation of FITC-, [125]I-Labeled SASD- and Sulfo-SANPAH-Labeled Peptides from Human Lactotransferrin and Localization of the Modified Lysine Residues on the Human Lactotransferrin Sequence

Since derivatization of human lactotransferrin with reagents such as fluorescein, iodinated SASD and sulfo-SANPAH can modify the behaviour of lactotransferrin binding to mitogen-stimulated human lymphocytes, it was of interest to identify the derivatized amino acids by isolating the labeled peptides from lactotransferrin. Pepsin was used for isolating [125]I-labeled SASD-peptides because of its ability to efficiently cleave the N-terminal moiety of human lactotransferrin and liberate the intact C-terminal lobe as a 40 kDa C-peptic fragment. The FITC and sulfo-SANPAH-labeled peptides were isolated after reduction and alkylation steps and tryptic hydrolysis.

The sequences of the FITC-, SASD- and sulfo-SANPAH-derivatized peptides and the localization of modified lysine residues on human lactotransferrin molecule are indicated in Fig. 2. The main SASD-derivatized amino acid was Lys-74. Lys-264 and Lys-278 bound 15%

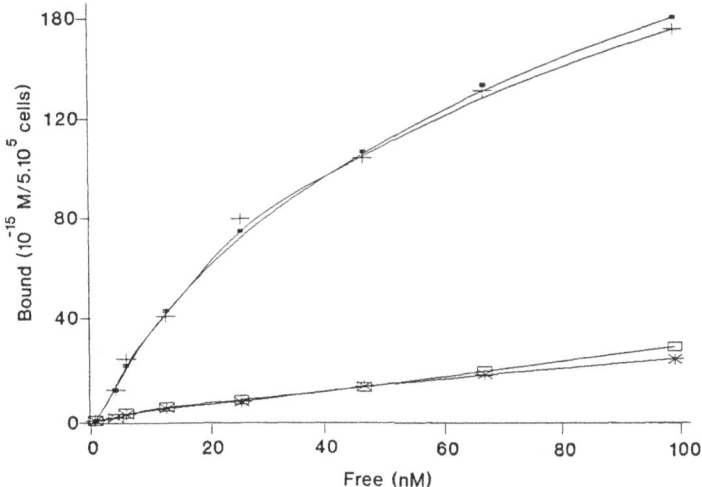

**Figure 1.** Binding assays of non-derivatized (■-■), FITC- (∗-∗), iodinated-SASD- (□-□) or sulfo-SANPAH- (+-+) derivatized diferric human lactotransferrin. All derivatized lactotransferrins used in the experiments possess one mol of reagent per mol of protein.

and 10% of total SASD, respectively. The sequence of the unique SANPAH-derivatized peptide was F-G-∗-D-K, indicating that SANPAH was exclusively bound to Lys-281. These results, in conjunction with the results previously obtained with FITC derivatization of human lactotransferrin (22) (Fig. 2), show that, in the N-terminal, domain I Lys-74 is the preferential binding site for SASD, Lys-264 for FITC and Lys-281 for sulfo-SANPAH.

## Isolation and Characterization of the Receptor-Binding Peptide from Human Lactotransferrin

The 30 kDa N-terminal tryptic fragment from human lactotransferrin, which was found to specifically bind to the lymphocyte receptor (28), was hydrolyzed by the *Staphylococcus aureus* V8 protease and 11 fractions were obtained (numbered from I to XI). Analysis of these peptide-containing fractions by 16% SDS-PAGE indicated that fractions I to IV contained unstained material, presumably salts or short peptides or amino acids. Fractions VI, VIII, IX and X contained a mixture of peptides with molecular masses ranging from 30 kDa (undigested N-tryptic fragment) to 1 kDa. The inhibition of $^{125}$I-labeled human lactotransferrin binding to

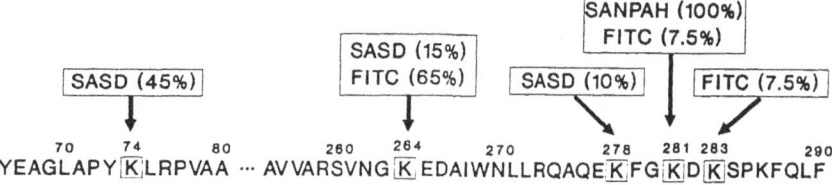

**Figure 2.** Localization of the FITC-, $^{125}$I-labeled-SASD- and sulfo-SANPAH-derivatized lysine residues on human lactotransferrin. Boxed amino acids on the human lactotransferrin peptide chain correspond to the derivatized lysine residues. The relative binding percentages of FITC, $^{125}$I-labeled-SASD and sulfo-SANPAH for each derivatized lysine residue are indicated in boxes.

PHA-activated human lymphocytes was assayed in the presence of a 100-molar excess of fractions III to XI, the 30 kDa N-tryptic fragment, the 50 kDa C-tryptic fragment and the 20 kDa N2-glycopeptide. The binding of radiolabeled lactotransferrin to mitogen-stimulated human lymphocytes was 55% and 33% inhibited by a 100-fold molar excesses of unlabeled lactotransferrin and 30 kDa N-tryptic fragment, respectively, whereas no more than 5% inhibition was noted for both the 20 kDa N2-glycopeptide and the 50 kDa C-tryptic fragment. A strong inhibition of radiolabeled lactotransferrin (45%) was obtained in the presence of fraction VII which contains a single 6 kDa peptide. The inhibition rate obtained by fraction VII rose from 45% to 55–60% when adding a 100-fold molar excess of unlabeled lactotransferrin, demonstrating that lactotransferrin and 6 kDa peptide bound to the same specific lymphocyte binding sites (data not shown).

Edman degradation of the peptide contained in fraction VII yielded the two following N-terminal sequences R-R-S-V-Q and A-T-K-C-F in the same molar ratio. Taking into account the two N-terminal sequences, the amino acid composition analysis of fraction VII (not shown) and the potential selective cleavage sites of lactotransferrin by V8-protease, the peptide was located on the human lactotransferrin sequence from residue 4 to 16 and from residue 17 to 52 (Fig. 1). The presence of the disulfide bridge 1 (Cys-10 and Cys-46) allows the two peptide chains to be covalently associated.

### Molecular Modeling Studies

The above-mentioned experimental results are unable to explain why the sulfo-SANPAH-derivatized lactotransferrin can bind to the cell receptor while FITC- or SASD-derivatized proteins cannot. This is why molecular modeling was used in order to determine where and how FITC, SASD and sulfo-SANPAH molecules fit into human lactotransferrin. SASD, FITC and sulfo-SANPAH, at the locations found (Lys-74, Lys-264 and Lys-281, respectively), were set so that they could interact with the receptor-binding domain from residues 4 to 52. Though FITC and SASD molecules are 190 amino acid residues away from each other on the polypeptide chain (from Lys-74 to Lys-264), they both cover a part of the receptor-binding peptide from residues 4 to 52 (Fig. 3). On the contrary, the sulfo-SANPAH molecule bound on Lys-281, which does not inhibit the binding of lactotransferrin on the activated lymphocyte receptor, is located at the opposite side of domain I.

### Effect of Lactotransferrin and Lactotransferrin-Derived Fragments on ADP-Induced Platelet Aggregation

Human lactotransferrin as well as its N-t tryptic fragment inhibited ADP-induced platelet aggregation. The measured inhibitory effect was concentration dependent down to a lactotransferrin concentration of 5 nM for lactotransferrin and of 2 μM for the N-t tryptic fragment. On the contrary, as expected, no inhibitory effect was observed with either C-t tryptic lactotransferrin or FITC-modified lactotransferrin. In order to demonstrate that lactotransferrin acts on platelet function through the binding to the lactotransferrin receptor, the effects of the two synthetic peptides, KRDS tetrapeptide and CFQWQRNMRKVRGPPVSC octodecapeptide, were investigated. Interestingly, an inhibitory effect was also observed for both the tetrapeptide and the octodecapeptide. The median inhibitory concentration calculated in our experimental conditions increased when the molecular mass of peptides decreased and was at maximum for the KRDS tetrapeptide. The median inhibitory concentration we have measured for the KRDS tetrapeptide (500 μM) was in good accordance with the concentration previously determined (350 μM) by Drouet et al. (29) and Mazoyer et al. (30). The inhibitory effect of the

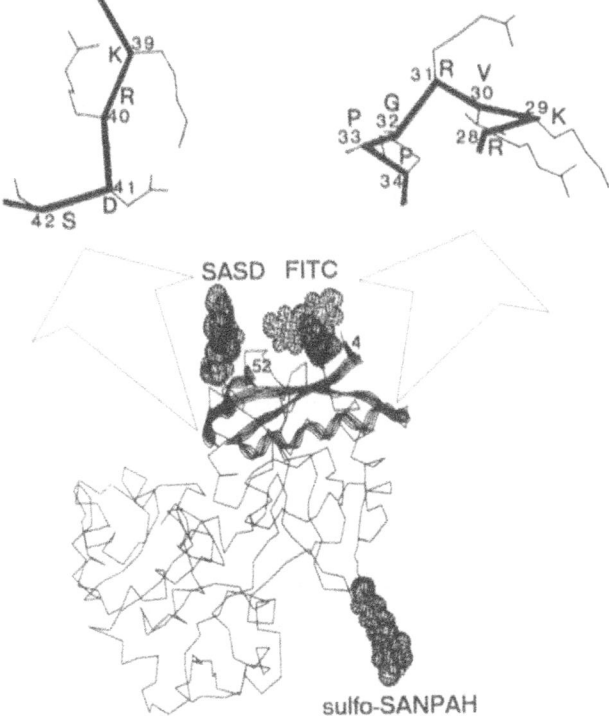

**Figure 3.** Masking effect of FITC, SASD and sulfo-SANPAH on the human lactotransferrin polypeptide chain. Backbone α-carbon tracing of the N-terminal lobe of human lactotransferrin (residues 1–330) showing the folding of (ribbon) the receptor-binding peptide from residue 4 to 52 and the relative positions of FITC, iodinated-SASD and sulfo-SANPAH molecules, represented with van-der-Waals surfaces

octodecapeptide is 10 times higher (20 μM) than for the KRDS sequence and it can be suggested that the accessible lactotransferrin loop 28–34 is mainly involved in the mechanism of the inhibition of platelet aggregation and therefore in the binding of lactotransferrin to cells.

## DISCUSSION

The aim of the present study was to go further in the characterization of the peptide sequence of human lactotransferrin which interacts with both the human PHA-activated lymphocyte and platelet receptor. The present results obtained by derivatizing human lactotransferrin show that SASD mainly conjugated to Lys-74, inhibits the binding of lactotransferrin to the lymphocyte receptor while sulfo-SANPAH conjugated to Lys-281, close to Lys-264, does not. Moreover, the isolation of a 6 kDa peptide able to inhibit the specific binding of human lactotransferrin to the lymphocyte receptor brings direct evidence that the amino acid residues involved in receptor binding are included between residues 4 and 52.

In fact, as shown by molecular modeling, sulfo-SANPAH conjugated to Lys-281 is unable to prevent molecular interactions between amino-acid residues 4 to 52 and the lymphocyte receptor. On the contrary, FITC and SASD molecules could mask two among the three main solvent-accessible areas from residues 4 to 52 : the area including residues 4-6 and residues 28–34 and the area from residues 38 to 45. The solvent-accessible amino acid residues of the α-helix from residues 16 to 28 are unlikely to be hidden by either FITC or SASD molecules.

Moreover, it is worth noting that FITC could only mask the solvent-accessible area comprising residues 4–6 and 28–34, while SASD could only mask part of the loop 38–45. Since lactotransferrin undergoes significant receptor-binding inhibition when derivatized by either FITC or SASD, the two above-mentioned regions could be involved in receptor binding. Moreover, the involvement of the two loop-containing regions has been corroborated by using peptide synthesis. The biological activity of lactotransferrin binding to platelet has been rediscovered with the two synthetic polypeptides KRDS and CFQWQRNMRKVRGPPVSC. Therefore, it can be assumed that the domain 4–52 is essential in the biological role of lactotransferrin.

## ACKNOWLEDGEMENTS

This work was supported in part by the Université des Sciences et Technologies de Lille, the Ministère de l'Education Nationale and the Centre National de la Recherche Scientifique (Unité Mixte de Recherche n° 111 du CNRS, Director: Prof. André Verbert).

We are grateful to Dr. E.N. Baker (Massey University, New-Zealand) and to Dr. P.F. Lindley (London University, Great Britain) for providing the X-ray crystallographic data of human lactotransferrin and rabbit serotransferrin, respectively, to Professor G. Vergoten who performed the molecular modeling, to Mrs O. Labiau and C. Brassart, CNRS technicians, for their skillful technical assistance and to Prof. A. Tartar for helpful advice. We thank Dr. D. Brévière from the Centre Régional de Transfusion Sanguine de Lille (Director: Dr. J.J. Huart) for providing us with human blood and Dr. C. Mazurier for measuring platelet aggregation.

## REFERENCES

1. Montreuil J, Mullet S (1960a) C R Acad Sci Paris 250:1736-1737.

2. Montreuil J, Tonnelat J , Mullet S (1960b) Biochim Biophys. Acta 45:413–421.

3. Biserte G, Havez R , Cuvelier R (1963) Exp Ann Bioch Med 25:85–120.

4. Masson PL, Heremans JF, Schonne E (1969) J Exp Med 130:643–658.

5. Masson PL (1970) in: La lactoferrine, protéine des sécrétions externes et des leucocytes neutrophiles, Editions Arsca, Bruxelles, p.93

6. Metz-Boutigue MH, Jolles J, Mazurier J, Schoentgen F, Legrand D, Spik G, Montreuil J, Jollès P (1984) Eur J Biochem 145:659–676.

7. Powell MJ , Ogden JE (1990) Nucl Acids Res 18:4013.

8. Rey MW, Woloshuk SL, deBoer HA, Pieper FR (1990) Nucl Acids Res 18:5288.

9. Anderson BF, Baker H, Dodson EJ, Norris GE, Rumball SV, Waters JM, Baker EN (1987) Proc Natl Acad Sci USA 84:1769–1773.

10. Anderson BF, Baker HM, Norris GE, Rice DW, Baker EN (1989) J Mol Biol 209:711–734.

11. Anderson BF, Baker HM, Norris GE, Rumball SV, Baker EN (1990) Nature (London) 344:784–787.

12. Baker EN, Rumball SV, Anderson BF (1987) Trends Biochem Sci 12:350–353.

13. Hu WL, Mazurier J, Montreuil J, Spik (1990) Biochemistry 29:535–541.

14. Retegui LA, Moguilevski N, Castragne CE, Masson PL (1984) Lab Invest 50:323–328.

15. Birgens HS, Karle H, Hansen NE, Kristensen L (1984) Scand J Haematol 33:275–280.

16. Moguilevski N, Retegui LA, Courtoy CE, Castracane CE, Masson PL (1984) Lab Invest 50:335–340.

17. Goavec M, Mazurier J, Montreuil J, Spik G (1985) C R Acad Sci Paris 301:689–694.

18. Mazurier J, Legrand D, Hu WL, Montreuil J, Spik G (1989) Eur J Biochem 179:481–487.

19. Leveugle B, Mazurier J, Legrand D, Mazurier C, Montreuil, J, Spik G (1993) Eur J Biochem 213:1205–1211.

20. Legrand D, Mazurier J, Metz-Boutigue MH, Jollès J, Jollès P, Montreuil J, Spik G (1984) Biochim Biophys Acta 787:90–96.

21. Legrand D, Mazurier J, Aubert JP, Loucheux-Lefebvre MH, Montreuil J, Spik G (1986) Biochem J 236:839–844.

22. Legrand D, Mazurier J, Maes P, Rochard E, Montreuil J, Spik G (1991) Biochem J 276:733–738.

23. Legrand D, Mazurier J, Elass A, Rochard E, Vergoten G, Maes P, Montreuil J, Spik G (1992) Biochemistry 31:9243–9251

24. Patscheke H, Wörner P (1978) Thrombos Res 12:485–496.

25. Spik G, Strecker G, Fournet B, Bouquelet S, Montreuil J, Dorland L, Van Halbeek H, Vliegenthart JFG (1982) Eur J Biochem 121:413–419.

26. Mazurier J, Lhoste JM, Montreuil J, Spik G (1983) Biochim Biophys Acta 745:44–49.

27. Sybyl, inventors (1988) Tripos Associates, assignee, St Louis, MO, USA 189–213.

28. Rochard E, Legrand D, Mazurier J, Montreuil J, Spik, G. (1989) FEBS Lett 255:201–204.

29. Drouet LO, Bal Dit Sollier C, Cisse MT, Pignaud G, Mazoyer E, Fiat AM, Jollès P, Caen JP (1990) Nouv Rev Fr Hematol 32:59–62.

30. Mazoyer E, Levy-Toledano S, Rendu F, Hermant L, Lu H, Fiat, AM, Jollès P, Caen J (1990) Eur J Biochem 194:43–49.

# PHYSICAL CHARACTERISTICS AND POLYMERIZATION DURING IRON SATURATION OF LACTOFERRIN, A MYELOPOIETIC REGULATORY MOLECULE WITH SUPPRESSOR ACTIVITY

Charlie Mantel,[†] Keisuke Miyazawa,[†] and Hal E. Broxmeyer[†,‡]

†Walther Oncology Center and Department of Medicine
   (Hematology/Oncology)
‡Department of Microbiology and Immunology
  Indiana University School of Medicine
  Indianapolis, Indiana 46202

## ABSTRACT

Lactoferrin (LF) has been implicated in normal regulation of myeloid blood cell production *in vitro* and *in vivo* and abnormalities in LF-cell interactions have been associated with progression of leukemia and other hematopoietic disorders. LF may be clinically useful and for this reason we studied selected biochemical characteristics of LF. Purified human milk LF was saturated with iron from solution and analyzed by gel electrophoresis, ion-exchange and gel filtration chromatography. The metalloprotein was found to contain several molecular weight species on polyacrylamide gels. High resolution ion-exchange chromatography demonstrated the binding of LF to both anionic and cationic media under identical conditions indicating a bipolar charge distribution. Gel filtration studies revealed a tetramerized form of LF, the formation and stability of which was dependent on the ionic strength of the solution.

## INTRODUCTION

Lactoferrin (LF) is an iron binding glycoprotein of approximately 76 kD molecular weight (1) found in milk, tears, semen and plasma (2–5); within the hematopoietic system it is found principally in the secondary granules of neutrophilic granulocytes (6). LF is structurally related to serotransferrin, the major plasma iron transport protein (1, 7, 8) and is believed to be involved in the prevention of bacterial colonization of the above mentioned secretions (2, 9). LF has been implicated in the negative regulation of myeloid blood cell production *in vitro* and *in vivo* (10–15) an effect mediated at least in part by LF decreasing the release of interleukin-1 (IL-l) from monocytes which in turn is capable of inducing the release of growth factors, such as hematopoietic colony stimulating factors (CSF), from other cell types (16). LF decreases proliferation (5, 10–15) and survival (17) of hematopoietic progenitor cells and on certain cell lines also decreases differentiation of cells (18). A number of molecules can abrogate the myelosuppressive effects of LF in vitro, including IL-l (16, 17), IL-6 (19) and bacterial lipopolysaccharide (LPS) (10, 11, 20). The LPS-abrogation effects occur through LPS binding to LF molecules and once an LPS-LF complex is formed, binding to LPS-receptors is favored over that of binding to LF-receptors (20). Abnormal interactions have been noted with cells from patients with leukemia and related disorders which reflect abnormal quantitative effects (such as decreased amounts of LF) and abnormal qualitative effects (such as inactivity) in the LF molecule, as well as decreased sensitivity of patient cells to the actions of normal LF molecules (5, 21, 22). This highlights a role for LF in myelopoietic regulation. LF has been shown to be beneficial in animal models of virally induced hematopoietic abnormalities (23), effects enhanced by pretreatment of mice with interferon-gamma (24) in a manner analogous to the responsiveness of leukemia cell lines to the suppressive effects of LF induced by interferon-gamma *in vitro* (25). Little is known about regulation of LF production, but it is of interest that LF-gene expression of normal human bone marrow is transcriptionally decreased by tumor necrosis factor-alpha (26), a molecule whose production from monocytes is decreased by LF (27).

LF may be useful in the clinic in such hyperproliferative disorders as chronic myelogenous leukemia by dampening proliferation of cells and thus potentially slowing disease progression. The recent expression of recombinant human lactoferrin (28) will no doubt hasten the testing of this possibility. This highlights the need to further understand the characteristics which make this molecule biologically active. The myelosuppressive effect of LF is lost when LF undergoes polymerization (29). Polymerization of LF is reported to be calcium dependent (30). Since many of the biological effects of LF are related to its capacity to bind iron (9–11), preparations of LF are saturated with iron either by donation from an iron chelate (such as nitrilotriacetic acid), or by direct binding of ferric ion from solution. In this communication we assess the migration of native and iron-saturated LF in different gels and demonstrate the polymerization of LF during iron saturation in solution in the absence of calcium and show that this self-association is related to the ionic strength of the solvent during iron saturation. Evidence is presented indicating a mechanism for this phenomenon. The data also suggest that LF has a "bipolar" charge density and the significance of this to its biological effects is discussed.

## MATERIALS AND METHODS

Lactoferrin (95% pure) was purchased from Sigma Chemical Co. (St. Louis, MO) in an iron poor form isolated from human milk (native LF; NLF). It was routinely saturated with iron by the method of Masson and Heremans (31) with modification (32). Briefly, NLF was dissolved in 0.01 M sodium phosphate, pH 7.4 plus 150 mM NaCl (PBS) at 10 mg/ml. Then a solution of L-ascorbic acid and ferric ammonium sulfate was added to make the final concentration 0.89 mM and 0.27 mM respectively. This was rotated at room temperature for two hours then dialyzed against PBS overnight with several changes of buffer in Spectrapore No. 1 dialysis tubing. After dialysis, iron saturated LF (FeLF) was sterilized by 0. 22 μm filtration and stored at 4°C. Iron saturation was confirmed spectrophotometrically by an increase in the absorbance at 460 nm (A460/A280 = 0. 040).

SDS-polyacrylamide gel electrophoresis (SDS-PAGE) was done on 7.5% or 12%, 0.75 mm thick polyacrylamide gels prepared according to the method of Laemmli (33), and run in a Protein II vertical gel electrophoresis apparatus (Bio-Rad, Richmond, CA) at 15 mA per gel for 3 to 6 hours. Molecular weight standards were from Bio-Rad. Protein bands were visualized either by Coomassie Brilliant Blue R-250 stain or by the Bio-Rad Silver Stain kit.

LF preparations were chromatographed on a Superose-12 HR 10/30 (Pharmacia, Piscataway, NJ) gel-filtration column in PBS or the indicated buffers at flow-rates of 0. 5 or 1.0 ml/min. Typically 100 μg of protein was loaded in 0. 2 ml and protein elution was continuously monitored by optical density (OD) at 280 mM. Alternatively, LF was chromatographed on a 1.6 × 60 cm Sephacryl S-300 (Pharmacia LKB, Piscataway, NJ) column (packed according to manufacturer's instructions) in the buffers indicated at a flow-rate of 0.5 column volume per hour. Columns were calibrated using the Bio-Rad Gel Filtration Calibration kit.

Ion exchange chromatography of LF was done on either a Mono-S or Mono-Q column (1 ml; Pharmacia) using a Fast Protein Liquid Chromatography system (FPLC; Pharmacia) to deliver the gradient. Typically, 100 µg protein was charged in 0.1 ml of 0.01 M sodium phosphate pH 7.4. The column was washed with 5 column volumes start buffer, and eluted with a linear NaCl gradient from 0 to 1 M over 10 column volumes at 1 ml/min. Protein elution was monitored as above.

## RESULTS AND DISCUSSION

### 1. Lactoferrin Characterization

Before and after iron saturation, LF was analyzed by SDS-PAGE. FeLF migrated as the major Coomassie Blue stained band at a MW of 76 kD when run unreduced on 12% gels (Fig. 1a, lane #2). After reduction, several other bands (66, 47.5 and 35 kD) could be visualized (Fig. 1a, lane #3). The two lower bands may correspond to the two "lobes" of LF (34,35). This would imply that there may exist inter-lobe disulfide bonds or very strong non-covalent association between the two lobes of some LF molecules. Such strong inter-lobe interactions have been reported (36).

When analyzed on 7% gels, FeLF migrated as two bands at 76 kD and 62 kD when unreduced, (Fig. 1b, lane #2). A third band was also found at 67 kD. Kijlstra et al. also observed an apparent lighter form of LF in tears (3). They showed that the smaller molecular weight form was iron saturated and the larger form was unsaturated. Bezwoda and Mansoor also observed this behavior in non- reduced samples (37). These smaller forms were not seen with NLF in this study (data not shown). After reduction, two bands of MW 47.5 and 35 kD appeared, while the major form was 76 kD (Fig. 1b, lane #3). These data suggest that there are considerable conformational changes in the LF molecule after iron binding. This conclusion has been suggested by others (3,7,38).

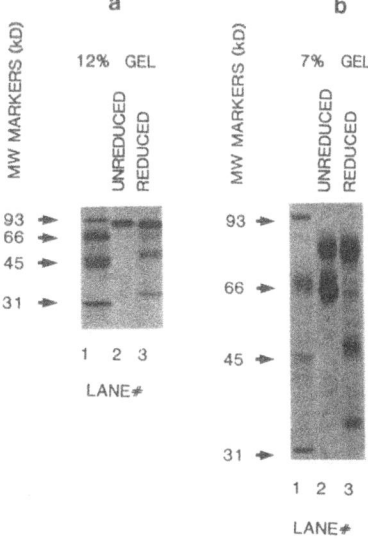

**Figure 1.** SDS-PAGE of iron saturated lactoferrin. a) 12% acry-lamide gel, lane 1 = MW markers, 2 = unreduced lactoferrin, 3 = reduced lactoferrin b) 7% acrylamide gel, lane 1 = MW markers, 2 = unreduced lactoferrin, 3 = reduced lactoferrin.

LF was next characterized by gel filtration. When NLF was chromatographed on a calibrated Superose-12 column in PBS it eluted as a single peak with an apparent relative MW of 51 kD (Fig. 2a). This implies that LF may be retarded by some minor ionic or hydrophobic interaction with this media. This is frequently observed with Superose. After iron saturation another peak could be observed eluting at 350 kD (Fig. 2b). As has already been observed by others (29, 30), this is most consistent with tetramerization.

LF was also characterized by ion-exchange chromatography. NLF could bind to a cation exchanger (Mono S) at pH 7.4 in 10 mM sodium phosphate. It was eluted with a linear gradient from 0 to 1 M sodium chloride as a single broad peak at 500 mM NaCl (Fig. 3a). FeLF had an identical elution profile (not shown). LF could also bind to an anion exchanger (Mono Q) at pH 7.4 in 10 mM sodium phosphate. When eluted with the same gradient, FeLF could be separated into at least two major components (Fig. 3b). Results were exactly the same when NLF was run (not shown). The first, eluting at about 230 mM NaCl, was a broad peak while

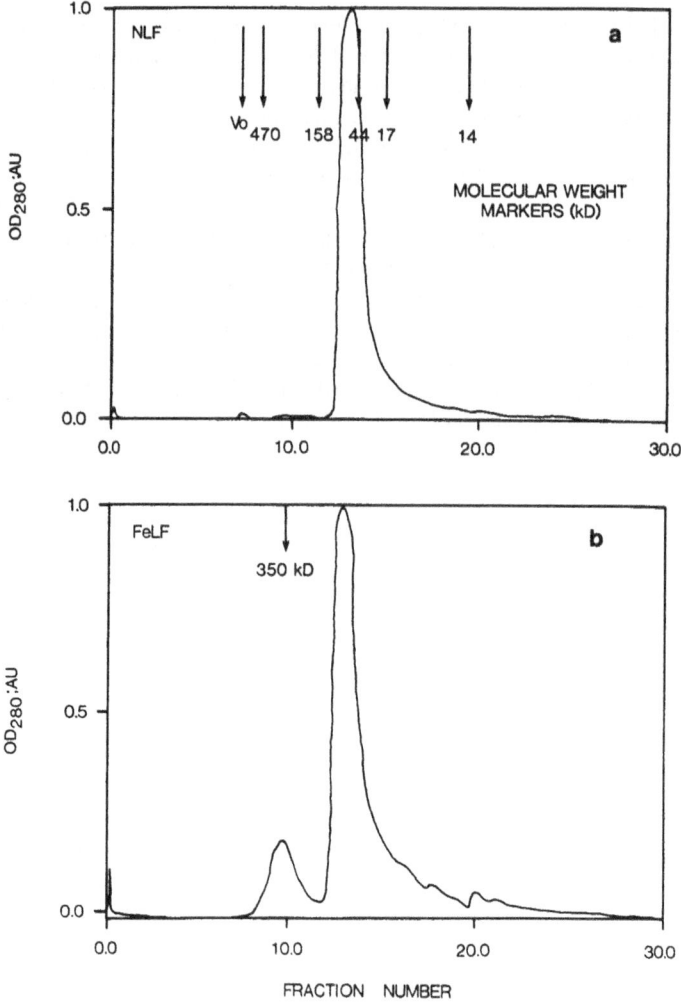

**Figure 2.** Gel filtration chromatography of native (a) and iron saturated (b) lactoferrin on Superose-12 in PBS.

the other was a sharper peak eluting at 325 mM NaCl. Another minor peak at 300 mM could also be observed. The fact that NLF and FeLF could bind to both a cation and anion exchanger in the same buffer at the same pH suggests that the molecule is composed of two well separated domains of charge density, one being predominantly positively charged at pH 7.4 and the other predominately negative. This notion is consistent with the reported structure of LF. Like serum transferrin, LF is composed of 2 major "lobes" (one each at the carboxyl and amino termini) (7). However, unlike serum transferrin, LF has two dissimilar lobes of differing iron binding affinity and size. Bluard-Decominck et al. (35) have reported that after enzymatically cleaving the two lobes, the lobes could be separated by ion-exchange chromatography because the amino terminal lobe was basic in nature while the carboxyl terminal lobe was acidic. These authors also reported the migration of the two fragments toward different electrodes in native agarose

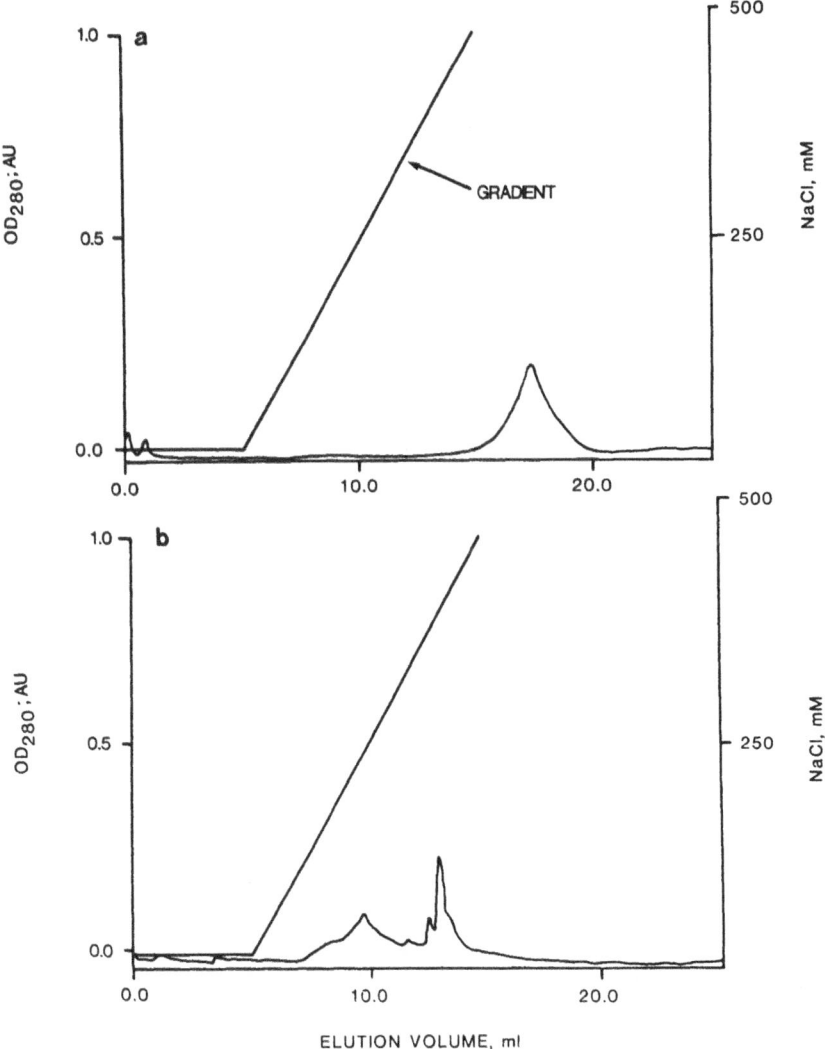

**Figure 3.** Ion-exchange chromatography of lactoferrin on Mono-Q anion exchanger (a) and mono-S cation exchanger (b).

electrophoresis gels. The data presented in this report strongly suggest that the charge density of LF is zwitterionic in nature. This may be important to both the observed physical properties of LF (self-association and association to other macromolecules (30, 39, 40) as well as to its biological properties such as cellular binding (13, 41, 42)).

## 2. Lactoferrin Polymerization

Since LF was shown to undergo polymerization during iron saturation (Fig. 2b), the kinetics of this phenomenon were studied. In as little as 1 minute after adding the ferric ascorbate solution (Fig. 4), polymer begins to form. After 60 minutes there was a gradual increase in the percentage of polymer generated, which appears to level off after one or two days. LF was usually treated for several hours in this fashion before dialysis, so the actual exposure time to the iron saturation solution may vary. Subsequently, the percentage of polymerization after 24 hours of dialysis under these conditions ranged from 15% to 25% from preparation to preparation. However, once formed, the polymer appears to be moderately stable

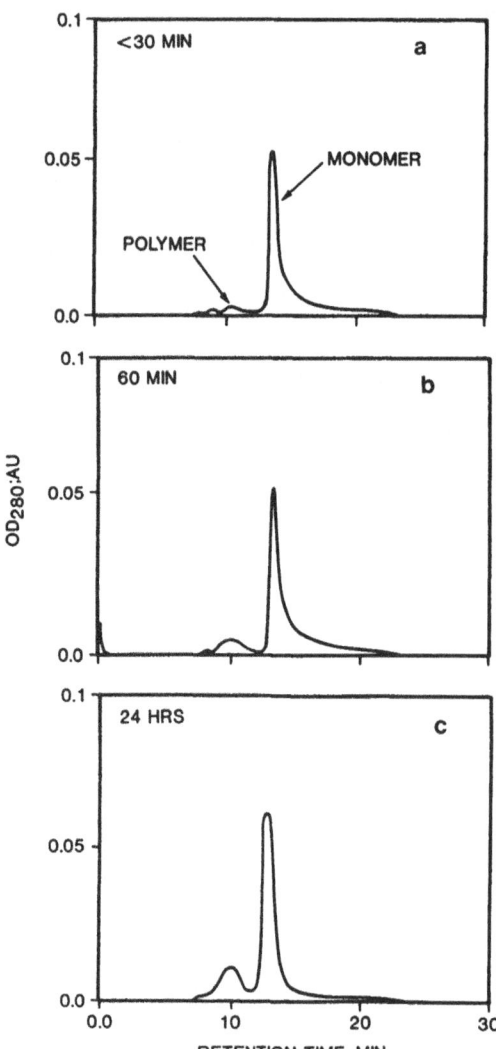

**Figure 4.** Effect of iron saturation time on polymerization of lactoferrin. Chromatography is on Superose-12 column in PBS.

when stored at concentrations from 1 to 10 mg/ml at 4°C in PBS. The polymer slowly decomposes into monomer but may take as long as one or two months to completely decompose. Interestingly, when the polymer is separated from the monomer by gel filtration in 10 mM sodium phosphate, pH 7.4 (PB), the decomposition is much more rapid and may be complete within several days. This process is highly dependent upon the ionic strength of the solvent. By increasing the ionic strength with NaCl, the decomposition time can be reduced to 60 min at 35 mM NaCl, 20 min at 75 mM, 10 min at 100 mM, and 5 min at 150 mM. This demonstrates a relationship between the stability of the polymer and the ionic strength of its solvent.

To more clearly understand the role of ionic strength on the chromatographic behavior of LF, native and FeLF was dialyzed into PB with or without added NaCl (75 mM) and chromatographed on a Superose 12 column equilibrated and previously calibrated in PB (the result for PBS; 150 mM NaCl, was shown in Fig. 2 a and b). The result for NLF (Fig. 5a) was the elution of the monomeric peak at a MW of 26 kD, nearly one half that observed for PBS. Also this peak developed a significant trailing shoulder. This suggests that the interaction of LF with this media may be ionic in nature and that the lower ionic strength of this buffer exacerbated this interaction. There was also a significant amount of LF "aggregates" (MW > $10^6$ kD) observed. When FeLF was chromatographed, there was a marked increase in retention

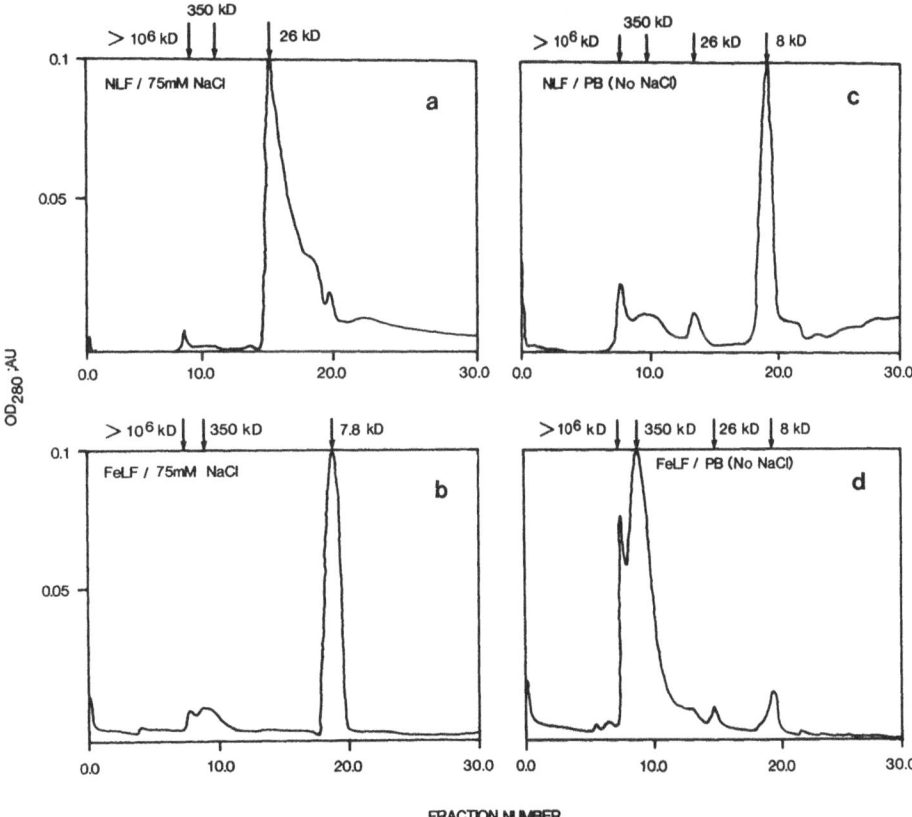

**Figure 5.** Effect of solvent ionic strength on elution profile of lactoferrin on Superose-12. a) native lactoferrin in 10 mM phosphate buffer, pH 7.4 (PB) plus 75 mM NaCl; b) iron saturated lactoferrin in PB plus 75 mM NaCl; c) native lactoferrin in PB alone; d) iron saturated lactoferrin in PB alone.

time compared to PBS (Fig. 5b). FeLF monomer eluted with an apparent MW of 7.8 kD, while the polymeric form was unchanged (350 kD). Again, there was an increase in the percentage of material in the void volume ($> 10^6$ kD). When PB alone was used (no NaCl added), NLF eluted as four peaks (Fig. 5c). The void volume peak increased even more than that observed with 75 mM NaCl. A polymeric peak developed at 350 kD, a small peak at 26 kD and a major peak at 8 kD. FeLF showed two minor peaks at 8 kD and 26 kD, while most of the LF was observed at 350 kD and void volume (Fig. 5d). When total protein recovery was determined, it was discovered that in the case of PB alone, sometimes as much as 80% of the total loaded protein was left on the column. This material could be recovered by washing the column with PB plus 1 M NaCl.

These data indicate that LF can self-associate in low ionic strength aqueous solvents. They also imply that this association is ionic in nature. This is not surprising in view of the amphoteric properties of the molecule's charge density. It can also be deduced from these data that after iron binding, LF may undergo considerable conformational changes which alters its hydrophobic/ ionic interactions with column packing such as Superose as well as enhances self-association.

To confirm these observations, another gel filtration media, Sephacryl S-300 (S-300) was used. Figure 6a shows the elution profile of FeLF in PBS on S-300. Compare this to Superose-12 (Figure 2b). When this column was equilibrated with PB only and an identical amount of FeLF (dialyzed into the same buffer) was applied (Figure 6b), only a small amount of the loaded protein eluted at the void volume. As with Superose, the protein could be recovered with 1 M NaCl.

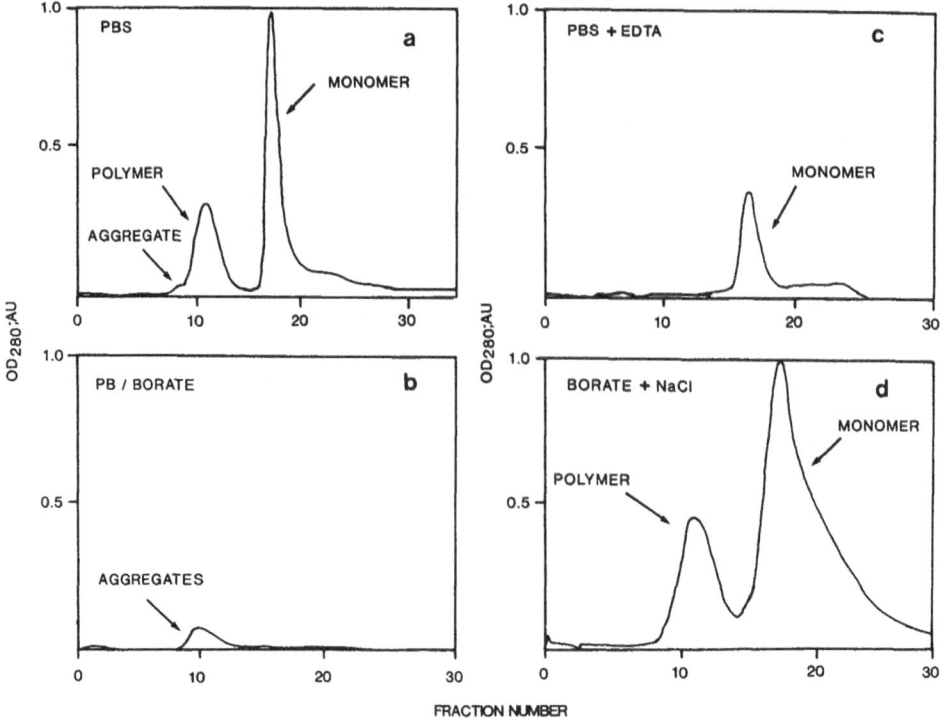

**Figure 6.** Effect of different buffers on elution profile of iron saturated lactoferrin on Sephacryl S-300 gel filtration column. a) PBS; b) PB only or borate buffer only; c) PBS plus 10 mM EDTA; d) borate buffer plus 150 mM NaCl.

Bennett et al. (30) previously showed that FeLF undergoes polymerization in calcium-containing buffers, while a buffer supplemented with EDTA maintained only monomeric LF. These results were interpreted as indicating that the polymerization was a direct result of an interaction between calcium and LF and that EDTA could prevent polymerization by chelation of calcium. To determine the relationship of this polymeric form of LF to the polymer generated during iron saturation, 10 mM EDTA or 10 mM $CaCl_2$ was added to the elution buffer and the S-300 column run with FeLF dialyzed into the same buffer. Figure 6c shows that the polymer was completely decomposed when EDTA was added. Since $CaCl_2$ cannot be added directly to phosphate buffers because of insoluble precipitates of calcium phosphate, and also because the original characterization of LF polymers by Bennet et al. (30) was done in a borate buffer, the S-300 column was run again with 0.2 M sodium borate, pH 7.4 plus 10 mM $CaCl_2$. The elution profile of FeLF was identical to that when run without $CaCl_2$ (Fig. 6b). Again, the protein remaining on the column could be removed with 1 M NaCl. Additionally, when the borate buffer was supplemented with 150 mM NaCl (Figure 6d), FeLF displayed both the polymeric and monomeric peaks, although somewhat broadened. These data demonstrate that LF may self-associate in the absence of calcium ion and that EDTA can rapidly decompose the polymer. A close examination of the retention time of the polymer generated in the lower ionic strength borate buffer suggested that it was not tetramer, but a higher MW polymer than that generated during iron saturation ("aggregate"). This may have significance because Bennett et al. (30) reported a loss of biological activity of this polymer. They proposed that LF tetramerization due to calcium could be important in hematopoietic regulation. While our data agree on the important point that LF does polymerize, it suggests however that the polymerization observed by us and others is the result of self-association in low ionic strength solvent. In these experiments, calcium concentrations above 10 mM did not promote polymerization in borate buffer supplemented with NaCl. However, when high concentrations of $CaCl_2$ (50 to 100 mM) were used it caused a complete decomposition of existing polymer. This was probably due to the increase in ionic strength.

Next, the amount of iron bound by the polymer was studied. To determine if the polymer had bound iron during the saturation process, the optical density of the Superose-12 column eluates were monitored with a dual wavelength monitor, simultaneously measuring OD at 280 and 460 nm. The modal absorbance ratios ($AU_{460}/AU_{280}$) for the polymer and monomer were 0.020 and 0.040, respectively. This suggested that the polymer peak contains fewer iron saturated molecules of LF than the monomeric peak. However, to confirm this result, NLF was iron saturated in the presence of $^{59}FeCl_2$. After dialysis FeLF was subjected to gel filtration on Superose-12 in PBS and 1 ml fractions collected and the radioiron counted in a gamma counter. The ratio of $^{59}Fe^{total}/OD_{280}^{total}$ (cpm/total AU) of the pooled peak fractions was 138 for polymer and 34 for monomer. This result was checked by measuring the protein content directly (Bio-Rad protein assay). Using this technique, the ratios were 143 for polymer and 38 for monomer. These data are in contrast to the extinction data and suggest there is four times more radioiron associated with the polymer than monomer. One interpretation of this is that the configuration of the polymer reduces the absorption of the visible wavelength-absorbing domains of the molecule. However, another interpretation is that the polymer has "trapped" ferric ions without actually binding iron in the iron-binding sites of the molecule. This could explain the high radioiron ratio of the polymer without changing the optical absorption properties of the protein.

The formation of polymer during iron saturation may be explained by exposure to high concentrations of ferric ion. Since it has been shown that there may be additional iron associated with the polymer other than that tightly bound, it may be that ferric ions act as an ionic bridge between LF molecules, promoting polymerization and forming an iron "core". To test this idea, semi-purified, dialyzed $^{56}$Fe-labelled polymer was decomposed with NaCl and subjected to

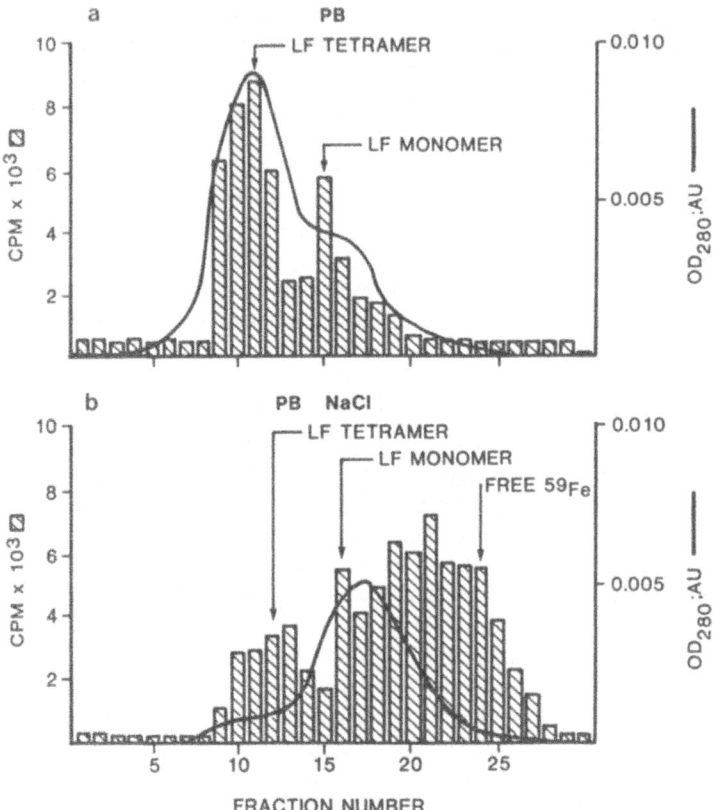

**Figure 7.** Gel filtration chromatography of $^{59}$Fe labelled polymerized lactoferrin on Superose-12. a) PB only; b) PB plus 150 mM NaCl.

gel-filtration on Superose-12 (Figure 7a and b). "Free" radioiron was released from the polymer, while the generated monomer retained its iron (Fig. 7b). These data agree with the notion of trapped iron during polymerization.

Our studies noted above add to the studies of others which have characterized the LF molecule. This information may be of relevance to the biological actions of LF.

## ACKNOWLEDGMENT

These studies were supported by U.S. Public Health Service Grants R37 CA36464 and RO1 HL49202 from the National Cancer Institute and the National Institutes of Health to HEB.

## REFERENCES

1. Metz-Boutique MH, Jolles J, Mazurier J, Schoentgen F, Legrand D, Spik G, Montreuil J, Jolles P (1984) Human lactoferrin: amino acid sequence and structural comparisons with other transferrins. Eur J Biochem 145:659–676.

2. Masson PL, Heremans JF, Dive C (1966) An iron-binding protein common to many external secretions. Clin Chim Acta 14:735–739.

3. Kijlstra A, Kuizenga A, van der Velde M, van Haeringen NJ (1989) Gel electrophoresis of human tears reveals various forms of tear lactoferrin. Curr Eye Res 8:581–588.

4. Bennett RM, Eddie-Quartey AC, Holt PJL (1973) Lactoferrin-an iron binding protein in synovial fluid. Arthritis Rheum 16:186–190.

5. Broxmeyer HE, Gentile P, Bognacki J, Ralph P (1983) Lactoferrin, transferrin and acidic isoferritins: Regulatory molecules with potential therapeutic value in leukemia. Blood Cells 9:83–105.

6. Baggiolini M, DeDuve C, Masson PL, Heremans JF (1970) Association of lactoferrin with specific granules in rabbit heterophil leukocytes. J Exp Med 131:559–570.

7. Anderson B, Baker HM, Dodson E, Orris GE, Rumbail SV, Waters JM, Baker EN (1987) Structure of human lactoferrin at 3.2-Å resolution. Proc Natl Acad Sci USA 84:1769–1773.

8. Aisen P, Listowsky I (1980) Iron transport and storage proteins. Annu Rev Biochem 49:357–393.

9. Arnold RR, Cole MF, McGhee JR (1977) A bactericidal effect for human lactoferrin. Science 197:263–265.

10. Broxmeyer HE, Smithyman A, Eger RR, Meyers P, DeSousa M (1978) Identification of lactoferrin as the granulocyte-derived inhibitor of colony-stimulating activity production. J Exp Med 148:1052–1067.

11. Broxmeyer, HE, DeSousa M, Smithyman A, Ralph, P Hamilton J, Kurland JI, Bognacki J (1980) Specificity and modulation of the action of lactoferrin, a negative feedback regulator of myelopoiesis. Blood 55:324–333.

12. Bagby GC, Riga VD, Bennett RM, Van denbark AA, Garewal HS (1989) Interaction of lactoferrin, monocytes, and T-lymphocyte subsets in the regulation of steady-state granulopoiesis in vitro. J Clin Invest 68:56–63.

13. Broxmeyer HE, Bicknell DC, Gillis S, Harris EL, Pelus LM, Sledge GW (1986) Lactoferrin: Affinity purification from human milk and polymorphonuclear neutrophils using monoclonal antibody (II 2C) to human lactoferrin, development of an immunoradiometric assay using II 2C, and myelopoietic regulation and receptor-binding characters. Blood Cells 11:429–446.

14. Gentile P, Broxmeyer HE (1983) Suppression of mouse myelopoiesis by administration of human lactoferrin in vivo and the comparative action of human transferrin. Blood 61:982–993.

15. Broxmeyer HE, Williams DE, Hangoc G, Cooper S, Gentile P, Shen RN, Ralph P, Gillis S, Bicknell DC (1987) The opposing action in vivo on murine myelopoiesis of purified preparations of lactoferrin and the colony stimulating factors. Blood Cells 13:31–48.

16. Zucali JR, Broxmeyer HE, Levy D, Morse C (1989) Lactoferrin decreases monocyte induced fibroblast production of myeloid colony stimulating activity by suppressing monocyte release of interleukin-l. Blood 74:1531–1536.

17. Hangoc G, Falkenburg JHF, Broxmeyer HE (1991) Influence of T-lymphocytes and lactoferrin on the survival-promoting effect of IL-I and IL-6 on human bone marrow granulocyte-macrophage and erythroid progenitor cells. Exp Hematol 19:697–703.

18. Okabe-Kado J, Hayashi M, Honma Y, Hozumi M (1985) Characterization of a differentiation-inhibitory activity from non-differentiating mouse myeloid leukemia cells. Cancer Res 45:4848–4852.

19. Gentile P, Broxmeyer HE (1991) Interleukin-6 ablates the accessory cell- mediated suppressive effects of lactoferrin on human hematopoietic progenitor cell proliferation in vitro. Ann NY Acad Sci 628;74–83.

20. Miyazawa K, Mantel C, Lu L, Morrison DC, Broxmeyer HE (1991) Lactoferrin- lipopolysaccharide interactions. Effect on lactoferrin binding to monocyte/macrophage-differentiated HL-60 cells. J Immunol 146:723–729.

21. Broxmeyer HE, Mendelson N, Moore MAS (1977) Abnormal granulocyte feedback of colony forming and colony stimulating activity-producing cells from patients with chronic myelogenous leukemia. Leukemia Res 1:3–12.

22. Broxmeyer HE, Bicknell DC, Cooper S, Sledge G Jr., Williams DE, McGuire WA, Coates TD (1991) Quantitative functional deficiency of affinity-purified lactoferrin from neutrophils of patients with chronic myelogenous leukemia, and lactoferrin/H-ferritin-cell interactions in a patient with lactoferrin-deficiency with normal numbers of circulating leukocytes. Pathobiol 59:26–35.

23. Lu L, Hangoc G, Chen LT, Shen RN, Oliff A, Broxmeyer HE (1987) The protective influence of lactoferrin on mice infected with the polycythemia-inducing strain of the Friend Virus Complex. Cancer Res 47:4184–4188.

24. Lu L, Shen RN, Zhou SZ, Srivastava C, Harrington MA, Miyazawa K, Wu B, Lin ZH, Ruscetti S, Broxmeyer HE (1991) Synergistic effect of human lactoferrin and recombinant murine interferon gamma on disease progression in mice infected with the polycythemia-inducing strain of the Friend virus complex. Int J Hematol 54:117–124.

25. Broxmeyer HE, Piacibello W, Juliano L, Platzer E, Berman E, Rubin BY (1986) Gamma interferon induces colony forming cells of the human monoblast cell line U937 to respond to inhibition by lactoferrin, transferrin, and acidic isoferritins. Exp Hematol 14:35–43.

26. Srivastava CH, Rado TA, Bauerle D, Broxmeyer HE (1991) Regulation of human bone marrow lactoferrin and myeloperoxidase gene expression by tumor necrosis factor-$\alpha$. J Immunol 146:1014–1019.

27. Crouch SPM, Slater KJ, Fletcher J (1992) Regulation of cytokine release from mononuclear cells by the iron-binding protein lactoferrin. Blood 80:235–240.

28. Ward PP, Lo JY, Duke M, May GS, Headon DR, Conneely OM (1992) Production of biologically active recombinant human lactoferrin in *Asperigillus Oryzae*. Biotechnology 10:784–789.

29. Bagby GC Jr, Bennett RM, Wilkinson B, Davis J (1982) Feedback regulation of granulopoiesis: polymerization of lactoferrin abrogates its ability to inhibit CSA production. Blood 60:108–112.

30. Bennett RM, Bagby GC, Davis J (1981) Calcium-dependent polymerization of lactoferrin. Biochem Biophys Res Commun 101: 88–95.

31. Masson PL, Heremans JF (1968) Metal-combining properties of human lactoferrin (Red Milk Protein). Eur J Biochem 6: 579–584.

32. Bates GW, Schlabach MR (1973) The reaction of ferric salts with transferrin. J Biol Chem 248: 3228–3232.

33. Laemmli UK (1970) Cleavage of structural proteins during the assembly of the head of bacteriophage. Nature 227: 680–685.

34. Legrand D, Mazurier J, Metz-Boutique M, Jolles J, Jolles P, Montreuil J, Spik G (1984) Characterization and localization of an iron-binding 18 kDa glycopeptide isolated from the N-terminal half of human lactotransferrin. Biochim Biophys Acta 787:90–96.

35. Bluard-Deconinck J, Williams J, Evans RT, Van Snick J, Osinski PA, Masson PL (1978) Iron-binding fragments from the N-terminal and C-terminal regions of human lactoferrin. Biochem J 171:321–327.

36. Legrand D, Mazurier J, Aubert JP, Loucheux-Lefebvre MH, Montreuil J, Spik G (1986) Evidence for interactions between the 30 KDa N- and 50 KDa C-terminal tryptic fragments of human lactotransferrin. Biochem J 236:839–844.

37. Bezwoda WR, Mansoor N (1986) Isolation and characterization of lactoferrin separated from human whey by adsorption chromatography using Cibacron Blue F3G-A linked affinity adsorbent. Clin Chim Acta 157:89–94.

38. Rosseneu-Motreff MY, Soetewey F, Lamote R, Peeters H (1971) Size and shape determination of apotransferrin and transferrin monomers. Biopolymers 10:1039–1048.

39. Erlanson-Albertsson C, Sternby B, Johannesson U (1985) The interaction between pancreatic carboxylesterhydrolase (bile-salt stimulated lipase of human milk) and lactoferrin. Biochim Biophys Acta 829:282–287.

40. Hekman A (1971) Association of lactoferrin with other proteins as demonstrated by changes in electrophoretic mobility. Biochim Biophys Acta 251:380–387.

41. Van Snick JL, Masson PL (1976) The binding of human lactoferrin to mouse peritoneal cells. J Exp Med 144:1568-1580.

42. Yamada Y, Amagasaki T, Jacobsen DW, Green R (1987) Lactoferrinbinding by leukemia cell lines. Blood 70:264–270.

# LACTOFERRIN AND THE INFLAMMATORY RESPONSE

Roy D. Baynes and Werner R. Bezwoda[†]

Division of Hematology
Department of Medicine
Kansas University Medical Center
Kansas City, Kansas 66160-7402

[†]Department of Medicine
University of the Witwatersrand
Johannesburg, South Africa

## ABSTRACT

Polyclonal antibodies were prepared to purified breast milk lactoferrin and used in an ELISA to measure plasma concentrations in investigations of various aspects of the inflammatory response. They were also used, *in situ*, to evaluate granulocyte lactoferrin content in disease states. The first series of studies addressed the putative role of lactoferrin in the pathogenesis of the hypoferremic, hyperferritinemic response to acute inflammation. Dissociation between the lactoferrin response and the iron related changes in rheumatoid arthritis and after α-interferon administration suggested that the relationship observed in acute and chronic bacterial infection may reflect coincidental effects of inflammatory cytokines. That lactoferrin does not mediate the inflammatory hypoferremic response was established by the finding that bone marrow transplant recipients, post-myeloablation, developed a hypoferremic response during septic episodes despite virtually undetectable plasma lactoferrin concentrations. The second series of investigations employed the plasma lactoferrin concentration as an index of granulocyte activation and function in a number of inflammatory conditions. Markedly increased initial plasma concentrations in acute pneumonia reflecting profound intravascular granulocyte activation were documented to predict sepsis related mortality. Plasma and granulocyte lactoferrin studies established that viral infection is associated with an acquired granulocyte lactoferrin deficiency. Plasma measurements indicated that asthmatics, even when clinically asymptomatic, have evidence of persistent granulocyte activation.

## INTRODUCTION

Lactoferrin is synthesized in two predominant cell types, namely those of the myeloid series and secretory epithelia. While the bacteriostatic properties of lactoferrin in various body secretions derived from secretory epithelia are well established, the biological significance of plasma lactoferrin derived from granulocytic cells is less well defined. Granulocytic lactoferrin is synthesized by maturing granulocytes at the myelocyte stage, appears to be largely complete by the early band stage (Rado et al. 1984) and is stored in the secondary or specific granule system. A simplified comparison of the primary or azurophilic granule system contents and the secondary granule composition is shown in Table 1. Given the localization of lactoferrin within the neutrophil granules, it has been inferred that lactoferrin is important in host defense. This view has been enhanced by clinical observation that the serum concentration is strongly influenced, not only by the size of the granulocyte pool but, in addition, by the host response to infection and inflammation (Baynes et al. 1986a). Prior data have suggested that the increased plasma concentration in such situations has physiological significance. Suggested biological roles for lactoferrin have included the major mediator of the hypoferremic, hyper-ferritinemic response to infection (Van Snick et al. 1974); a factor modulating neutrophil adherence and migration (Boxer et al. 1982); a major enhancer of free radical production in its iron loaded state while functioning as a major defender against potential free radical injury in its iron free or apo-state (Britigan & Edeker, 1991); an inhibitor or promotor of intracellular killing again dependent on its saturation with iron (Byrd & Horwitz, 1991); a modulator of the primary antibody response to infection (Duncan & McArthur, 1981) and a feedback regulator of granulopoiesis by regulating cytokine release from mononuclear cells (Crouch et al. 1992). One aspect of this manuscript will focus on the role of lactoferrin in the inflammatory hypoferremic, hyperferritinemic response. The association between lactoferrin and neutrophil adhesion and migration appears increasingly to be an indirect one insofar as the secondary granule contains both lactoferrin and a large number of receptors for various structural proteins including those for laminin, fibrinogen, fibronectin, and vitronectin. Fusion of the secondary granule limiting membrane affects expression of this series of adhesion receptor molecules (Singer et al. 1989). Thus, lactoferrin reflects rather than influences activity of the secondary granule or "adhesome". Aspects which will not be covered in this manuscript include the effect of lactoferrin on phagocytic cell performance or antibody production although convincing evidence for a biological *in vivo* significance is somewhat lacking. Likewise, free radical metabolism will not be addressed although this may yet prove highly significant particularly in the setting of granulocyte participation in sepsis related morbidity and mortality. Lactoferrin and regulation of granulopoiesis will be covered in another section of this book. Given lactoferrin's concomitant release with "adhesome" incorporation to surface membrane it does provide a useful assay for neutrophil activation and functional integrity. This property was

**Table 1.** Simplified Comparison of Neutrophil Granules and Their Contents

| Granules | Primary | Secondary |
|---|---|---|
| Anitmicrobial Agents | Defensins<br>Myeloperoxidase<br>Lysozyme<br>Cationic proteins | Lysozyme |
| Proteinases | Elastase<br>Cathepsin G | Collagenase |
| Other |  | Lactoferrin<br>$B_{12}$ binding protein<br>Laminin receptor (67 K)<br>CD11b/CD18 (C3bi)<br>Fibronectin receptor $\alpha$<br>Vitronectin recptor $\alpha$ |

utilized in this manuscript to address neutrophil performance in septic death, viral infection, and bronchial asthma.

Since the nature of the planned studies was essentially clinical, a prerequisite was the development of a reproducible assay. For this purpose, lactoferrin was purified from human breast milk, polyclonal rabbit antibodies raised to this, and an enzyme-linked immunosorbent assay (ELISA) established, the details of which have been fully published elsewhere (Bezwoda et al. 1985). In the process, an interesting finding was the variable electrophoretic migration of lactoferrin on SDS-PAGE dependent on its iron saturation (Bezwoda & Mansoor, 1986).

The meaningful interpretation of the assay data was further dependent on knowledge of the clearance rate of lactoferrin. Initial cross-species data indicated a rapid clearance with initial $T_{1/2}$ between 3 and 33 min (Van Snick et al. 1975; Karle et al. 1979; Ziere et al. 1992). Clearance of homologous lactoferrin in human subjects does, however, appear to be moderately slower. Clearance studies of isotopically labelled lactoferrin yielded an initial $T_{1/2}$ of 1.25 hr (Bennett & Kokocinski, 1979). In observations of serum concentrations after infusion of bone marrow containing high concentrations of lactoferrin to recipients previously preparatively myeloablated the $T_{1/2}$ appeared to be on the order of 2.2 hr (Baynes et al. 1986b). Given this rapid clearance, it is apparent that the serum concentration reflects both granulocytic mass and granulocytic activation and that the lactoferrin:granulocyte ratio provides a useful measure for comparisons between subjects (Rosenmund et al. 1988).

## LACTOFERRIN AND THE HYPOFERREMIC RESPONSE

A consistent finding in inflammation is a reduction in serum iron associated with an increase in serum ferritin (Lee, 1983). Teleologically, it has been argued that this shift from exchangeable iron to a more sequestered storage form reduces that available for microorganisms and as such, is of survival advantage (Weinberg, 1984). The early observation that this hypoferremic response could be abrogated in animals rendered neutropenic by exhibition of cytostatic agents led to the conclusion that a granulocyte derived factor is responsible for this response. Experiments in which injections of supraphysiological amounts of human lactoferrin (5–15 mg) induced hypoferremia in rats were thought to support the notion that lactoferrin was the granulocyte derived mediator (Van Snick et al. 1974). The thesis was that inflammation associated cytokines stimulate neutrophil degranulation releasing apo-lactoferrin which competes with transferrin particularly in low pH environments such as the inflammatory nidus for exchangeable iron. The rapid clearance of the diferric lactoferrin by reticuloendothelial elements and hepatocytes and its non-availability to erythroid precursors would explain enhanced iron stores and iron limited erythropoiesis.

While this hypothesis appears mechanistically sound, there are a number of theoretical objections. These include the following (Taylor et al. 1987). Lactoferrin has never been documented to have a role in internal iron exchange. Its concentration in human sera is several orders of magnitude lower than that of transferrin. The difference in relative affinities of these two iron binding proteins is maximal at pH's well below 7.0. Finally, this hypothesis does not accord well with prior dual label kinetic data in inflammation. In these rat studies, incorporation of two isotopes of iron, one from a reticuloendothelial label (heat damaged erythrocytes) and one from an erythropoietic label (tagged transferrin) indicated that the major abnormality in inflammation was one of sequestration of the reticuloendothelial label (Hershko et al. 1974). Studies in mice also revealed that inflammation is associated with a normal plasma iron turnover but with hold up of iron derived from heat damaged red cells in reticuloendothelial ferritin (Letendre & Holbein, 1983; Letendre & Holbein, 1984). Unpublished studies on the fate of injected $^{59}$Fe as ferric citrate indicate no difference in the organ distributions of isotope in rats with or without inflammation as induced by turpentine abscess production. This accords well with organ distribution studies as conducted in mice with and without acute Neisseria meningitidis infections (Letendre & Holbein, 1983).

To resolve the status of lactoferrin in the hypoferremic response, we conducted a number of clinical observations in various pathological settings to define the relationships between iron related-, inflammation related-, and lactoferrin measurements. In acute pyogenic pneumonia and chronic pulmonary tuberculosis, time dependent changes in these parameters after institution of appropriate antibacterial therapy were followed. In these situations, plasma lactoferrin, serum ferritin (a measure of storage iron) and serum C-reactive protein (an acute phase reactant) do appear to decline in parallel while transferrin saturation shows an inverse relationship. However, time dependent studies of subjects starting gold therapy for rheumatoid arthritis revealed a dissociation between ferritin and lactoferrin measurements.

Similar observations were made in normal volunteers who received a single dose of $\alpha$-interferon at a concentration of $2 \times 10^6$ U/m$^2$. The rationale for this study was that while viral illness is known to result in the characteristic hypoferremic response, we have observed that such illness is associated with reduced plasma and neutrophil lactoferrin contents (Baynes et al. 1986c; Baynes et al. 1988). Interferon therapy is well-known to produce as a side effect, symptoms of a viral infection. The administration of $\alpha$-interferon resulted in the prompt development of hypoferremia with hyperferritinemia but with the expected reduction in both lactoferrin and neutrophils (Baynes et al. 1990).

The critical clinical observations were obtained in subjects who were neutropenic after chemotherapeutic ablation as preparation for bone marrow transplantation. Plasma measurements in this situation reveal that lactoferrin is virtually absent. Due to suppressed erythropoiesis, iron utilization in this setting is markedly reduced such that serum iron, transferrin saturation, and serum ferritin concentrations are markedly elevated. When sepsis supervenes, as it often does in this setting and as it did in the subjects studied, a febrile response is observed along with an increase in acute phase proteins. Hypoferremia does occur but without any increase in the plasma concentrations of lactoferrin.

These clinical observations showing dissociation between hypoferremia and hyperferritinemia, on the one hand, and plasma lactoferrin concentration, on the other, strongly suggest that the lactoferrin hypothesis, as the mechanism of production of hypoferremia, is not tenable. These data provide the clinical corroboration for the doubts raised about this hypothesis based on quantitative, prior kinetic, and proportional considerations.

## PLASMA LACTOFERRIN AS AN INDICATOR OF GRANULOCYTE ACTIVITY AND FUNCTION

### Plasma Lactoferrin as a Predictive Indicator of Sepsis Related Morbidity and Mortality

It was recognized relatively early that plasma lactoferrin concentration showed prognostic value in certain clinical settings. For example, in subjects with severe burns, the initial plasma lactoferrin concentration appeared to predict the development of neutropenia and pulmonary infiltrates (Wolach et al. 1984). The plasma concentration was also shown to be a good predictor of the development of the adult respiratory distress syndrome in subjects who were at risk (Hallgren et al. 1984). It is also well-known that subjects undergoing cardiopulmonary bypass frequently develop multi-organ dysfunction including impaired alveolar-arterial gas exchange, renal dysfunction, and subtle impairment of cerebral function. This has been shown to be the consequence of complement activation and neutrophil degranulation in response to the extracorporal perfusion surfaces and oxygenators. Increases in plasma lactoferrin concentrations have shown a direct proportion to the severity of the post-perfusion syndrome (Nilsson et al. 1988; 1990a,b).

A clinical study was conducted to evaluate plasma lactoferrin concentrations and the lactoferrin:leukocyte ratio as predictors of outcome in subjects presenting with acute lobar pneumonia. Twenty-one subjects were studied with acute uni- or multilobar pneumonia. Eight subjects cultured positive. Three had *Streptococcus pneumonia*, 3 *Klebsiella pneumonia*, 1 *Pseudomonas pyocyaneus*, and 1 *Haemophilus influenza*. Patients received appropriate antibiotic therapy, intensive care treatment when indicated and life support when indicated. Initial lactoferrin concentrations were significantly higher (geometric mean 2115 g/L) in the five subjects who eventually died as compared with the 16 subjects who survived (geometric mean 1075 μg/L) (t = 2.8; p < 0.03). There were no differences in initial neutrophil count. The initial lactoferrin:leukocyte ratio was significantly higher in the group that succumbed. The nature of demise in those subjects who succumbed was worsening adult respiratory distress syndrome along with multi-organ failure.

These initial observations (Baynes et al. 1986a) have been confirmed and extended in subsequent studies. Worsening pulmonary status and multi-organ failure in this setting is now thought to represent "malignant intravascular inflammation" with one of the major contributors being profound intravascular granulocyte activation leading to oxygen mediated tissue injury (Holman & Saba, 1988; Nuytinck et al. 1988). In addition to oxidative tissue injury, there is also evidence that other granulocyte products may be profoundly cytotoxic to human pulmonary tissue and endothelium. More particularly, the relatively recently described small molecular weight microbicidal peptides, named defensins, and constituents of the primary granule system, appear to be highly cytotoxic (Okrent et al. 1990). In a recent investigation it was confirmed that a poor outcome in sepsis was associated with intense intravascular stimulation of granulocytes, with mean plasma lactoferrin concentrations roughly twice that in subjects with a better outcome. This was associated with reduced granule content of the neutrophils, increased superoxide production, decreased chemotaxis, and reduced granulocyte delivery to a skin blister window (Tellado & Christou, 1991). In another recent study, other markers of granulocyte activation were also found to predict for the outcome of sepsis. Concentrations of elastase-$\alpha_1$ antitrypsin complexes were higher in 27 patients dying of sepsis than in 21 patients surviving. Employing an arbitrary cut-off value of 10 nM for this measurement, subjects below this level had a 37% mortality whereas those above this level had a mortality of 81% (Nuijens et al. 1992).

The nature of the activating stimulus resulting in this fatal hyperstimulation of granulocytes is as yet incompletely defined. Tumor necrosis factor (TNF)-$\alpha$ has been well documented to stimulate granulocyte adherence to fibrinogen matrices with concomitant mobilization of secondary granule contents (Hanlon et al. 1991). Injection of recombinant TNF to normal subjects resulted in significant increases in elastase-$\alpha_1$ antitrypsin complex concentrations, plasma lactoferrin concentrations, and interleukin-6 concentrations (van der Poll et al. 1992). In the previously mentioned study of neutrophil activation markers as a predictor of sepsis survival there were direct correlations observed between elastase-$\alpha_1$ antitrypsin complex concentrations and the concentrations of interleukin 6 and C3a (Nuijens et al. 1992).

These data all indicate that neutrophil activation and degranulation, induced by multiple cytokine and activated complement agonists, contribute to the development of fatal complications in patients with sepsis.

## Plasma and Granulocyte Lactoferrin Content in Viral Infection

Clinical wisdom indicates that a fairly frequent complication of viral infection is the development of bacterial superinfection. In an attempt to address factors which might be contributory, we undertook studies of plasma and neutrophil lactoferrin in subjects with viral infections. As already mentioned, plasma lactoferrin and the lactoferrin:neutrophil ratio provide a useful indicator of granulocyte function and activation.

In the initial investigation, 67 young male subjects with viral illness were studied. Their illnesses included varicella (15), arbovirus (5), Epstein–Barr virus (7), viral hepatitis (20 hepatitis A, 2 hepatitis B), measles (5), rubella (12), and aseptic meningitis (1). The diagnosis was based on compatible clinical and laboratory findings along with positive serological investigations. In the vast majority of these subjects, plasma lactoferrin concentrations were markedly reduced (with a mean value of 136.3. µg/L (SD 68.4) as compared with a mean value in normal control subjects of 279.5 µg/L (SD 171.8). The lactoferrin:granulocyte ratios were also markedly different being 81.8 µg/$10^9$ cells (SD 37.9) in normals and 47.5 µg/$10^9$ cells (SD 36.2) in subjects with viral illness. These data established that viral illness appears to be associated with either an acquired abnormality of neutrophil granule composition or of granulocyte activation.

A further investigation was conducted in an attempt to define the level of the acquired abnormality. In this study, neutrophil lactoferrin content was evaluated by means of an *in situ* immunoperoxidase method employing a semiquantitative visual scoring. Fixed peripheral blood smears were stained. Under the fixing and staining conditions employed, the lactoferrin had a nuclear staining pattern. Twenty-six young adults with viral illness and 9 normal young adults were studied. The viral illnesses included varicella (11), measles (3), rubella (8), Epstein–Barr virus (2), and 1 each with hepatitis A and hepatitis B. Each subject was evaluated during the acute disease presentation. Using the visual scoring index, normal subjects had a mean score of $196 \pm 4$ units/100 neutrophils. By contrast, the subjects with viral illness had a mean score of $98 \pm 38$ units/100 neutrophils. This difference was highly significant (p < 0.0005). These data suggest that viral illness is associated with an acquired defect of lactoferrin production. Whether this finding is peculiar to lactoferrin or reflective of a more generalized defect in secondary granule biology merits further study. The mechanism whereby these observed changes in viral infection (Baynes et al. 1986c; 1988) are produced is as yet unclear. There are early data beginning to appear on factors modulating lactoferrin gene expression. A recent report indicates that TNFα results in a 70% reduction in transcription of the lactoferrin gene in low density human bone marrow cells (Srivastava et al. 1991 ). Further work is clearly indicated in regard to other cytokines and inflammatory mediators.

## Plasma Lactoferrin and Asthma

It has been well appreciated that acute asthmatic attacks are associated with the presence of significant numbers of granulocytes in the airways and interstitium. Whether they are present merely as a response to chemotactic mediators or contribute directly to the acute inflammatory process is incompletely defined. It is also uncertain as to whether these cells are involved only in acute exacerbations of the illness or whether chronic granulocyte activation is present in asthmatics. In the current investigation, a longitudinal assessment over a 24 hr period of plasma lactoferrin and lactoferrin:neutrophil ratios were compared between normal subjects and 17 patients with mild asthma. Diagnosis was based on airway obstruction with $FEV_1$ < 70% of predicted, a response to $\beta_2$ sympathomimetic agonists and/or a positive exercise bronchoprovocation test (Kallenbach et al. 1992). Patients had, on account of mildness of their disease, not received corticosteroid therapy and other medications had been discontinued 12 hr prior to study. Serial plasma lactoferrin measurements, neutrophil counts, and measurements of peak expiratory flow rates were made during the next 24 hrs. Plasma lactoferrin concentrations were elevated at all times in the asthmatic subjects while there were no differences in neutrophil count between the groups. Consequently, the lactoferrin:neutrophil ratio was elevated at all times in the asthmatic subjects. This is reflected in the time weighted average for the ratio in the asthmatic group being 1.4 times that in normal subjects. When the asthmatic group was divided into those with "morning dipping" (8) and those without (9), there was a consistent mild increase in the lactoferrin:neutrophil ratio in the "dippers." This did not, however, reach statistical significance. Morning dipping was defined by a greater than 16% fall in peak expiratory flow rate between 1600 hr and 0400 hr and is thought to reflect more severe asthma. It should, however, be remembered that these were all mildly affected largely asymptomatic asthmatics. These data suggest strongly that persistent low grade neutrophil activation is a feature of bronchial asthma. Given the finding that neutrophil derived secretory products are toxic to pulmonary cells and endothelium (Okrent et al. 1990), further studies of this nature are indicated in asthma to better define the pathogenetic significance of the neutrophil in this disease. The inclusion of more severe, symptomatic asthmatics is likely to accentuate the differences observed in such measurements.

## Future Applications

In recent years with the clinical introduction of hemopoietic growth factors, accelerated granulocyte count recovery after myeloablation and improvement in neutropenic conditions can be readily achieved. Preliminary studies are beginning to indicate that while agents such as GMCSF or GCSF increase circulating granulocyte counts, this is not reflected in a proportional increase in tissue delivery of granulocytes. The use of lactoferrin measurements in serum and in granulocytes will be of great value in establishing the functional integrity and, more particularly, that of the specific granules in such stimulated granulopoiesis.

## CONCLUSIONS

The development of reproducible immunological assays for human lactoferrin by us and other workers has firmly established the relationship between plasma concentrations of this protein and the response to inflammation. The biological functions of the significant amounts of lactoferrin in plasma are as yet incompletely understood.

In the current manuscript, the data are summarized refuting a significant mechanistic role for lactoferrin in the pathogenesis of the inflammation associated hypoferremic response. That plasma lactoferrin provides a very useful assay for neutrophil activation and more particularly secondary granule content release is well established. In the current manuscript, this approach was used to indicate the role of the neutrophil as contributing to multi-organ failure in sepsis, the possibility of an acquired neutrophil defect in many viral infections and the apparent persistent activation of neutrophils in subjects with bronchial asthma. This approach should prove useful for the future definition of the role of the granulocyte in other pathological processes.

## ACKNOWLEDGEMENTS

We would like to acknowledge our collaborators J. Kallenbach, Q. Khan, N. Mansoor, D. Dajee and the assistance of Ms. L. Kuharich in the preparation of this manuscript. Supported by AID Cooperative Agreement DAN-5115-A-00-7908-00.

## REFERENCES

Baynes, R.D., Bezwoda, W.R., Bothwell, T.H., Khan, Q. & Mansoor, N. (1986a) The non-immune inflammatory response: serial changes in plasma iron, iron-binding capacity lactoferrin, ferritin and C-reactive protein. Scand J Clin Lab Invest, 46:695-704.

Baynes, R.D., Bezwoda, W.R., Khan, Q. & Mansoor, N. (1986b) Relationship of plasma lactoferrin content to neutrophil regeneration and bone marrow infusion. Scand J Haematol, 36:79–84.

Baynes, R.D., Bezwoda, W.R., Khan, Q. & Mansoor, N. (1986c) Plasma lactoferrin content: Differential effect of steroid administration and infective illnesses: Lack of effect of ambient temperature at which specimens are collected. Scand J Haematol, 37:353–359.

Baynes, R.D., Bezwoda, W.R. & Mansoor, N. (1988) Neutrophil lactoferrin content in viral infections. Am J Clin Path, 89:225–228.

Baynes, R.D., Bezwoda, W.R., Dajee, D., Lamparelli, R.D. & Bothwell, T.H. (1990) Effects of alpha-interferon on iron-related measurements in human subjects. S Afr Med J, 78:627–628.

Bennett, R.M. & Kokocinski, T. (1979) Lactoferrin turnover in man. Clin Sci, 57:453–460.

Bezwoda, W.R., Baynes, R.D., Khan, Q. & Mansoor, N. (1985) Enzyme linked immunosorbent assay for lactoferrin. Plasma and tissue measurements. Clin Chim Acta, 151:61–69.

Bezwoda, W.R. & Mansoor, N. (1986) Isolation and characterisation of lactoferrin separated from human whey by adsorption chromatography using Cibacron Blue F3G-A linked affinity adsorbent. Clin Chim Acta, 157:89–93.

Boxer, L.A., Björksten, B., Björk, J., Yang, H.H., Allen, J.M. & Baehner, R.L. (1982) Neutropenia induced by systemic infusion of lactoferrin. J Lab Clin Med, 99:866–872.

Britigan, B.E. & Edeker, B.L. (1991) Pseudomonas and neutrophil products modify transferrin and lactoferrin to create conditions that favor hydroxyl radical formation. J Clin Invest, 88:1092–1102.

Byrd, T.F. & Horwitz, M.A. (1991) Lactoferrin inhibits or promotes Legionella pneumophila intracellular multiplication in nonactivated and interferon gamma-activated human monocytes depending upon its degree of iron saturation. Iron-lactoferrin and nonphysiological iron chelates reverse monocyte activation against Legionella pneumophila. J Clin Invest, 88:1103–1112.

Crouch, S.P.M., Slater, K.J. & Fletcher, J. (1992) Regulation of cytokine release from mononuclear cells by the iron-binding protein lactoferrin. Blood, 80:235–240.

Duncan, R.L., Jr. & McArthur, W.P. (1981) Lactoferrin-mediated modulation of mononuclear cell activities. I. Suppression of the murine in vitro primary antibody responses. Cellular Immunol, 63:308–320.

Hallgren, R., Borg, T., Venge, P. & Modig, J. (1984) Signs of neutrophil and eosinophil activation in adult respiratory distress syndrome. Critical Care Med, 12:14–18.

Hanlon, W.A., Stolk, J., Davies, P., Humes, J.L., Mumford, R. & Bonney, R.J. (1991) rTNF-alpha facilitates human polymorphonuclear leukocyte adherence to fibrinogen matrices with mobilization of specific and tertiary but not azurophilic granule markers. J Leukocyte Biol, 50:43–48.

Hershko, C., Cook, J.D. & Finch, C.A. (1974) Storage iron kinetics. VI. The effect of inflammation on iron exchange in the rat. Br J Haematol, 28:67–75.

Holman, J.M. & Saba, T.M. (1988) Hepatocyte injury during post operative sepsis. Activated neutrophils as potential mediators. J Leukocyte Biol, 43:193–20:

Kallenbach, J., Baynes, R., Fine, B., Dajee, D. & Bezwoda, W. (1992) Persistent neutrophil activation in mild asthma. J Allergy Clin Immunol, 90:272–274.

Karle, H., Hansen, N.E., Malmquist, J., Karl, A.K. & Larsson, I. (1979) Turnover of human lactoferrin in the rabbit. Scand J Haematol, 23:303–312.

Lee, G.R. (1983) The anemia of chronic disease. Semin Hematol, 20:61-80.

Letendre, E.D. & Holbein, B.E. (1983) Turnover in the transferrin iron pool during the hypoferremic phase of experimental Neisseria meningitidis infection in mice. Infect Immun, 39:50–59.

Letendre, E.D. & Holbein, B.E. (1984) Mechanism of impaired iron release by the reticuloendothelial system during the hypoferremic phase of experimental Neisseria meningitidis infection in mice. Infect Immun, 44:320–325.

Nilsson, L., Brunnkvist, S., Nilsson, U., Nyström, S.-O., Tyden, H., Venge, P. & Åberg, T. (1988) Activation of inflammatory systems during cardiopulmonary bypass. Scandi J Thor Cardiovasc Surg, 22:51–53.

Nilsson, L., Nilsson, U., Venge, P., Johansson, O., Tyden, H., Åberg, T. & Nyström, S.-O. (1990a) Inflammatory system activation during cardiopulmonary bypass as an indicator of biocompatibility: A randomized comparison of bubble and membrane oxygenators. Scand J Thor Cardiovasc Surg, 24:53–58.

Nilsson, L., Tyden, H., Johansson, O., Nilsson, U., Rönquist, G., Venge, P., Åberg, T., Nyström, S.-O. (1990b) Bubble and membrane oxygenators—comparison of postoperative organ dysfunction with special reference to inflammatory activity. Scand J Thor Cardiovasc Surg, 24:59–64.

Nuijens, J.H., Abbink, J.J., Wachtfogel, Y.T., Colman, R.W., Eerenberg, A.J.M., Dors, D., Kamp, A.J.M., Strack van Schijndel, R.J.M., Thijs, L.G. & Hack, C.E. (1992) Plasma elastase alpha 1-antitrypsin and lactoferrin in sepsis: Evidence for neutrophil mediators in fatal sepsis. J Lab Clin Med, 119:159–168.

Nuytinck, H.K., Offermans, X.J., Kubat, K. & Goris, J.A. (1988) Whole body inflammation in trauma patients. An autopsy study. Arch Surg, 123:1519–1524.

Okrent, D.G., Lichtenstein, A.K. & Ganz, T. (1990) Direct cytotoxicity of polymorphonuclear leukocyte granule proteins to human lung-derived cells and endothelial cells. Am Rev Resp Dis, 141:179–185.

Rado, T.A., Bollekens, J., St. Laurent, G., Parker, L. & Benz, E.J. (1984) Lactoferrin biosynthesis during granulocytopoiesis. Blood, 64:1103–1109.

Rosenmund, A., Friedli, J., Bebie, H. & Straub, P.W. (1988) Plasma lactoferrin and the plasma lactoferrin/neutrophil ratio. A reassessment of normal values and of the clinical relevance. Acta Haematol, 80:40–48.

Singer, I.I., Scott, S., Kawka, D.W. & Kazazis, D.M. (1989) Adhesomes: Specific granules containing receptors for laminin, C3bi/fibrinogen, fibronectin, and vitronectin in human polymorphonuclear leukocytes and monocytes. J Cell Biol, 109:3169–3182.

Srivastava, C.H., Rado, T.A., Bauerle, D. & Broxmeyer, H.E. (1991) Regulation of human bone marrow lactoferrin and myeloperoxidase gene expression by tumor necrosis factor. J Immunol, 146:1014–1019.

Taylor, C., Rogers, G., Goodman, C., Baynes, R.D., Bothwell, T.H., Bezwoda, W.R., Kramer, F. & Hattingh, J. (1987) Hematologic, iron-related, and acute-phase protein responses to sustained strenuous exercise. J Appl Physiol, 62:464–469.

Tellado, J.M. & Christou, N.V. (1991) Critically ill anergic patients demonstrate polymorphonuclear neutrophil activation in the intravascular compartment with decreased cell delivery to inflammatory foci. J Leukocyte Biol, 50:547–553.

van der Poll, T., van Deventer, S.J.H., Hack, C.E., Wolbink, G.J., Aarden, L.A., Buller, H.R. & ten Cate, J.W. (1992) Effects of leukocytes after injection of tumor necrosis factor into healthy humans. Blood, 79:693–698.

Van Snick, J.L., Masson, P.L. & Heremans, J.F. (1974) The involvement of lactoferrin the hyposideremia of acute inflammation. J Exp Med, 1:1068–1084.

Van Snick, J.L., Masson, P.L. & Heremans, J.F. (1975) The affinity of lactoferrin for the reticuloendothelial system (RES) as the molecular basis for the hyposideremia of inflammation. Proteins of Iron Storage and Transport in Biochemistry and Medicine (ed., P.R. Crichton), p.433. North Holland Publishing Co., Amsterdam.

Weinberg, E.D. (1984) Iron withholding: a defense against infection and neoplasia. Physiol Rev, 64:65–102.

Wolach, B., Coates, T.D., Hugli, T.E., Baehner, R.L. & Boxer, L.A. (1984) Plasma lactoferrin reflects granulocyte activation via complement in burn patients. J Lab Clin Med, 103:284–293.

Ziere, G.J., van Dijk, M.C.M., Bijsterbosch, MK. & van Berkel, T.J.C. (1992) Lactoferrin uptake by the rat liver. Characterization of the recognition site and effect of selective modification of arginine residues. J Biol Chem, 267:11229–11235.

# THE ROLE OF LACTOFERRIN AS AN ANTI-INFLAMMATORY MOLECULE

Bradley E. Britigan,[†][*] Jonathan S. Serody,[‡] and Myron S. Cohen[‡]

[†]Research Service and Department of Internal Medicine
VA Medical Center
and Department of Internal Medicine
University of Iowa College of Medicine
Iowa City, Iowa

[‡]Departments of Medicine and Microbiology and Immunology
University of North Carolina at Chapel Hill
Chapel Hill, North Carolina

## SUMMARY

The formation of hydroxyl radical via the iron catalyzed Haber–Weiss reaction has been implicated in phagocyte-mediated microbicidal activity and inflammatory tissue injury. The fact that neutrophils contain lactoferrin and mononuclear phagocytes have the capacity to acquire exogenous iron has suggested that iron bound to lactoferrin may influence the nature of free radical products generated by these cells. Over the years the iron-lactoferrin complex has been heralded as both a promoter and inhibitor of hydroxyl radical formation. This manuscript is intended to provide an overview of work performed to date related to this controversy and to present results of a number of preliminary studies which shed further light on the role of lactoferrin in inflammation.

## INTRODUCTION

Lactoferrin is an 76.4 kDa glycoprotein found in neutrophil specific granules and at a variety of mucosal surfaces (1,2) with the capacity to bind ferric iron ($Fe^{+3}$). Over the last two decades the importance of the generation of cytotoxic oxygen-centered free radical species in a variety of pathologic processes has been demonstrated (reviewed in (3)). Formation of one of these species, hydroxyl radical, often requires the presence of a catalytic iron chelate. Accordingly, we have had a long-standing interest in the role of lactoferrin in free radical

---

[*] Address correspondence to Bradley E. Britigan, M.D., Department of Internal Medicine, University of Iowa, SW54, GH, Iowa City, IA 52242.

biology. This article is intended to: provide an overview of work performed to date; present results of a number of preliminary studies; and outline fruitful areas for future research.

## THE NATURE OF REDUCTION PRODUCTS OF MOLECULAR OXYGEN AND THEIR TOXICITY FOR BIOLOGICAL SYSTEMS

Much of the normal metabolism of human cells involves biochemical reactions in which molecular oxygen ($O_2$) serves as the terminal electron acceptor. In general these reactions are tightly regulated such that $O_2$ is reduced directly to its four electron reduction product, $H_2O$. This is important from the standpoint of cell integrity and function because the oxygen intermediates which would result if this process occurs in a univalent fashion are toxic for biologic systems. Sequential univalent reduction of $O_2$ yields the superoxide free radical ($O_2^-\cdot$), hydrogen peroxide ($H_2O_2$), hydroxyl radical ($\cdot OH$), and $H_2O$.

At physiologic pH, the formation of $O_2^-\cdot$ automatically leads to $H_2O_2$ generation as $O_2^-\cdot$ will rapidly react ($k_{obs} = 2 \times 10^5$ $M^{-1}s^{-1}$) with itself (dismute) as shown below:

$$O_2^-\cdot + O_2^-\cdot + 2H^+ \rightarrow H_2O_2 + O_2$$

Hydroxyl radical is the one electron reduction product of $H_2O_2$. Stoichiometrically this is not a favorable reaction. Nevertheless, there has been considerable interest in $\cdot OH$ formation in recent years because its highly reactive nature (see below). Although $O_2^-\cdot$ can directly reduce $H_2O_2$ to $\cdot OH$, the rate of this reaction is slow. Thus, it is not considered to be of major importance in human biology. However, in the presence of certain transition metal chelates the reaction proceeds much more rapidly. Iron and to a lesser extent copper chelates have been evaluated to the greatest extent in this regard. The resulting reaction sequence, referred to as the Haber–Weiss reaction, is outlined below (4):

$$O_2^-\cdot + Fe^{+3} \rightarrow O_2 + Fe^{+2}$$

$$H_2O_2 + Fe^{+2} \rightarrow \cdot OH + OH^- + Fe^{+3}$$

$$\overline{O_2^-\cdot + H_2O_2 \rightarrow \cdot OH + OH^- + O_2}$$

In this reaction $O_2^-\cdot$ functions as both the source of $H_2O_2$ and as the reducing agent for the generation of $Fe^{+3}$ from $Fe^{+2}$. Although $O_2^-\cdot$ and $H_2O_2$ are cytotoxic, it is felt that the role of $O_2^-\cdot$ and/or $H_2O_2$ formation in many diseases is to serve as the substrates for the generation of $\cdot OH$ (5). As will be discussed later, not all iron or copper complexes are able to efficiently catalyze this reaction (5).

## PHAGOCYTES, OXIDANTS, AND INFLAMMATORY TISSUE DAMAGE

Even though normal cellular metabolism of $O_2$ is tightly regulated it is estimated that up to 1% of $O_2$ reduction which takes place within cells occurs univalently. Since reactive oxygen species such as $O_2^-\cdot$, $H_2O_2$, and $\cdot OH$ are potentially capable of damaging almost every type of biological compound—e.g., protein, DNA, lipid, and carbohydrate, most cells contain a variety of enzymatic and other mechanisms designed to prevent the accumulation of these species (6, 7). In contrast, phagocytic white blood cells (neutrophils, monocytes, macrophages, and eosinophils) purposefully generate $O_2^-\cdot$ via a plasma membrane associated NADPH-dependent oxidase complex as part of their microbicidal armamentarium (8, 9). Once generated, $O_2^-\cdot$ spontaneously dismutes to form $H_2O_2$ ($k = 2 \times 10^5$ $M^{-1}s^{-1}$). In the case of neutrophils, monocytes, and eosinophils (but not macrophages) which contain myeloperoxidase (MPO) or eosinophil peroxidase, $H_2O_2$ is converted to other microbicidal oxidants such as hypochlorous acid (HOCl) or other hypohalous acids and chloramines (10, 11).

Formation of $O_2^-$· and other oxidants is critical to the optimal microbicidal activity of human phagocytes as evidenced by the recurrent infections suffered by individuals with chronic granulomatous disease of childhood (CGD) whose phagocytes are unable to generate $O_2^-$· (12). Congenital MPO deficiency is relatively common (1:2000–1:4000), yet these individuals are not generally observed to be at an increased risk of infection (13).

The same process of phagocyte oxidant production that offers protection from infectious diseases has also been linked to the pathogenesis of a variety of pathologic processes (e.g. ARDS, ischemia/reperfusion injury, arthritis) in which activated phagocytes, particularly neutrophils, damage host tissue. Considerable *in vitro* data support the ability of phagocyte-derived oxidants to damage cells from a variety of tissues (14).

## PHAGOCYTE-ASSOCIATED HYDROXYL RADICAL PRODUCTION

The iron-catalyzed formation of ·OH has been implicated in both phagocyte microbicidal activity and inflammatory tissue injury (5, 14). Killing of some microorganisms by phagocytes or their products ($H_2O$) are prevented by the iron chelator deferoxamine which binds iron in a form incapable of acting as a ·OH catalyst. In animal models of acute lung injury, damage is prevented by: 1) depletion of neutrophils prior to the initiating insult (15, 16); 2) simultaneous presence of ·OH "scavenging agents," (17); or 3) pretreatment of the animals with deferoxamine (18, 19). In addition, examination of joint fluid obtained from individuals with rheumatoid arthritis who were receiving salicylates as part of their therapy revealed the presence of salicylate oxidation products which would be generated via interaction of this compound with ·OH (20).

Oxidation products which result from the interaction of ·OH with a number of substrates have been detected following neutrophil stimulation *in vitro* with some (but not all) ·OH scavengers inhibiting detection of these products (21–28). These data implied that ·OH was a "physiologic product" of neutrophil stimulation. This hypothesis was supported by Ambruso and Johnston who reported that iron bound to lactoferrin, which is released from neutrophil specific granules during activation, was capable of catalyzing the Haber–Weiss reaction (25).

However, in the above studies no attempt was made to exclude the possibility that iron contaminating the buffers used was responsible for any ·OH formed; and the ·OH detection systems employed have been criticized for their lack of specificity (29). A further confounding observation was that in some of these studies, azide, the inhibitor of heme containing enzymes such as myeloperoxidase, inhibited these events. In recent years improved techniques for detecting ·OH generation in biologic systems have been developed. Using a variety of different spin trapping techniques we have found no evidence for the capacity of human neutrophils, monocytes, and macrophages to generate ·OH via the Haber–Weiss reaction in the absence of an exogenous iron catalyst (30–32). Similar conclusions have been reached by other investigators employing alternative techniques for ·OH detection (33–36). Recent data from our laboratories suggesting that neutrophils and monocytes generate small amounts of ·OH via the reaction of $O_2^-$· with MPO-derived HOCl (37) may also explain some of the earlier observations.

## LACTOFERRIN AND THE HABER–WEISS REACTION

The above data indicate that neutrophils and other phagocytes require the availability of a transition metal catalyst in order to generate ·OH via the Haber–Weiss reaction. Since lactoferrin is present in neutrophil specific granules, at mucosal surfaces, and within inflammatory exudates there has been considerable interest as to the capacity for iron chelated to lactoferrin to participate in this reaction. Two opposing view points have been presented: 1) iron bound to lactoferrin promotes $\cdot OH$ generation, microbicidal activity, and inflammation; or 2) lactoferrin binds iron in a form which is incapable of acting as a Haber–Weiss catalyst thereby allowing it to serve as an antioxidant.

As noted above, using an assay involving the detection of ethylene formed from 2-keto-4-thiomethylbutyric acid (KMB) Ambruso and Johnston reported evidence that lactoferrin which was partially or fully saturated with iron catalyzed the formation of $\cdot OH$ by stimulated neutrophils or cell-free $O_2^-/H_2O_2$ generating systems (25). Work from two other laboratories employing the KMB assay (38) and/or spin trapping (38, 39) supported this conclusion. However, spin trapping studies such as these which utilize only 5,5 dimethyl-pyrroline-l-oxide (DMPO) as the trapping agent are not reliable indicators of $\cdot OH$ generation (40). Furthermore, the KMB assay also lacks specificity for $\cdot OH$ (29). In contrast to these studies, investigators using a variety of $\cdot OH$ detection systems have been unable to demonstrate that iron-lactoferrin chelates are capable of acting as an $\cdot OH$ catalyst (41–43). This appeared to be the case even under conditions of low pH in which the ability of lactoferrin to bind iron would be decreased (43). Winterbourn suggested (42) that the discrepancy between her results and others (25, 38) may have related to the presence of residual nitrilotriacetic acid (NTA) used in the lactoferrin purification procedure by other investigators which could have provided the catalytic iron chelate responsible for $\cdot OH$ generation. We and others have also shown that the addition of apo-lactoferrin to cell-free $\cdot OH$ generating systems comprised of sources of $O_2^-/H_2O_2$ and other catalytic iron chelates such as $Fe^{+3}$-NTA inhibits the generation of $\cdot OH$ (43–45) and its sequelae (46). The sum of these individual studies suggests that $Fe^{+3}$ bound to lactoferrin is not an efficient catalyst of the Haber–Weiss reaction. Similar conclusions have been reached with regard to transferrin-$Fe^{+3}$ complexes (41, 43, 47). Recently, Klebanoff and Waltersdorph presented data suggesting that apotransferrin and/or apolactoferrin accelerates the autooxidation of $Fe^{+2}$ at acid pH with apparent generation of $\cdot OH$ via an $H_2O_2$ intermediate (48). The likelihood of such conditions existing under *in vivo* conditions is unclear however.

## EFFECT OF LACTOFERRIN RELEASE ON NEUTROPHIL-MEDIATED OXIDANT INJURY

Studies assessing the effect of lactoferrin on the formation of $\cdot OH$ by cellular systems provide additional evidence that lactoferrin may serve an important antioxidant role *in vivo*. Using spin trapping techniques we demonstrated that release of lactoferrin from cytoplasmic granules by PMA-stimulated neutrophils prematurely terminates formation of $\cdot OH$ when they are stimulated in the presence of an exogenous iron catalyst (49, 50). These data suggest that release of lactoferrin from activated neutrophils, two-thirds of which is secreted outside the phagosome (2), serves to limit the potential for the generation of extracellular $\cdot OH$ which could be potentially damaging to adjacent tissue. Consistent with this observation, Molloy and Winterbourn demonstrated that neutrophil-derived lactoferrin binds iron released from *E. coli* ingested and killed by these phagocytes (51). They too suggested that this process could prevent inflammatory tissue injury (51). In addition, infusion of apolactoferrin has been shown to prevent neutrophil-mediated lung injury in an animal model (18). However, not all studies of cell injury agree with an antioxidant role for neutrophil-derived lactoferrin. Vercelotti and colleagues have argued that lactoferrin promotes neutrophil-mediated injury to erythrocytes via enhancement of $\cdot OH$ generation (52).

A few patients have been described whose neutrophils do not contain secondary granules and therefore lactoferrin (53). These patients suffer from recurrent infections (53). As assessed by spin trapping, the neutrophils from one such individual were reported (in contrast to those from a normal control) to lack the ability to generate $\cdot$OH (54). However, the spin trapping system employed lacks specificity for $\cdot$OH (29) and multiple other defects in the function of neutrophils with secondary granule deficiency have been described (53, 54). Furthermore, other studies on neutrophil-like cells deficient in lactoferrin do not support these conclusions. In the presence of DMSO, the promyelocytic HL-60 cell line differentiates to a form which resembles a mature neutrophil except that these cells lack secondary granules and therefore lactoferrin (55, 56). Alternatively, human neutrophils from which both primary and secondary granules had been removed (cytoplasts) are another model of a lactoferrin-deficient neutrophil. Using a spin trapping system with improved specificity for $\cdot$OH, we found that both neutrophilic HL-60 cells and neutrophil cytoplasts actually have a greater propensity to generate $\cdot$OH when they were stimulated in the presence of exogenous iron (49).

Thus, although the data are not entirely in agreement, the bulk of evidence suggests that iron bound to lactoferrin is an extremely poor catalyst of $\cdot$OH generation and that the presence of lactoferrin at sites of inflammation *in vivo* likely provides protection from the local consequences of untoward $\cdot$OH production. The fact that most lactoferrin present *in vivo* does not contain iron (57), a state necessary for anti-oxidant activity, would allow it to readily perform such a function. However, the ability of lactoferrin to serve such an antioxidant role *in vivo* will depend on the relative ratio of iron and lactoferrin present. This most likely explains the results of a recent study in which iron-dependent peroxidation of membranes by neutrophils stimulated in the presence of ferritin was unaffected by neutrophil lactoferrin release (58). In a less complicated system we have also shown the ability of high concentrations of iron to overwhelm the protective effects of lactoferrin (49). Under these circumstances the inhibitory effects of MPO on the Haber–Weiss reaction (36, 50) may take on greater importance.

## BINDING OF LACTOFERRIN TO MONONUCLEAR PHAGOCYTES PREVENTS HYDROXYL RADICAL-MEDIATED INJURY

Although monocytes and MDM do not contain lactoferrin, they and other myeloid cells appear to take up exogenous lactoferrin via a lactoferrin receptor (59–66). We recently confirmed that incubation of monocytes or monocyte-derived macrophages (MDM) with human milk apo- or diferric lactoferrin resulted in a similar concentration-dependent increase in cell-associated lactoferrin (67). This process was unaltered by transferrin but the fucose polymer fucoidan was markedly inhibitory (67), likely due to formation of an irreversible lactoferrin-fucoidan complex (68). $\gamma$-interferon did not alter MDM lactoferrin uptake (67). Monocytes/MDM incubated at 4°C with $^{125}$I-apo-lactoferrin showed saturable binding (67). Assuming a 1:1 binding ratio, Scatchard analysis indicated a single lactoferrin receptor (Kd = $3.56 \times 10^{-6}$ with $3.4 \times 10^7$ binding sites/cell). These data suggesting a low affinity high density receptor are similar to earlier lactoferrin receptor studies (61, 69). However, as previously noted (70), the propensity of lactoferrin to form multimeric complexes allowing a single lactoferrin receptor to bind more than one lactoferrin molecule could result in an overestimation of lactoferrin receptor density. Others have reported lower lactoferrin receptor density (62, 65).

At the highest lactoferrin concentration employed, cell-associated lactoferrin was threefold greater than could be accounted for by the number of lactoferrin receptor predicted by Scatchard data. Although this is consistent with lactoferrin internalization and receptor recycling (64), other binding mechanisms of the lactoferrin receptor could be involved. When lactoferrin-loaded cells were incubated in lactoferrin free buffer for 30 min, 40% of was released as intact lactoferrin (both SDS-PAGE and immunoblot analysis) (67).

Relative to control cells, apo-lactoferrin loaded monocytes and MDM showed a 43% and 42% decrease, respectively in $\cdot$OH generation (2-deoxyribose oxidation assay) upon stimulation with PMA in the presence of Fe-NTA (67). No decrease was seen with the substitution of apo-transferrin or diferric lactoferrin for apo-lactoferrin (67). Similarly, monocyte membrane peroxidation (malonaldehyde formation) decreased 28% in apo-lactoferrin-loaded cells relative to controls following PMA stimulation in the presence of Fe-NTA (67). Surprisingly, monocytes loaded with diferric lactoferrin also were protected. Conceivably this could relate to iron removal from lactoferrin during the 1.5 hour incubation procedure. None of these results could be explained on the basis of differences in $O_2^-/H_2O_2$ generation among the cell populations. These data were generated with monocytes/MDM loaded with relatively large amounts of lactoferrin; this was necessary because of the high concentrations of iron required for detection of $\cdot OH$ and lipid peroxidation by the assay systems employed. At the lower iron concentration to which these cells would be exposed *in vivo* it seems likely that lesser amounts of cell-associated lactoferrin would be similarly protective. These data suggest that surface binding of lactoferrin to mononuclear surface membranes may serve to prevent damage to both the phagocyte and surrounding tissue by iron-dependent oxidant formation. Oxidants may induce DNA damage in phagocytic cells as they do in other cells (71, 72), possibly via site specific $\cdot OH$ injury. Lactoferrin can bind to DNA and has been reported to localize in part to euchromatin shortly after binding to the surface of mononuclear phagocytes (73). These observation raise the possibility that lactoferrin might protect the cell therefore from $\cdot OH$-mediated DNA damage resulting from $O_2^-/H_2O_2$ generation during the "respiratory burst".

Most studies of lactoferrin uptake by monocytes/MDM have used milk lactoferrin. Since uptake of milk lactoferrin can be prevented with the fucose polymer fucoidan and human neutrophil lactoferrin does not contain fucose residues (74), it seemed possible monocytes/MDM might not take up neutrophil-derived lactoferrin. Accordingly, neutrophil secondary granules were obtained by subcellular fractionation (75, 76) and their lactoferrin content (2–4 mg/donor) determined by ELISA. Lactoferrin uptake by monocytes incubated with lysed neutrophil secondary granules was 43% (n = 4) of that observed with the same concentration of milk lactoferrin. Consistent with previous results (69), however, preliminary data suggests no difference in the Kd or receptor number for milk or neutrophil lactoferrin by standard Scatchard analysis. Perhaps other secondary granule contents present could have inhibited lactoferrin uptake. Nevertheless, our data point out the ability of monocytes/MDM to take up large quantities of lactoferrin in the presence of other neutrophil granule proteins.

Consistent with the *in vivo* relevance of the above studies, immunoblot and ELISA analysis revealed the association of lactoferrin (1–2 $\mu g/10^7$ cells) with human pulmonary alveolar macrophages harvested from non-inflamed airways (67). These lactoferrin concentrations are those expected given the amount of macrophage uptake of lactoferrin we have previously observed (67) and the concentration of lactoferrin we detect (0.0025 mg/mg protein) in airway secretions (77). Although not measured, alveolar macrophage-associated lactoferrin should increase with inflammation, as we find airway levels of lactoferrin are 10–100 fold greater under such conditions (77).

## FURTHER CHARACTERIZATION OF THE MONONUCLEAR PHAGOCYTE LACTOFERRIN RECEPTOR

In a series of experiments, we have examined the potential for utilizing myeloid cell lines for further study of the monocyte/MDM lactoferrin receptor. We found that the quantities of both milk and neutrophil lactoferrin taken up by undifferentiated HL-60 cells are similar to monocytes/MDM whereas lactoferrin uptake by DMSO-differentiated (neutrophilic) HL-60 is slightly less. In contrast, both undifferentiated monocytoid U937 cells as well as monocytic (TPA differentiated) U937 cells exhibit similar lactoferrin uptake.

In order to provide additional evidence that lactoferrin binds to a specific surface receptor, we have attempted to cross-link (4°C) biotinylated milk apo-lactoferrin to the monocytes/MDM lactoferrin receptor with disuccinimidyl suberate (78–80). The samples were subjected to SDS-PAGE and transferred to nitrocellulose. Biotinylated apolactoferrin was then visualized by incubating the blots in strepavidin-linked horseradish peroxidase (HPO), $H_2O_2$, and tetramethyl-benzidene. A band of 80 kDa corresponding to free lactoferrin and another distinct band of about 190 kDa have been variably detectable. A third species at the stacker/gel interface is also sometimes detected which could be due a multimeric lactoferrin complex or lactoferrin bound to DNA which is unable to enter the gel. Nearly identical results have been obtained with non-biotinylated apo-lactoferrin in which polyclonal rabbit anti-lactoferrin followed by [125]I Staph protein A or peroxidase conjugated antirabbit IgG was used to identify cross-linked lactoferrin.

We have also used ligand-blotting (81, 82) to identify the size of the lactoferrin receptor. Monocytes, MDM or U937 were subjected to non-reducing SDS-PAGE, transferred to nitrocellulose, following which the blot was incubated in [125]I apo- or diferric lactoferrin. [125]I lactoferrin binding, which is inhibitable by "cold" lactoferrin, to a 100–120 kDa band has been occasionally detected. More prominent have been bands of 37 kDa and 16 kDa. These may be subunits of the lactoferrin receptor or represent proteolysis products. Similar results have been obtained with human monocytes. No lactoferrin binding has been seen using equal numbers of erythrocytes.

## PURIFICATION OF A LACTOFERRIN RECEPTOR FROM U937 CELLS

Using an approach similar to that used to isolate the lymphocyte lactoferrin receptor (83), U937 cells were incubated in lactoferrin, solubilized, and applied to a Sephadex column to which anti-lactoferrin antibody had been linked. Subsequent elution of bound material with high molar urea yielded a protein of 110 kDa which retained the ability to bind apo-lactoferrin. In the presence of DTT, this protein migrates at about 37 kDa. Sequential analysis of the 37 kDa subunit of the putative U937 lactoferrin receptor has been undertaken utilizing Edman degradation. Amino acid analysis of the 37 kDa subunit revealed a similar percentage of all the amino acids with the exception of an increase in the number of glycines as compared to the intestinal brush border lactoferrin receptor (84). Using cyanogen bromide cleavage of the putative receptor subunit, we have isolated several fragments which we have sequenced. Currently, we are attempting to purify this sequence to homogeneity as a first step in cloning the gene for the receptor. Work is also currently ongoing to determine whether a similar approach can be used for purification of a lactoferrin receptor from human monocytes.

Although these data are not definitive, in conjunction with the cross-linking and ligand blotting results, they suggest that monocytes, MDM, and U937 cells contain a 110 kDa protein (possibly composed of 37 kDa subunits) which is a strong candidate for the lactoferrin receptor. This would make the monocytes/MDM lactoferrin receptor similar in size to the lymphocyte and intestinal lactoferrin receptor (83, 84).

## EFFECTS OF OXIDANTS, PROTEASES, AND LIPOPOLYSACCHARIDE ON THE ANTIINFLAMMATORY PROPERTIES OF LACTOFERRIN

Although, as discussed above purified forms of lactoferrin are capable of inhibiting the Haber–Weiss-mediated formation of ·OH, under *in vivo* conditions the protein would likely encounter a variety of factors which could decrease its ability to act as an oxidant. The release of myeloperoxidase (MPO) from neutrophil and monocyte granules during activation results in the generation of hypohalous acids (HOX where $X = Cl$, Br, or I) and other oxidant species (85). These MPO-derived products have been shown to oxidize a large number of proteins (85), including lactoferrin (86). Previous work by other investigators (86, 87) had shown that iodination of transferrin or lactoferrin by MPO did not alter the ability of the protein to either take up $Fe^{+3}$ nor result in release of previously bound $Fe^{+3}$. However, these observations did not directly address the redox properties of the $Fe^{+3}$ bound to lactoferrin after exposure to HOX or MPO. Therefore we recently examined the effect of exposure to HOCl on the potential for diferric lactoferrin to act as a catalyst of the Haber–Weiss reaction as well as the ability of apo-lactoferrin to inhibit ·OH formation catalyzed by another iron chelate, Fe-NTA. We found no evidence that diferric lactoferrin previously exposed to HOCl had been converted into a form which now supported ·OH production (44). In addition, HOCl treated apo-lactoferrin was equal to the untreated control in its ability to inhibit ·OH formation by a Fe-NTA supplemented hypoxanthine/xanthine oxidase $O_2^-/H_2O_2$ generating system (44). These results were different than with transferrin in which exposure of the apo form of the protein to HOCl decreased its ability to inhibit ·OH production (44).

At sites of inflammation, lactoferrin is also likely exposed to a variety of proteases. Numerous investigators had previously demonstrated the ability of various human and bacterial proteases to cleave lactoferrin and transferrin (88–93). Attesting to the clinical relevance of these findings, using a highly sensitive chemiluminescence immunoblot system we have consistently (n = 22) detected transferrin cleavage products in bronchoalveolar lavage (BAL) from *P. aeruginosa*-infected cystic fibrosis patients but not normal individuals (n = 7) or those with bacterial bronchiectasis (n = 11). 20/21% of cystic fibrosis patient BAL contain lactoferrin cleavage products with no evidence for such a process in the other patients studied (93a).

Accordingly, we examined the impact of such cleavage of diferric and apo-lactoferrin in the context of the Haber–Weiss reaction. We cleaved (SDS-PAGE confirmed) diferric-trans-ferrin or diferric-lactoferrin with physiologic concentrations of pseudomonas elastase, human neutrophil elastase, pseudomonas alkaline protease, or trypsin (44). ·OH was detected (spin trapping and the deoxyribose oxidation assay) upon addition of pseudomonas elastase cleaved diferric transferrin to a reaction mixture of hypoxanthine and xanthine oxidase or PMA stimulated neutrophils (44). Diferric transferrin cleavage by a combination of pseudomonas elastase and neutrophil elastase doubled ·OH generation compared to that seen with diferric transferrin exposed to pseudomonas elastase alone (44). Except for a slight amount of ·OH detected when diferric lactoferrin treated with a high concentration pseudomonas elastase treated was employed, ·OH formation was not detected with any of the other combinations (44). Cleavage of apo-transferrin by pseudomonas elastase, but not the other three proteases, also inhibited apo-transferrin's ability to prevent ·OH generation by Fe-NTA supplemented hypoxanthine/xanthine oxidase (44). Apo-lactoferrin's ability to inhibit this reaction was unaffected by any of the single or multiple protease exposures (44).

Binding of bacterial lipopolysaccharide (LPS) to lactoferrin has been implicated in the microbicidal mechanism of the protein for some gram negative bacteria (94). Miyazawa et al. have recently demonstrated (66) that the binding of LPS to lactoferrin alters the mechanism of lactoferrin binding to a myeloid cell line. Given the high likelihood that lactoferrin would encounter considerable amounts of LPS at sites of gram negative infections we examined the impact of LPS binding to lactoferrin on the ability of lactoferrin to inhibit the Haber–Weiss reaction. Regardless of whether experiments were performed at pH 7.4 or 4.5 binding of *E. coli* LPS to the protein had no effect on its ability to inhibit ·OH formation resulting from an iron-supplemented xanthine/xanthine oxidase system as assessed by the deoxyribose oxidation assay (45).

LPS induces a large number of cellular responses, many of which have been linked to the pathophysiology of septic shock and other consequences of gram negative infection. One of the effects of LPS is increase the magnitude of neutrophil $O_2^-\cdot$ production upon exposure to various stimuli of the "respiratory burst" such as the chemotactic peptide n-formyl-methionyl-leucyl-phenylalanine (FMLP) (95, 96). This process, termed "priming" has been suggested to enhance both neutrophil microbicidal activity and inflammatory tissue damage. In recent experiments we found that binding of LPS to lactoferrin inhibits the subsequent ability of that LPS to "prime" neutrophils for enhanced $O_2^-\cdot$ production in response to FMLP (45). Results were identical regardless of the extent to which the protein was iron-loaded. However, these results were only demonstrable if the lactoferrin was first exposed to a chelating column to remove the large amount of LPS (as detected by limulus assay) which regularly each of the commercial preparations examined contained. The ability to inhibit LPS "priming" was not unique to lactoferrin. Identical results were obtained with transferrin and albumin (45) which are also able to bind LPS (94, 97, 98). Our findings that lactoferrin inhibits LPS priming of neutrophils are somewhat in conflict with the recent report of Gahr et al. which indicated that lactoferrin by itself has the capacity to prime neutrophils (99). However, the concentration of lactoferrin employed was fifty-fold greater than we employed.

The mechanism whereby LPS binding to lactoferrin decreases its priming effect on neutrophils is not yet clear. Possibilities include a decrease in affinity for the LPS receptor or simultaneous alteration in the signal transduction mechanism which leads to priming. Nevertheless our data suggest the possibility that binding of LPS to lactoferrin could provide a means of decreasing the proinflammatory events which occur in the setting of septic shock. Consistent with this possibility lactoferrin has been reported to decrease mortality in a mouse model of *E. coli*-induced septic shock (100).

## SUMMARY AND CONCLUSION

In summary, the current data strongly suggests that among its potential roles *in vivo*, lactoferrin may serve as an antioxidant defense mechanism at least two levels. First by binding any catalytic iron which may be generated during the course of cell destruction. During an inflammatory response it may serve to prevent hydroxyl-radical mediated tissue injury associated with neutrophil-oxidant production. The ability of lactoferrin to inhibit $\cdot OH$ formation can also be extended to those phagocytes which do not contain lactoferrin (i.e., monocytes and macrophages) via the ability of these cells to bind lactoferrin through a specific surface receptor for the protein. The capacity of lactoferrin to function as an inhibitor of the Haber–Weiss reaction appears to be quite resistant to alteration by the effects of compounds such as HOCl, proteases, or LPS to which the protein would likely be exposed at sites of inflammation. The second anti-inflammatory mechanism exhibited by lactoferrin is its ability to bind LPS with a resultant decrease in LPS bioactivity. Given evidence that plasma levels of LPS increase in response to LPS and during septic shock (101, 102), lactoferrin may potentially serve to ameliorate LPS-induced toxicity. Clearly the relative role of lactoferrin in endogenous host antiinflammatory mechanisms as well as its potential as a therapeutic agent for diseases involving oxidant-mediated tissue injury are worthy of continued study.

## ACKNOWLEDGEMENTS

This work was supported by the VA Research Service, NIH awards HL44275, AI28412, AI92959, a Grant-In-Aid from the American Heart Association, and an award from the American Council for Tobacco Research. This work was performed during the tenure of Dr. Britigan as a Research Associate of the VA Research Service. Dr. Serody is a recipient of a National Research Service Award from NIH. We acknowledge the important contributions of our colleagues Gerald Rosen and Sovitj Pou in much of the work we have described and we thank Naomi Erickson for her help with preparation of the manuscript.

## REFERENCES

1. Masson, PL, Heremans, JF, Dive, CH. (1966) Studies on lactoferrin, an iron-binding protein common to many external secretions. Clin. Chim.Acta 14:735–739.

2. Wang-Iverson, P, Pryzwansky, KB, Spitznagel, JK, Cooney, MH. (1978) Bactericidal capacity of phorbol myristate acetate treated human polymorphonuclear leukocytes. Infect. Immun. 22:945–955.

3. Cross, CE, Halliwell, B, Borish, ET, Pryor, WA, Saul, RL, McCord, JM, Harman, D. (1987) Oxygen radicals and human disease. Ann. Intern. Med. 107:526–545.

4. Haber, F, Weiss, J. (1934) The catalytic decomposition of hydrogen peroxide by iron salts. Proc. R. Soc. Lond. Math. Phys. Soc. 147:332–351.

5. Halliwell, B, Gutteridge, JMC. (1986) Oxygen radicals and iron in relation to biology and medicine: some problems and concepts. Arch. Biochem. Biophys. 246:501–514.

6. Bannister, JV, Bannister, WH, Rotilio, G. (1987) Aspects of the structure, function, and applications of superoxide dismutase. CRC Crit. Rev. Biochem. 22:111–180.

7. Fridovich, I. (1978) The biology of oxygen radicals: the superoxide radical is an agent of oxygen toxicity; superoxide dismutases provide an important defense. Science 201:875–880.

8. Root, RK, Cohen, MS. (1981) The microbicidal mechanisms of human neutrophils and eosinophils. Rev. Infect. Dis. 3:565–598.

9. Clark, RA. (1990) The human neutrophil respiratory burst oxidase. J. Infect. Dis. 161:1140–1147.

10. Klebanoff, S J, Hamon, CB,. (1972) Role of myeloperoxidase-mediated antimicrobial systems in intact leukocytes. J. Reticuloendothel. Soc. 12:170–196.

11. Weiss, S J, Lampert, MD, Test, ST. (1983) Long-lived oxidants generated by human neutrophils: characterization and bioactivity. Science 222:625–628.

12. Tauber, AI, Borregaard, N, Simons, E, Wright, J. (1983) Chronic granulomatous disease: a syndrome of phagocyte oxidase deficiencies. Medicine (Baltimore) 62:286–308.

13. Nauseef, WM. (1990) Myeloperoxidase deficiency. Hematol. Pathol. 4: 165-178.

14. Weiss, SJ. (1986) Oxygen, ischemia and inflammation. Acta Physiol. Scand. (suppl)548:9–37.

15. Till, GO, Johnson, KJ, Kunkel, R, Ward, PA. (1982) Intravascular activation of complement and acute lung injury: dependency on neutrophils and toxic oxygen metabolites. J. Clin. Invest. 69:1126–1135.

16. Shasby, DM, Vanbenthuysen, KM, Tate, RM, Shasby, SS, McMurthry, I, Repine, JE. (1982) Granulocytes mediate acute edematous lung injury in rabbits and in isolated rabbit lungs perfused with phorbol myristate acetate: role of oxygen radicals. Am. Rev. Respir. Dis. 125:443–447.

17. Fox, RB. (1984) Prevention of granulocyte mediated lung injury in rats by a hydroxyl radical scavenger, dimethylthiourea. J. Clin. Invest. 74: 1456–1464.

18. Ward, PA, Till, GO, Kunkel, R, Beauchamp, C. (1983) Evidence for the role of hydroxyl radical in complement and neutrophil-dependent tissue injury. J. Clin. Invest. 72:789–801.

19. Till, GO, Hatherill, JR, Tourtellotte, WW, Lutz, MJ, Ward, PA. (1985) Lipid peroxidation and acute lung injury after thermal trauma to skin. Am. J. Pathol. 119:376–384.

20. Grootveld, M, Halliwell, B. (1986) Aromatic hydroxylation as a potential measure of hydroxyl radical formation in vivo. Biochem. J. 237:499–504.

21. Weiss, S J, Rustagi, PK, LeBuglio, AF. (1978) Human granulocyte generation of hydroxyl radical. J. Exp. Med. 147:3 16–323.

22. Tauber, AI, Babior, BM. (1977) Evidence for hydroxyl radical production by human neutrophils. J. Clin. Invest. 60:374–379.

23. Repine, JE, Eaton, JW, Anders, MW, Ohidal, JR, Fox, RB. (1979) Generation of hydroxyl radical by enzymes, chemicals, and human phagocytes in vitro. J. Clin. Invest. 64:1642–1651.

24. Sagone, AL, Jr., Decker, MA, Wells, RM, Democko, C. (1980) A new method for the detection of hydroxyl radical production by phagocytic cells. Biochim. Biophys. Acta 628:90–97.

25. Ambruso, DR, Johnston, RB, Jr. (1981) Lactoferrin enhances hydroxyl radical production by human neutrophils, neutrophil particulate fractions and an enzymatic generating system. J. Clin. Invest. 67:352–360.

26. Weiss, S J, King, GW, LoBuglio, AF. (1977) Evidence for hydroxyl radical generation by human monocytes. J. Clin. Invest. 60:370–373.

27. Speer, CP, Ambruso, DR, Grimsley, J, Johnston, RB, Jr.. (1985) Oxidative metabolism in cord blood monocytes and monocyte-derived macrophages. Infect. Immun. 50:919–921.

28. Hume, DA, Gordon, S, Thornalley, PJ, Bannister, JV. (1983) The production of oxygen-centered radicals by Bacillus-Calmette-Guerin-activated macrophages: an electron paramagnetic resonance study of the response to phorbol myristate acetate. Biochim. Biophys. Acta 763:245–250.

29. Cohen, MS, Britigan, BE, Hassett, DJ, Rosen, GM. (1988) Do human neutrophils form hydroxyl radical? Evaluation of an unresolved controversy. Free Radic. Biol. Med. 5:81–88.

30. Britigan, BE, Rosen, GM, Chai, Y, Cohen, MS. (1986) Do human neutrophils make hydroxyl radical? Detection of free radicals generated by human neutrophils activated with a soluble or particulate stimulus using electron paramagnetic resonance spectrometry. J. Biol. Chem. 261:4426–4431.

31. Pou, S, Cohen, MS, Britigan, BE, Rosen, GM. (1989) Spin trapping and human neutrophils: limits of detection of hydroxyl radical. J. Biol. Chem. 264:12299–12302.

32. Britigan, BE, Coffman, TJ, Buettner, GR. (1990) Spin trap evidence for the lack of significant hydroxyl radical production during the respiration burst of human phagocytes using a spin adduct resistant to superoxide mediated destruction. J. Biol. Chem. 265:2650–2656.

33. Thomas, MJ, Shirley, PS, Hedrick, C, DeChatelet, LR. (1986) Role of free radical processes in stimulated human polymorphonuclear leukocytes. Biochemistry 25:8042–8048.

34. Kaur, H, Fagerheim, Z, Grootveld, M, Puppo, A, Halliwell, B. (1988) Aromatic hydroxylation of phenylalanine as an assay for hydroxyl radicals: application to activated neutrophils and heme protein leghemoglobin. Anal. Biochem. 172:360–367.

35. Greenwald, RA, Rush, SW, Mark, SA, Weitz, Z. (1989) Conversion of superoxide generated by polymorphonuclear leukocytes to hydroxyl radical: a direct spectrophotometric detection system based on degradation of deoxyribose. Free Radic. Biol. Med. 6:385–392.

36. Winterbourn, CC. (1986) Myeloperoxidase as an effective inhibitor of hydroxyl radical production: implications for the oxidative reactions of neutrophils. J. Clin. Invest. 78:545–550.

37. Ramos, CL, Pou, S, Britigan, BE, Cohen, MS, Rosen, GM. (1992) Spin trapping evidence for myeloperoxidase-dependent hydroxyl radical formation by human neutrophils and monocytes. J. Biol. Chem. 267:8307-8312.

38. Bannister, JV, Bannister, WH, Hill, HAO, Thornalley, PJ. (1982) Enhanced production of hydroxyl radicals by the xanthine-xanthine oxidase reaction in the presence of lactoferrin. Biochim. Biophys. Acta 715:116–120.

39. Nakamura, M. (1990) Lactoferrin-mediated formation of oxygen radicals by NADPH- cytochrome P-450 reductase system. J. Biochem. 107:395–399.

40. Britigan, BE, Cohen, MS, Rosen, GM. (1987) Detection of the production of oxygen-centered free radicals by human neutrophils using spin trapping techniques: a critical perspective. J. Leukocyte Biol. 41:349–362.

41. Baldwin, DA, Jenny, ER, Aisen, P. (1984) The effect of human serum transferrin and milk lactoferrin on hydroxyl radical formation from superoxide and hydrogen peroxide. J. Biol. Chem. 259:13391–13394.

42. Winterbourn, CC. (1983) Lactoferrin-catalyzed hydroxyl radical production: additional requirements for a chelating agent. Biochem. J. 210:15–19.

43. Aruoma, OI, Halliwell, B. (1987) Superoxide-dependent and ascorbate-dependent formation of hydroxyl radicals from hydrogen peroxide in the presence of iron: are lactoferrin and transferrin promoters of hydroxyl-radical generation? Biochem. J. 241:273–278.

44. Britigan, BE, Edeker, BL. (1991) *Pseudomonas* and neutrophil products modify transferrin and lactoferrin to create conditions that favor hydroxyl radical formation. J. Clin. Invest. 88:1092–1102.

45. Cohen, MS, Mao, J, Rasmussen, GT, Serody, JS, Britigan, BE. (1992) Interaction of lactoferrin and lipopolysaccharide: effects on the antioxidant property of lactoferrin and thlactoferrinlactoferrinlactofelactoferrie ability of lipopolysaccharides to prime human neutrophils for enhanced superoxide formation. J. Infect. Dis. 166:1375–1378.

46. Gutteridge, JMC, Paterson, SK, Segal, AW, Halliwell, B. (1981) Inhibition of lipid peroxidation by the iron-binding protein lactoferrin. Biochem. J. 199:259–261.

47. Buettner, GR. (1987) The reaction of superoxide, formate radical, and hydrated electron with transferrin and its model compound, Fe(III)-ethylenediamine-N,N′-bis [2-(2-hydroxyphenyl) acetic acid] as studied by pulse radiolysis. J. Biol. Chem. 262:11995–11998.

48. Klebanoff, S J, Waltersdorph, AM. (1990) Prooxidant activity of transferrin and lactoferrin. J. Exp. Med. 172:1293–1303.

49. Britigan, BE, Rosen, GM, Thompson, BY, Chai, Y, Cohen, MS. (1986) Stimulated neutrophils limit iron-catalyzed hydroxyl radical formation as detected by spin trapping techniques. J. Biol. Chem. 261: 17026–17032.

50. Britigan, BE, Hassett, D J, Rosen, GM, Hamill, DR, Cohen, MS. (1989) Neutrophil degranulation inhibits potential hydroxyl radical formation: differential impact of myeloperoxidase and lactoferrin release on hydroxyl radical production by iron supplemented neutrophils assessed by spin trapping. Biochem. J. 264:447–455.

51. Molloy, AL, Winterbourn, CC. (1990) Release of iron from phagocytosed *Escherichia coli* and uptake by neutrophil lactoferrin. Blood 75:984–989.

52. Vercellotti, GM, van Asbeck, BS, Jacob, HS. (1985) Oxygen radical-induced erythrocyte hemolysis by neutrophils: critical role of iron and lactoferrin. J. Clin. Invest. 76:956–962.

53. Gallin, JI. (1985) Neutrophil specific granule deficiency. Ann. Rev. Med. 36:263–274.

54. Boxer, LA, Coates, TD, Haak, RA, Wolach, JB, Hoffstein, S, Baehner, RL. (1982) Lactoferrin deficiency associated with altered granulocyte function. N. Engl. J. Med. 307:404–410.

55. Harris, P, Ralph, P. (1985) Human leukemic models of myelomonocytic development: a review of the HL-60 and U937 cell lines. J. Leukocyte Biol. 37:407–422.

56. Thompson, BY, Sivam, G, Britigan, BE, Rosen, GM, Cohen, MS. (1988) The $O_2$ Metabolism of the HL-60 cell line: comparison of the effects of monocytoid and neutrophilic differentiation. J. Leukocyte Biol. 43:140–147.

57. van Snick, JL, Masson, PL, Heremans, JF. (1974) The involvement of lactoferrin in the hyposideremia of acute inflammation. J. Exp. Med. 140: 1068–1084.

58. Winterbourn, CC, Monteiro, HP, Galilee, CF. (1990) Ferritin-dependent lipid peroxidation by stimulated neutrophils: Inhibition by myeloperoxidase-derived hypochlorous acid but not by endogenous lactoferrin. Biochim. Biophys. Acta Mol. Cell Res. 1055:179–185.

59. Yamada, Y, Amagasaki, T, Jacobsen, DW, Green, R. (1987) Lactoferrin binding by leukemia cell lines. Blood 70:264–270.

60. Bennett, RM, Davis, J, Campbell, S, Portnoff, S. (1983) Lactoferrin binds to cell membrane DNA: association of surface DNA with an enriched population of B cells and monocytes. J. Clin. Invest. 71:611–618.

61. Campbell, EJ. (1982) Human leukocyte elastase, cathepsin G, and lactoferrin: family of neutrophil granule glycoproteins that bind to an alveolar macrophage receptor. Proc. Natl. Acad. Sci. USA 79:6941–6945.

62. Birgens, HS, Hansen, NE, Karle, H, Kristensen, LO. (1983) Receptor binding of lactoferrin to human monocytes. Br. J. Haematol. 54:383–391.

63. Birgens, HS, Kristensen, LO. (1990) Impaired receptor binding and decrease in isoelectric point of lactoferrin after interaction with human monocytes. Eur. J. Haematol. 45:31–35.

64. Birgens, HS, Kfistensen, LO, Borregaard, N, Karle, H, Hansen, NE. (1988) Lactoferrin-mediated transfer of iron to intracellular ferritin in human monocytes. Eur. J. Haematol. 41:52–57.

65. Lima, MF, Kierszenbaum, F. (1985) Lactoferrin effects on phagocytic cell function. I. Increased uptake and killing of an intracellular parasite by murine macrophages and human monocytes. J. Immunol. 134:4176–4183.

66. Miyazawa, K, Mantel, C, Lu, L, Morrison, DC, Broxmeyer, HE. (1991) Lactoferrin-lipopolysaccharide interactions: Effect on lactoferrin binding to monocyte/macrophage-differentiated HL-60 cells. J. Immunol. 146:723–729.

67. Britigan, BE, Serody, JS, Hayek, MB, Charniga, LM, Cohen, MS. (1991) Uptake of lactoferrin by mononuclear phagocytes inhibits their ability to form hydroxyl radical and protects them from membrane autoperoxidation. J. Immunol. 147:4271–4277.

68. Imber, MJ, Pizzo, SV. (1983) Clearance and binding of native and defucosylated lactoferrin. Biochem. J. 212:249–257.

69. Moguilevsky, N., Courtoy, P.J. and Masson, P.L. Study of lactoferrin-binding sites at the surface of blood monocytes. In: Proteins of Iron Storage and Transport, edited by Spik, G., Montreuil, J., Crichton, R.R. and Mazurier, J. Amsterdam: Elsevier Science Publishers, B.V., 1985, p. 199–202.

70. Bennett, RM, Davis, J. (1981) Lactoferrin binding to human peripheral blood cells: interaction with a B-enriched population of lymphocytes and a subpopulation of adherent mononuclear cells. J. Immunol. 127:1211–1216.

71. Schraufstatter, IU, Hinshaw, DB, Hyslop, PA, Spragg, RG, Cochrane, CG. (1986) Oxidant injury of cells: DNA strand breaks activate polyadenosine diphosphate-ribose polymerase and lead to depletion of nicotinamide adenine dinucleotide. J. Clin. Invest. 77:1312–1320.

72. Birnboim, HC. (1982) DNA strand breakage in human leukocytes exposed to a tumor promoter, phorbol myristate acetate. Science 215:1247–1249.

73. Steinmann, G, Broxmeyer, HE, de Harven, E, Moore, MAS. (1982) Immuno-electron microscopic tracing of lactoferrin, a regulator of myelopoiesis, into a subpopulation of human peripheral blood monocytes. Br.J. Haematol. 50:75–84.

74. Derisbourg, P, Wieruszeski, J-M, Montreuil, J, Spik, G. (1990) Primary structure of glycans isolated from human leucocyte lactotransferrin: absence of fucose residues questions the proposed mechanism of hyposideraemia. Biochem. J. 269:821–825.

75. Borregaard, N, Heiple, JM, Simons, ER, Clark, RA. (1983) Subcellular localization of the b-cytochrome component of the human neutrophil microbicidal oxidase. Translocation during activation. J. Cell Biol. 97:52–61.

76. Thomas, RM, Nauseef, WM, Iyer, SS, Peterson, MW, Stone, PJ, Clark, RA. (1991) A cytosolic inhibitor of human neutrophil elastase and cathepsin G. J. Leukocyte Biol. 50:568–579.

77. Howell, DR, Britigan, BE, Fick, RB, Jr., Cox, CD. (1990) Levels of iron and iron-binding proteins in bronchoalveolar lavage fluids of cystic fibrosis subjects. Clin. Res. 38:274A. (Abstract)

78. Roiron, D, Amouric, M, Marvaldi, J, Figarella, C. (1989) Lactoferrin-binding sites at the surface of HT29-D4 cells: comparison with transferrin. Eur. J. Biochem. 186:367–373.

79. Morgan, CL, Stanley, ER. (1984) Chemical cross linking of the mononuclear phagocyte specific growth factor CSF-1 to its receptor at the cell surface. Biochem. Biophys. Res. Commun. 119:35–41.

80. Pilch, PF, Czech, MN. (1979) Interaction of cross-linking agents with the insulin effector system of isolated fat cells. Covalent linkage of 125I-insulin to a plasma membrane protein of 140,000 daltons. J. Biol. Chem. 254:3375–3381.

81. Mazurier, J, Montreuil, J Spik, G. (1985) Visualization of lactotransferrin brush-border receptors by ligand-blotting. Biochim. Biophys. Acta 821:453–460.

82. Thaler, CJ, Vanderpuye, OA, McIntyre, JA Faulk, WP. (1990) Lactoferrin binding molecules in human seminal plasma. Biol. Reprod. 43:712–717.

83. Mazurier, J, Legrand, D, Hu, WL, Montreuil, J Spik, G. (1989) Expression of human lactotransferrin receptors in phytohemagglutinin-stimulated human peripheral blood lymphocytes: isolation of the receptors by antiligand affinity chromatography. Eur. J. Biochem. 179:481–487.

84. Kawakami, H, Lönnerdal, B. (1991) Isolation and function of a receptor for human lactoferrin in human fetal intestinal brush-border membranes. Am.J. Physiol. Gastrointest. Liver Physiol 261:G841–G846.

85. Hurst, JK, Barrette, WC, Jr.. (1989) Leukocyte oxygen activations and microbicidal oxidative toxins. CRC Crit. Rev. Biochem. Molec. Biol. 24:271–328.

86. Winterbourn, CC, Malloy, AL. (1988) Susceptibilities of lactoferrin and transferrin to myeloperoxidase-dependent loss of iron-binding capacity. Biochem. J. 250:613–616.

87. Clark, RA Pearson, DW. (1989) Inactivation of transferrin iron binding capacity by the neutrophil myeloperoxidase system. J. Biol. Chem. 264:9420–9427.

88. Doring, G, Pfestorf, M, Botzenhart, K Abdallah, MA. (1988) Impact of proteases on iron uptake of *Pseudomonas aeruginosa* pyoverdin from transferrin and lactoferrin. Infect. Immun. 56:291–293.

89. Brines, RD Brock, JH. (1983) The effect of trypsin and chymotrypsin on the *in vitro* antimicrobial and iron-binding properties of lactoferrin in human milk and bovine colostrum: unusual resistance of human apolactoferrin to proteolytic digestion. Biochim. Biophys. Acta 759:229–235.

90. Line, WF, Sly, DA, Bezkorovainy, A. (1976) Limited cleavage of human lactoferrin with pepsin. Int. J. Biochem. 9:203–208.

91. Bluard-Deconinck, J-M, Williams, J, Evans, RW, van Snick, J, Osinski, PA Masson, PL. (1978) Iron-binding fragments from the N-terminal and C-terminal regions of human lactoferrin. Biochem. J. 171:321–327.

92. Evans, RW, Williams, J. (1978) Studies of the binding of different iron donors to human serum transferrin and isolation of iron-binding fragments from the N- and C-terminal regions of the protein. Biochem. J. 173:543–552.

93. Esparza, I Brock, JH. (1980) The effect of trypsin digestion on the structure and iron-donating properties of transferrins from several species. Biochim. Biophys. Acta 622:297–307.

93a. Britigan, BE, Hayek, MB, Doebbeling, BN, and Fick, RB, Jr. (1993) Transferrin and lactoferrin undergo proteolytic cleavage in the *Pseudomonas aeruginosa*-infected lungs of patients with cystic fibrosis. Infect. Immun. 61:5049–5055.

94. Ellison, RT, III Giehl, TJ. (1991) Killing of Gram-negative bacteria by lactoferrin and lysozyme. J. Clin. Invest. 88:1080–1091.

95. Guthrie, LA, McPhail, LC, Henson, PM Johnston, RB,Jr. (1984) The priming of neutrophils for enhanced release of oxygen metabolites by bacterial lipopolysaccharide: Evidence for increased activity of the superoxide-producing enzyme. J. Exp. Med. 160:1656–1671.

96. Vosbeck, K, Tobias, P, Mueller, H, Allen, RA, Arfors, K-E, Ulevitch, RJ Sklar, LA. (1990) Priming of polymorphonuclear granulocytes by lipopolysaccharides and its complexes with lipopolysaccharide binding protein and high density lipoprotein. J. Leukocyte Biol. 47:97–104.

97. Wollenweber, H-W Morrison, DC. (1985) Synthesis and biochemical characterization of a photoactivable, iodinatable, cleavable bacterial lipopolysaccharide derivative. J. Biol. Chem. 260: 15068–15074.

98. Berger, D Berger, HG. (1987) Evidence for endotoxin binding capacity of human Gc- globulin and transferrin. Clin. Chim.Acta 163:289–299.

99. Gahr, M, Speer, CP, Damerau, B Sawatzki, G. (1991) Influence of lactoferrin on the function of human polymorphonuclear leukocytes and monocytes. J. Leukocyte Biol. 49:427–433.

99a. Britigan, BE, Hayek, MB, Doebbeling, BN, and Fick, RB, Jr. (1993) Transferrin and lactoferrin undergo proteolytic cleavage in the pseudomonas aeruginosa infected lungs of patients with cystic fibrosis. Infect. Immun. 61:5049–5055.

100. Zagulski, T, Lipinski, P, Zagulska, A, Broniek, S Jarzabek, Z. (1989) Lactoferrin can protect mice against a lethal dose of *Echerichia coli* in experimental infection *in vivo*. Br.J. Exp. Path. 70:697–704.

101. Gutteberg, TJ, Osterud, B, Volden, G Jorgensen, T. (1990) The production of tumour necrosis factor, tissue thromboplastin, lactoferrin and cathepsin C during lipopolysaccharide stimulation in whole blood. Scand. J. Clin. Lab. Invest. 50:421–427.

102. Nuijens, JH, Abbink, JJ, Wachtfogel, YT, Colman, RW, Eerenberg, AJM, Dors, D, Kamp, AJM, Strack van Schijndel, RJM, Thijs, LG Hack, CE. (1992) Plasma elastase, $\alpha_1$-antitrypsin and lactoferrin in sepsis: Evidence for neutrophils as mediators in fatal sepsis. J. Lab. Clin. Med. 119:159–168.

# INTERACTION OF LACTOFERRIN WITH
# MONONUCLEAR AND COLON CARCINOMA CELLS

Jeremy H. Brock, Maznah Ismail, and Lourdes Sánchez

University Department of Immunology,
Western Infirmary
Glasgow G11 6NT, Scotland, United Kingdom

## SUMMARY

Lactoferrin is known to bind to macrophages/monocytes and intestinal mucosal cells, but the nature and function of these interactions is not clear. We have therefore examined the interaction of lactoferrin in vitro with the promonocytic cell line U937 and with differentiated human colon carcinoma cells. U937 cells bound more lactoferrin than transferrin, although most of the lactoferrin binding was non-specific. Uptake of iron from transferrin was rapid, but uptake from lactoferrin was slow, and may have been due to prior transfer of iron to transferrin in the culture medium as a result of labilisation of iron from membrane-bound lactoferrin. Unlike transferrin, lactoferrin was not internalised by U937 cells. Lactoferrin significantly reduced uptake of non-transferrin-bound iron by the cells, but had no effect on uptake of transferrin-bound iron. Transport of lactoferrin-bound iron across monolayer cultures of differentiated Caco-2 cells in bicameral chambers was similar to that of ferric citrate, while transport of transferrin-bound iron was lower. Lactoferrin and transferrin themselves were not transported, although some proteolytically degraded material did cross the monolayer. Thus lactoferrin, unlike transferrin, is not an important iron donor to monocytic cells, but may instead serve to regulate iron uptake from other sources. It does not seem to enhance iron transport across mucosal cells.

## INTRODUCTION

Despite its close structural similarity to transferrin, the role of lactoferrin in iron transport and uptake remains unclear. The first suggestion that lactoferrin might play such a role came from the work of Van Snick et al. (1974) who demonstrated that lactoferrin could bind to peritoneal macrophages and transfer its iron to these cells. They proposed that lactoferrin might contribute to the hypoferraemia of inflammation by capturing transferrin-bound iron and diverting it to the reticuloendothelial system. Furthermore, lactoferrin may enhance the growth of certain cell lines in vitro (Amouric et al, 1984; Azuma et al. 1989), suggesting that like transferrin, it can supply iron required for cell division.

However, direct evidence that lactoferrin acts as an iron donor is scarce. Uptake of iron from transferrin depends upon expression of a well-characterised receptor which allows

*Lactoferrin: Structure and Function*
Edited by T.W. Hutchens *et al.*, Plenum Press, New York, 1994

endocytosis of transferrin and intracellular iron release to occur (Dautry-Varsat, 1986). No such mechanism has been described for lactoferrin. Furthermore, although there is evidence for the existence of specific lactoferrin receptors on some cells (Birgens, 1991), the nature of the membrane molecule(s) involved and their interaction with lactoferrin remain largely unknown, though a putative receptor molecule has been identified in activated lymphocytes (Mazurier et al, 1989) and a few other cells (see Spik, this volume). However, there is no evidence that such a receptor is itself regulated by cellular iron requirements, or that it preferentially binds the iron-saturated form of the protein, as occurs with transferrin.

Another area of controversy involves the role of lactoferrin in iron absorption. A lactoferrin receptor has been identified on intestinal brush border membranes (Davidson and Lonnerdal, 1988; Hu et al. 1990), suggesting a possible role for lactoferrin in mediating iron transport across the gut mucosa. However, in vivo studies have so far provided no clear evidence that lactoferrin enhances iron absorption (Fransson et al. 1983a,b; Davidson et. al. 1990) and indeed there is some evidence, reviewed by Brock (1980) that it may exert an inhibitory effect.

In order to address these questions, we have investigated the interaction of lactoferrin with the human promonocytic cell line U937, and with Caco-2, a human colon carcinoma cell line capable of differentiating and forming a mucosal cell monolayer in vitro which displays many features of the small intestine epithelial layer (Hidalgo et al, 1989).

## MATERIALS AND METHODS

### Cells

The human promonocytic cell line U937 was cultured as described previously (Iturralde et al, 1992). Log-phase cells, obtained 24 h after subculture, were used in all experiments. The human colon carcinoma cell line Caco-2 was kindly provided by Dr R.I. Freshney, Dept of Medical Oncology, Glasgow University, and cultured as described below.

### Protein Binding and Iron Uptake by U937 Cells

Human lactoferrin, prepared from colostrum (Johansson, 1969) and rendered largely iron-free by dialysis against EDTA at pH 4 (Mazurier and Spik, 1980), and human apotransferrin (Behringwerke, Hounslow, UK) were labelled with 59Fe and/or 125I as required, as described previously (Oria et al, 1988). Following trace labelling with 59Fe, the proteins were fully saturated by addition of unlabelled iron. This procedure prevents any slight excess of iron giving rise to unbound 59Fe. Specific and non-specific binding of proteins to cells was determined using 125I-labelled Fe-saturated lactoferrin or transferrin in the absence or presence of an excess of unlabelled protein and the results subjected to Scatchard analysis, as described previously (Iturralde et al, 1992). Iron uptake was determined by incubating the cells for appropriate periods with 59Fe-labelled lactoferrin or transferrin (50 µg/ml) in serum-free RPMI 1640 medium containing 1 mg/ml human serum albumin. To ensure identical conditions throughout, unlabelled lactoferrin was added to cultures containing labelled transferrin and *vice versa*. Cells and supernatants were separated by centrifugation and the cells washed twice, the suspension being transferred to a fresh tube prior to the last wash to minimise problems of isotope binding to the plastic tubes. In some experiments double-labelled proteins (59Fe and 125I) were used.

To determine the effect of lactoferrin on iron uptake from other sources, cells were incubated as above in the presence of 50 µg/ml 59Fe-transferrin or 8 µM 59Fe-nitrilotriacetate

(NTA) in the presence or absence of 50µg/ml of unlabelled apolactoferrin or Fe-lactoferrin, and iron uptake determined as above.

## Internalisation of Lactoferrin and Transferrin

U937 cells ($2 \times 10^6$ in 0.2 ml serum-free medium) were incubated at 4°C or 37°C for 30 min with 125I-labelled lactoferrin or transferrin (50 µg/ml), then spun down and washed rapidly at 4°C. Surface and intracellular distribution of the labelled proteins was determined by exposing the cells to 0.25 M acetate-0.5 M NaCl, pH 2.3, for 5 sec, and cells and supernatant separated by centrifugation through Versilube F50 oil (Alfa, Wokingham, UK). To determine whether endocytosed proteins were recycled back to the cell membrane, the cells were incubated for 30 min at 37°C in the presence of 3 mM primaquine (Sigma) following removal of unbound labelled proteins. The cells were then subjected to acid-washing as above.

## Transport of Iron and Proteins across Caco-2 Monolayers

Caco-2 cells were grown in Transwell bicameral chambers (Costar, High Wycombe, UK) in DMEM medium (Gibco, Paisley, UK) containing 10% fetal calf serum (Northumbria, Cramlington, UK), 1% non-essential amino acids (Flow, Rickmansworth, UK) and 1 µg/ml bovine insulin (Sigma). Confluent cultures of differentiated cells were normally obtained after 15–20 days, and integrity was checked by measuring exclusion of phenol red (Halleux and Schneider, 1991) and transepithelial electrical resistance (Alvarez-Hernandez et al, 1991). The medium in both the upper and lower compartments was then removed and replaced with a similar medium but without serum. To the upper chamber was added labelled lactoferrin or transferrin (50 µg/ml) or a similar amount of 59Fe-citrate, while apotransferrin (1 mg/ml) was added to the lower chamber as an iron acceptor. Medium from the lower chamber was removed at intervals for counting and replaced with fresh medium. At the end of the experiment the cells were dissolved in 2% (w/v) sodium dodecyl sulphate and assayed for radioactivity. Integrity of 125I-lactoferrin or transferrin in the lower compartment was determined by precipitation with 10% (w/v, final concentration) trichloroacetic acid.

## RESULTS

### Iron Uptake and Binding of Lactoferrin and Transferrin by U937 Cells

The cells readily acquired iron from transferrin, and there was apparently a slower but clearly detectable uptake from lactoferrin (Fig 1). However, when the experiment was repeated using doubly-labelled (59Fe, 125I) proteins it was evident that much of the apparent iron uptake from lactoferrin was due to the much greater binding of this protein to the cells, rather than to net iron uptake (Table 1). Binding of lactoferrin after 2 h was much greater than binding of transferrin and increased with time, whereas binding of transferrin remained fairly constant—after 2 h the molar ratio of cell-associated iron:lactoferrin was only 2.4, compared with 12.7 for transferrin, while after 18 h the figures were 5.9 and 66 respectively. Since each protein is theoretically capable of delivering 2 iron atoms to the cell in each endocytic cycle, this represents only about 3 cycles in 18 h for lactoferrin, as against 33 for transferrin.

A more detailed study of binding to U937 cells confirmed that total binding of lactoferrin to the cells was about 10 times greater than total binding of transferrin (Figs 2a, c). However, most of the lactoferrin binding was non-specific, although Scatchard

**Table 1.** Interaction of 59Fe, 125I-Transferrin and Lactoferrin with U937 Cells.

| Protein | Transferrin | | Lactoferrin | |
|---|---|---|---|---|
| Time | 2h | 18h | 2h | 18h |
| Amount of cell-associated: | | | | |
| Protein | 0.61 | 0.78 | 2.95 | 5.90 |
| Iron | 7.74 | 51.50 | 7.06 | 35.30 |
| Molar ratio Fe:protein | 12.7 | 66.0 | 2.4 | 5.9 |

Results are expressed as pmol/$10^6$ cells, and are from a single representative experiment.
Details as in Figure 1, except that doubly-labelled proteins were used.

analysis revealed that there was nevertheless a specific element, the number of sites per cell ($3 \times 10^6$) being similar to that for transferrin ($1.9 \times 10^6$), though the affinity was approximately 4-fold lower (Figs 2b, d). In other experiments (data not shown) it was found that binding of lactoferrin was not inhibited by transferrin (or vice versa) and that binding of lactoferrin was not inhibited by 5% fucose, nor by treatment of the cells with heparitinase, which cleaves acidic sugar residues. Furthermore lactoferrin did not inhibit binding of monoclonal antibodies to the cell-surface markers CD3, CD4, CD11b, CD14, CD16, CD25, CD45 and HLA-DR.

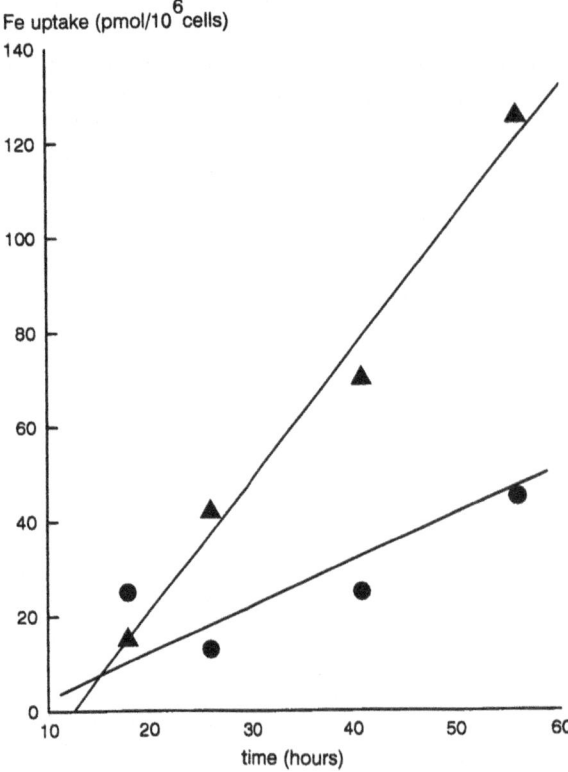

Fe uptake (pmol/$10^6$ cells)

time (hours)

**Figure 1.** Uptake of iron by U937 cells from lactoferrin (●) and transferrin (▲). The cells were incubated with 50 μg/ml 59Fe-labelled iron-saturated transferrin and 50 μg/ml unlabelled lactoferrin, or *vice versa*.

**Table 2.** Internalisation of Lactoferrin by U937 Cells. U937 cells were incubated with 50 µg/ml of labelled protein for 30 min and the proportion of surface-bound protein determined by washing with 0.25M acetate at pH 2.3

|  | % of bound protein retained after acid-washing | |
| --- | --- | --- |
| Incubation temperature: | 4°C | 37°C |
| Lactoferrin | 16 ± 1 | 19 ± 1 |
| Transferrin | 35 ± 1 | 59 ± 6 |

Figures are mean ± s.d., n=3

## Endocytosis and Recycling of Lactoferrin and Transferrin

As shown in Table 2, most lactoferrin could be removed from the cells by acid washing, irrespective of whether incubation was carried out at 4°C or 37°C, whereas with transferrin the majority of the protein was resistant to acid-washing following incubation at 37°C, indicative of internalisation. When cells were incubated with the recycling inhibitor primaquine (Reid and Watts, 1990) prior to acid-washing, 80% of transferrin was retained (Table 3), as against 40% in the absence of the inhibitor, indicating that endocytosed protein was being prevented from recycling back to the cell membrane. In contrast primaquine had no effect on the susceptibility of cell-associated lactoferrin to acid-washing, almost all the protein remaining surface-bound, which further confirms the lack of endocytosis of lactoferrin.

## Release of Iron from Cell-Associated Lactoferrin

Although binding of lactoferrin to U937 cells was largely non-specific, there was nevertheless some net accumulation of iron from lactoferrin, albeit much less than from transferrin (cf. Fig 1 and Table 1).

Since there was no evidence of any endocytosis of lactoferrin it seemed possible that iron uptake was occurring by an indirect mechanism involving exchange with transferrin in the medium. It was found that during an 18 h incubation of U937 cells in medium containing both transferrin and lactoferrin (50% iron-saturated in each case), 16.6% of lactoferrin-bound iron was transferred to transferrin, almost 3 times the amount transferred in the opposite direction (5.9%).

When cells previously incubated with 59Fe,125I-lactoferrin or transferrin were washed and reincubated at 4°C in fresh medium containing neither protein, but with 1 mM desferrioxamine, there was no release of iron previously acquired from transferrin (Fig 3a), even though there was a partial release of transferrin itself. In contrast, cells previously incubated with

**Table 3.** Effect of Primaquine on Internalisation of Lactoferrin and Transferrin by U937 Cells.

|  | % of bound protein retained after acid-washing | |
| --- | --- | --- |
| Primaquine: | + | − |
| Lactoferrin | 11 ± 2 | 13 ± 2 |
| Transferrin | 83 ± 5 | 40 ± 4 |

Figures are mean ± s.d., n=3
Cells were treated as in Table 1, except that cells were incubated at 37°C with 3 mM primaquine for 30 min prior to acid-washing.

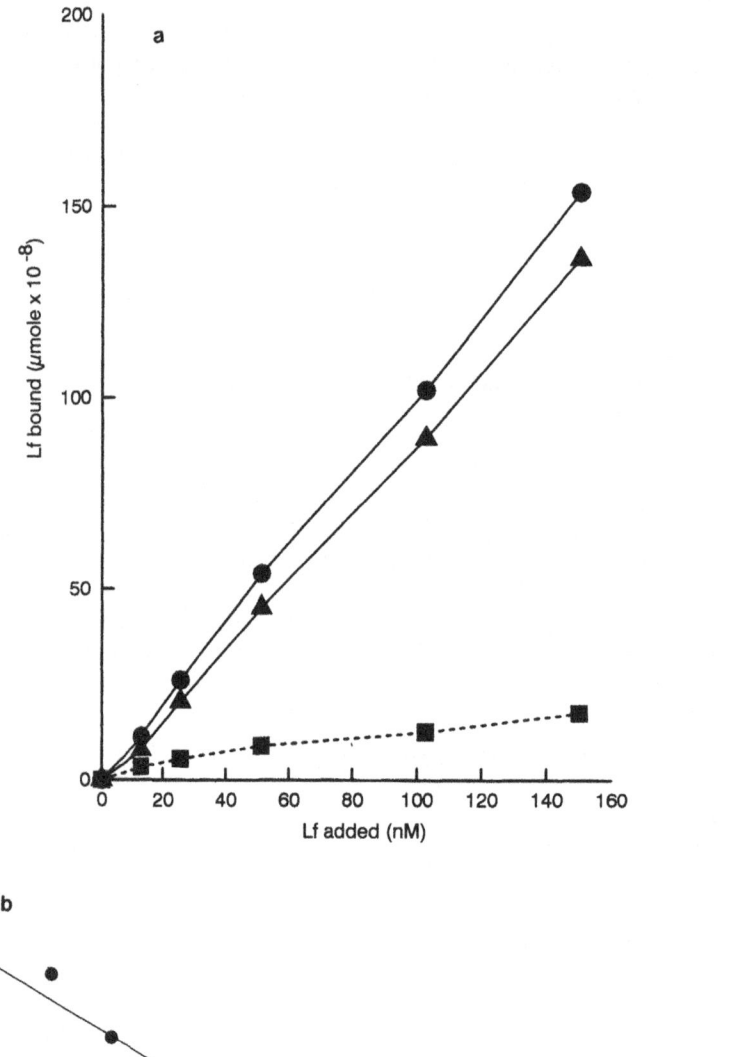

**Figure 2.** Binding of 125I-labelled lactoferrin (a, b) and transferrin (c, d) to U937 cells. Graphs (a) and (c) show total binding (—●—), non- specific binding (determined in the presence of a 200-fold excess of unlabelled protein) (—▲—), and specific binding (calculated by difference) (-■-). Graphs (b) and (d) show Scatchard analysis of the specific binding.

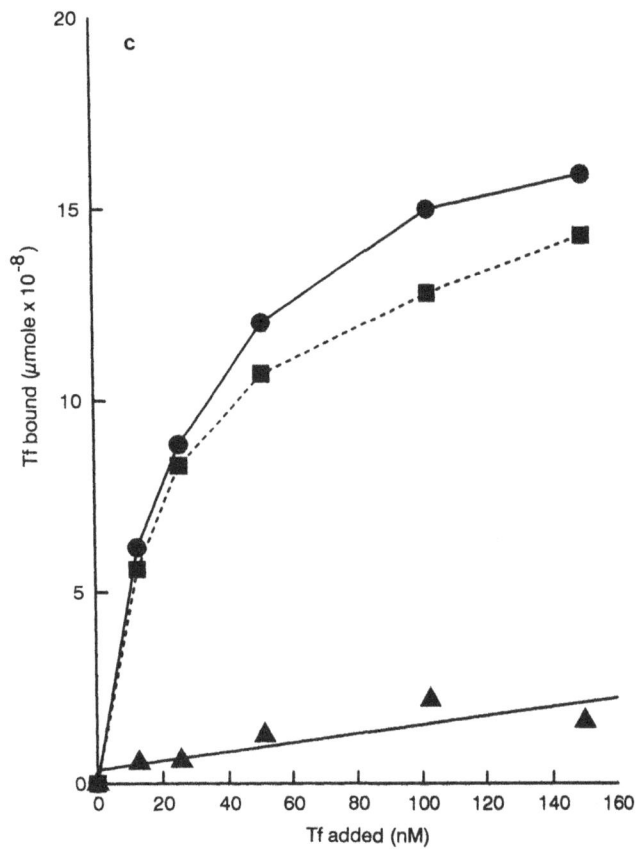

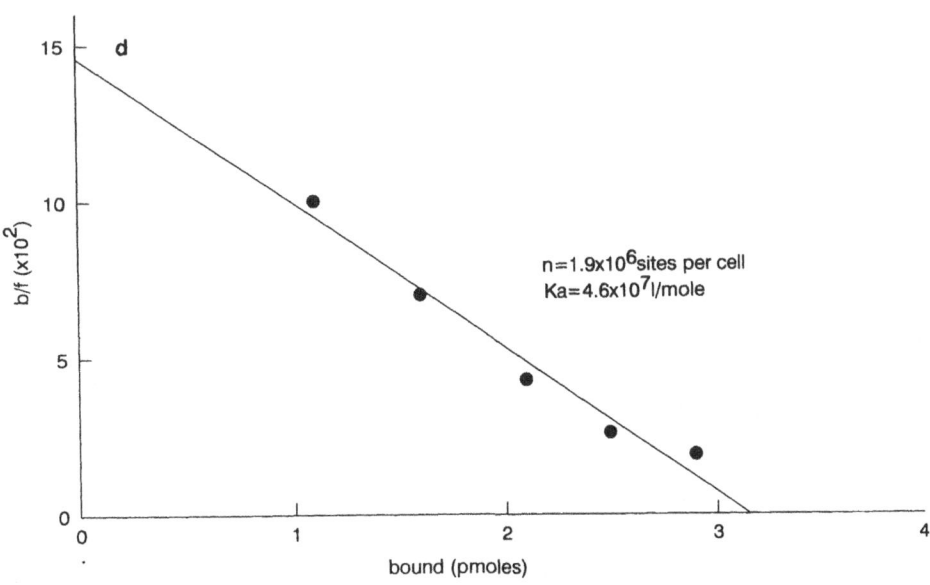

**Figure 2.** (cont'd)

lactoferrin released iron more rapidly than lactoferrin itself (Fig 3b), indicating that iron was being liberated from membrane-bound lactoferrin. It seems unlikely that desferrioxamine itself was responsible for this release, as similar results were obtained when apotransferrin was used instead of desferrioxamine (data not shown), and exchange of iron between transferrins and desferrioxamine is extremely slow under physiological conditions (Pippard, 1989).

### Effect of Lactoferrin on Uptake of Iron from other Sources

Both apo- and Fe-lactoferrin significantly reduced uptake of non- transferrin bound iron in the form of FeNTA, but neither had any effect on uptake of iron from transferrin (Fig 4).

### Transport of Iron, Lactoferrin and Transferrin across Caco-2 Monolayers

Transport of iron across Caco-2 monolayers was greater when iron was present as Fe-citrate or as Fe-lactoferrin than as Fe-transferrin (Fig 5)—indeed transport of transferrin-bound iron was almost undetectable. When 125I-labelled proteins were used to determine whether lactoferrin or transferrin themselves were transported, the results were similar for both proteins: about 4% of the radioactivity was transported across the monolayer in 24 h, but about 80% of this was in a low molecular weight form, suggesting that there was very little transport of intact lactoferrin or transferrin.

In contrast to the results for transport across the monolayer, there tended to be a greater amount of iron retained by the cells when the metal was initially bound to transferrin rather than lactoferrin, though the greatest retention occurred with Fe-citrate (Table 4).

## DISCUSSION

From the results reported here, it appears that the interaction of lactoferrin with the human promonocytic cell line U937 does not result in significant iron donation to the cells. Although the cells bound much more lactoferrin than transferrin, most of the binding was non-specific, and the amount of specific binding was similar for both proteins. The nature of the interaction of lactoferrin with U937 cells was not elucidated. Unlike the binding of human lactoferrin to human alveolar macrophages (Goavec et al, 1985), it does not seem to be mediated by the fucose moiety of the glycan chain, as fucose did not inhibit binding, nor did a purely electrostatic interaction appear to be involved, as treatment of the cells with heparitinase to remove acidic sugar residues did not affect binding. Furthermore, the binding site does appear to involve any of the surface molecules defined by various monocyte-reactive monoclonal antibodies. Nevertheless, the interaction is not totally non-specific as we have previously shown that binding of bovine lactoferrin to U937 cells can be inhibited by bovine but not human lactoferrin (Oria et al, 1992). Possibly a molecule similar to the putative lactoferrin receptor of activated lymphocytes (Mazurier et al, 1989, and Spik, this volume) is involved in the specific element of the binding of human lactoferrin.

Lactoferrin was not internalised to any measurable extent by U937 cells. Previous workers have suggested that uptake of iron from lactoferrin by macrophages (Van Snick et al, 1974) and monocytes (Birgens, 1991) may be significant, and contribute to the hypoferraemia of inflammation. However, the rate of uptake in those studies was in fact comparable to the apparent slow uptake reported here, and the present work suggests an alternative mechanism may have been responsible. It was found that some release of iron to the extracellular medium occurred from lactoferrin bound to U937 cell membranes. Since serum-containing culture medium must by definition also contain transferrin, it seems likely that if iron is released from lactoferrin bound at the cell surface it would be acquired by

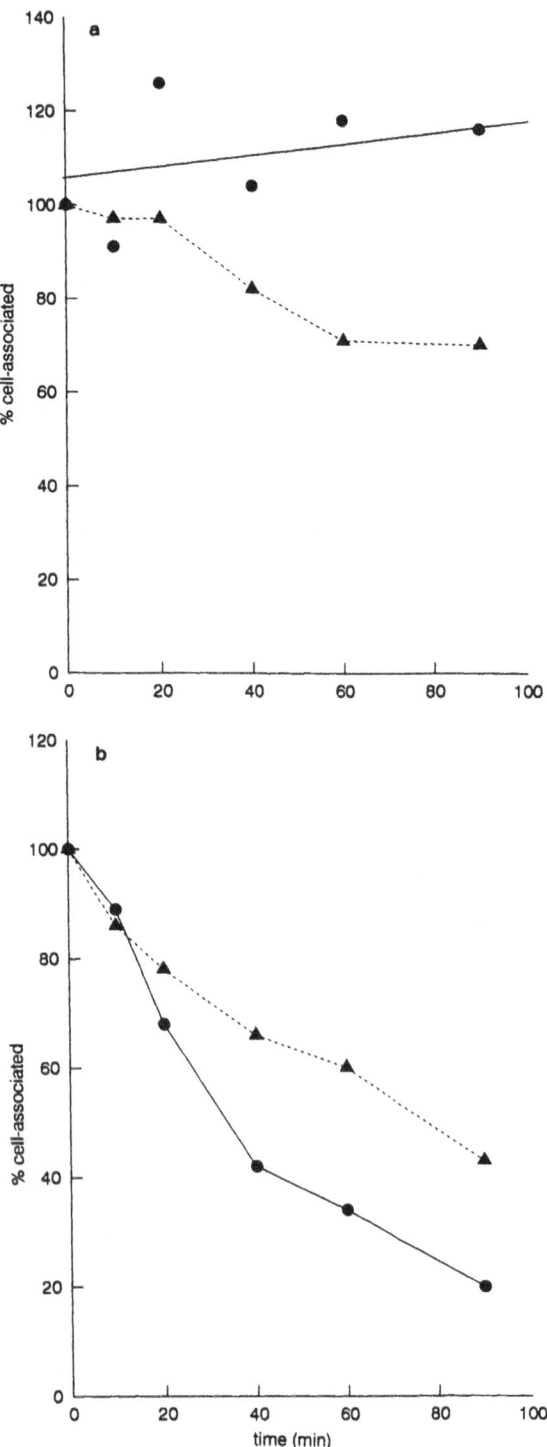

**Figure 3.** Dissociation at 4°C of iron (——●——) and the carrier protein (-▲-) from U937 cells previously incubated with 59Fe,125I-transferrin (a) or lactoferrin (b).

**Table 4.** Iron Uptake by Caco-2 Cell Monolayers.

| Source of Fe | % of added Fe incorporated by the cells |
|---|---|
| Fe-citrate | 8.7 + 1.4 |
| Fe-lactoferrin | 2.6 + 1.0 |
| Fe-transferrin | 3.6 ± 0·8 |

Figures are mean + s.d., n = 6.
Cells were grown in bicameral chambers as described in the text, and iron uptake from 50 μg/ml of 80%-saturated Fe-transferrin, Fe-lactoferrin, or an equivalent amount of Fe-citrate determined after 23 h of culture.

transferrin in the medium, which could donate the iron to the cells. Macrophages express transferrin receptors (Hamilton et al, 1984) while monocytes, although initially transferrin receptor-negative, express this receptor upon culture (Andreesen et al. 1984), so both could therefore readily acquire iron from transferrin. Only in the case of hepatocytes is there clear evidence for endocytosis of lactoferrin (McAbee and Esbensen 1991), and even this seems to be rather slow *in vivo* (Ziere et al, 1992).

Lactoferrin not only showed little ability to donate iron to U937 cells; it also inhibited uptake of iron from the chelate FeNTA, though not from transferrin. In this experiment the concentration of the iron added as FeNTA was 6 times higher than that capable of being bound by the lactoferrin present, and furthermore apolactoferrin and Fe-lactoferrin had similar effects. Thus inhibition was not simply due to binding of all the iron by lactoferrin in the medium. Rather, it appears that lactoferrin in some way regulates uptake of non-transferrin-bound iron at the cell membrane, which otherwise may enter in an uncontrolled manner and impair cell function (Djeha and Brock, 1992). The ability of membrane-bound lactoferrin to release iron to the extracellular medium could provide a mechanism whereby lactoferrin can regulate iron uptake even when potentially supersaturating concentrations are present, as may occur, for example, in inflammatory lesions where moribund cells have released their intracellular iron.

Release of iron from membrane-bound lactoferrin may account for the fact that transport of lactoferrin-bound iron by Caco-2 cell monolayers was comparable to transport of iron presented as Fe-citrate, and greater than that of iron from transferrin. If iron transport across the cells, rather than release from lactoferrin, is the rate-limiting step, then Fe-lactoferrin and

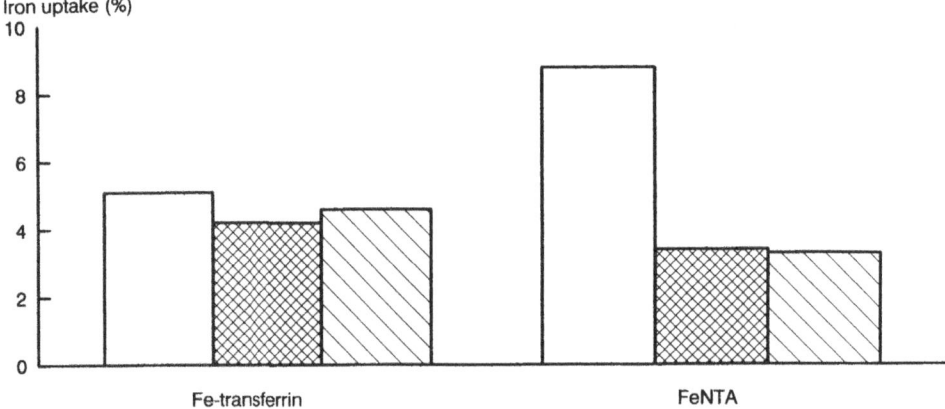

**Figure 4.** Iron uptake by U937 cells incubated with 50 μg/ml 59Fe-transferrin (1.5 μM Fe) or 8 μM 59Fe-nitrilo-triacetate (FeNTA) in the presence of 50 μg/ml apolactoferrin (☐) or Fe-lactoferrin (▨), or without lactoferrin (control) (▩).

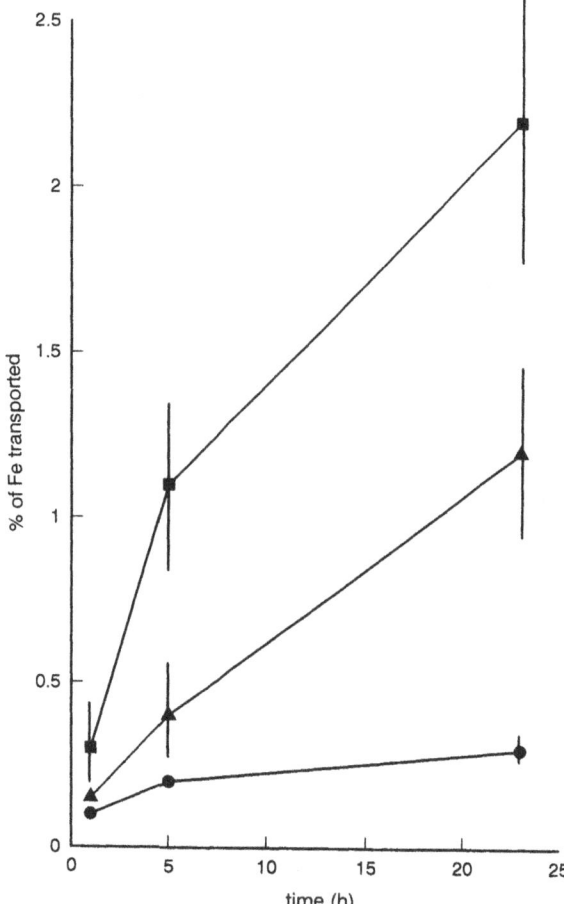

**Figure 5.** Transport of iron across differentiated Caco-2 cell monolayers in bicameral chambers. Iron was added to the upper chamber as Fe-transferrin (●), Fe-lactoferrin (▲), or Fe-citrate (■). The proteins were 80% saturated with iron.

Fe-citrate would be expected to behave similarly. In contrast, since Caco-2 cells express transferrin receptors (unpublished data), iron bound to transferrin may be taken up by endocytosis of the transferrin-iron complex and used for cell metabolism. The fact that rather more iron from transferrin than from lactoferrin was found within the cells, despite the minimal transport of transferrin-iron across the monolayer, tends to support this hypothesis, though proof would require examination of the intracellular distribution of iron. There was little evidence that either lactoferrin or transferrin themselves are transported, as most of the radiolabel found in the lower compartments was TCA-soluble, indicating protein degradation rather than transport of intact molecules.

These results are in line with *in vivo* studies showing that lactoferrin neither inhibits nor enhances absorption of inorganic iron (Fransson et al. 1983a,b; Davidson et al. 1990). It may nevertheless serve to bind iron in the lumen and prevent it enhancing growth of microbial pathogens, while at the same time not seriously impairing its bioavailability to the host.

Overall, therefore, these results suggest that lactoferrin-cell interactions may serve to regulate rather than facilitate iron uptake, particularly in conditions where binding by transferrin is ineffective. By doing so it could prevent uptake of unwanted iron by uncontrolled mechanisms, reduce potentially damaging free radical generation, and prevent uncontrolled growth of iron- requiring pathogenic microorganisms.

## REFERENCES

Alvarez-Hernandez X, Nichols GM, Glass J. (1991) Caco-2 cell line: s system for studying intestinal iron transport across epithelial cell monolayers. Biochim Biophys Acta 1070:205–208.

Amouric M, Marvaldi J, Pichon J, Bellot F, Figarella C. (1984) Effect of lactoferrin on the growth of a human adenocarcinoma cell line - comparison with transferrin. In Vitro 20:543–548.

Andreesen R, Osterholz J, Bodemann H, Bross KJ, Costabel U, Lohr GW. (1984) Expression of transferrin receptors and intracellular ferritin during terminal differentiation of human monocytes. Blut 49:195–202.

Azuma N, Mori H, Kaminogawa S, Yamauchi K. (1989) Stimulatory effect of lactoferrin on DNA synthesis in BALB/c 3T3 cells. Agric Biol Chem 53:31–35.

Birgens HS. (1991) The interaction of lactoferrin with human monocytes. Dan Med Bull 38:244–252.

Brock JH. (1980) Lactoferrin in human milk: its role in iron absorption and protection against enteric infection in the newborn infant. Arch Dis Child 55:417–421.

Dautry-Varsat A. (1986) Receptor-mediated endocytosis: the intracellular journey of transferrin and its receptor. Biochimie 68:375–381.

Davidson LA, Litov RE, Lonnerdal, B. (1990) Iron retention from lactoferrin-supplemented formulas in infant rhesus monkeys. Pediat Res 27:170–180.

Davidson LA, Lonnerdal, B. (1988) Specific binding of lactoferrin to brush border membrane: ontogeny and effect of glycan chain. Am J Physiol 254:G580–G585.

Djeha A, Brock JH. (1992) Effect of transferrin, lactoferrin and chelated iron on human T-lymphocytes. Br J Haematol 80:235–241.

Fransson GB, Keen, CL, Lonnerdal B. (1983a) Supplementation of milk with iron bound to lactoferrin using weanling mice. I. Effects on hematology and tissue iron. J Pediat Gastroenterol Nutr 2:693–700.

Fransson GB, Thoren-Tolling K, Jones B, Hambraeus L, Lonnerdal B. (1983b) Absorption of lactoferrin iron in suckling pigs. Nutr Res 3:373–384.

Goavec M, Mazurier J, Montreuil J, Spik G. (1985) Role des glycannes dans la fixation de la serotransferrine et de la lactotransferrine humaines sur les macrophages alveolaires humains. C R Seances Acad Sci (III) 301:689–694.

Halleux C, Schneider Y-J. (1991) Iron absorption by intestinal epithelial cells: 1. CaCo2 cells cultivated in serum-free medium, on polyethyleneterephthalate microporous membranes, as an in vitro model. In Vitro Cell Dev Biol 27A:293–302.

Hamilton TA, Weiel JA, Adams DO (1984) Expression of the transferrin receptor in murine peritoneal macrophages is modulated in different stages of activation. J Immunol 132:2285–2290.

Hidalgo IJ, Raub TJ, Borchardt RT (1989) Characterization of the human colon carcinoma cell line (Caco-2) as a model system for intestinal epithelial permeability. Gastroenterology 96:736–749.

Hu WL, Mazurier J, Montreuil J, Spik G. (1990) Isolation and partial characterisation of a lactotransferrin receptor from mouse intestinal brush border. Biochemistry 29:535–541.

Iturralde M, Vass JK, Oria R, Brock JH. (1992) Effect of iron and retinoic acid on the control of transferrin receptor and ferritin in the human promonocytic cell line U937. Biochim Biophys Acta 1133:241–246.

Johansson BG (1969) Isolation of crystalline lactoferrin from human milk. Acta Chem Scand 23:683–684.

Mazurier J, Legrand D, Hu WL, Montreuil J, Spik G. (1989) Expression of human lactotransferrin receptors in phytohemagglutinin-stimulated human peripheral blood lymphocytes. Isolation of the receptors by antiligand-affinity chromatography. Eur J Biochem 179:481–487.

Mazurier J, Spik G. (1980) Comparative study of the iron-binding properties of human transferrins. I. Complete and sequential iron saturation and desaturation of the lactotransferrin. Biochim Biophys Acta 629:399–408.

McAbee DD, Esbensen K. (1991) Binding and endocytosis of apo- and holo-lactoferrin by isolated rat hepatocytes. J Biol Chem 266:23624–23631.

Oria R, Alvarez-Hernandez X, Liceaga J, Brock JH (1988) Uptake and handling of iron from transferrin, lactoferrin and immune complexes by a macrophage cell line. Biochem J 252:221–225.

Oria R, Ismail M, Sanchez L, Calvo M, Brock JH. (1993) The effect of heat treatment and other milk proteins on the interaction of lactoferrin with monocytes. J Dairy Res 60:363–369.

Pippard MJ. (1989) Clinical use of iron chelation. In Iron in immunity, cancer and inflammation (ed. De Sousa M and Brock JH), John Wiley, Chichester, UK, pp361–392.

Reid PA, Watts C. (1990) Cycling of cell-surface MHC glycoproteins through primaquine-sensitive intracellular compartments. Nature 346:655–657.

Van Snick JL, Masson PL, Heremans, J. (1974) The involvement of lactoferrin in the hyposideremia of acute inflammation. J Exp Med 140:1068–1084.

Ziere GJ, Van Dijk MCM, Bijsterbosch MK, Van Berkel TJC. (1992) Lactoferrin uptake by the rat liver. Characterization of the recognition site and effect of selective modification of the arginine residues. J Biol Chem, 267:11229–11235.

# LACTOFERRIN RECEPTORS IN INTESTINAL BRUSH BORDER MEMBRANES

Bo Lönnerdal

Departments of Nutrition and Internal Medicine
University of California
Davis, California 95616

## SUMMARY

Lactoferrin from milk may have a physiological effect on the neonate by stimulating iron acquisition and/or mucosal growth. We have hypothesized that in order to achieve such an effect(s), lactoferrin will bind to a specific receptor located on the mucosal surface of the enterocyte. We have studied the presence of lactoferrin receptors in the brush border membrane from infant rhesus monkey intestine and from fetal and infant human intestine. The receptor exhibits saturation kinetics and the binding is specific for human and monkey lactoferrin—bovine lactoferrin or human transferrin do not bind to the receptor or compete with the binding of the primate lactoferrins. Enzymatic deglycosylation does not affect the binding of human lactoferrin to its receptor, suggesting that the glycan(s) is not needed for receptor recognition. Competitive binding experiments showed that holo-lactoferrin was more effective than less Fe-saturated forms of lactoferrin with regard to receptor binding. Mn-lactoferrin bound to the receptor, while we were unable to prepare Zn-lactoferrin in any physiological buffer. The human lactoferrin receptor was isolated and found to have a MW of ~110 kDa. This receptor has now been cloned and is being sequenced.

## INTRODUCTION

The high concentration of lactoferrin in human milk and its capacity to bind a major part of breast milk iron early led to the suggestion that it may have a physiological function in the newborn infant. This hypothesis was supported by experiments *in vitro*, which showed that Lf can resist the proteolytic action of several digestive enzymes, particularly when the pH for pepsin digestion is not very low (as in infants) and when it is fully iron-saturated (Brock et al., 1976; Brines & Brock, 1983). While the first suggested function for lactoferrin in the gastrointestinal tract was to provide a mechanism of bacteriostasis that can protect the infant against pathogens, several other functions have since been proposed (Sánchez et al., 1992). The role of a bacteriostatic agent was suggested to be exerted by lactoferrin, which is present in human milk with a very low degree of iron-saturation, binding the low quantities of available iron with a high affinity. Thus, iron-requiring bacteria would be deprived of this essential nutrient and consequently fail to proliferate. Most of this evidence was obtained *in vitro* and it is possible that this inhibitory effect on bacterial growth in fact may have been the result of a more direct bactericidal effect (see other chapters in this book). In any case, this potential effect is largely considered to be exerted within the gut lumen. Several of the other proposed physiological functions of lactoferrin include a direct interaction with the intestinal mucosa. This chapter will examine the support for the presence of such structures, or lactoferrin receptors, in the small intestine.

## PRESENCE OF INTESTINAL LACTOFERRIN RECEPTORS

The first report supporting the hypothesis of lactoferrin receptors in the small intestine was by Cox et al. (1979) who incubated biopsies from adult human small intestine with $^{59}$Fe- or $^{125}$I-labelled human lactoferrin. The biopsies showed increased binding of $^{59}$Fe-lactoferrin with time and that human serum transferrin did not bind. In contrast, iron was taken up by reticulocytes from transferrin, but not from lactoferrin. Experiments with $^{125}$I-lactoferrin suggested that lactoferrin did not enter the cell, while iron did. A lack of human intestinal material from infants and limitations on what is permissible in human infants directed the attention to different animal models. Although several species, including cows, goats and rats, have very little or no lactoferrin in their milk, some species like primates, pigs, rabbits and mice have significant concentrations of lactoferrin in their milk (Masson & Heremans, 1971).

Mazurier et al. (1985) incubated brush border membranes from rabbit small intestine and used ligand blotting utilizing $^{125}$I-lactoferrin to visualize the presence of structures on the brush border membrane that bind lactoferrin. A molecular weight of ~100 kD was estimated for the lactoferrin receptor and kinetic studies showed specific, saturable and calcium-dependent binding of lactoferrin. No specific binding of rabbit serum transferrin was observed. However, the lactoferrin used was of human origin and it is uncertain to what degree human and rabbit lactoferrin are similar. It should be cautioned that both human and bovine lactoferrin can bind specifically to rat brush border membrane vesicles although rat milk contains no lactoferrin (Kawakami et al., 1990), making experiments with heterologous lactoferrin difficult to evaluate. For example, it was found that brush border membranes from rats specifically bound bovine lactoferrin to its transferrin receptors (Kawakami et al., 1990), thereby possibly explaining a positive effect of bovine lactoferrin on iron absorption in rats (Kawakami et al., 1988). We have used the newborn rhesus monkey as a model for human infants. We first purified rhesus monkey milk lactoferrin and found that its biochemical properties are very similar to those of human lactoferrin; in fact, polyclonal antibodies against human lactoferrin cross-react with rhesus lactoferrin, but not with bovine lactoferrin (Davidson & Lönnerdal, 1985). In a kinetic study, using labelled rhesus lactoferrin and rhesus infant brush border membrane vesicles and a rapid filtration technique, we were able to show that binding was time-dependent and saturable (Davidson & Lönnerdal, 1988). Competitive binding experiments with excess unlabeled lactoferrin showed that the binding was specific; monkey and human lactoferrin effectively inhibited binding of rhesus lactoferrin at 50-fold excess, while a similar excess of bovine lactoferrin or human transferrin had no effect on binding. Together, these results provided support for an intestinal lactoferrin receptor in infant rhesus monkeys. Similar kinetic studies in brush border membranes from fetal, adolescent, and adult monkey intestine showed the presence of the lactoferrin receptor throughout the life cycle. Although the binding affinity for lactoferrin ($K_d$) was similar at all stages of development studied, the largest number of binding sites per mg of membrane protein was found during infancy, suggesting a higher need for these receptors during this age.

In a subsequent study, we explored the possibility of different variants of lactoferrin binding to the intestinal lactoferrin receptor. Although it is known that human lactoferrin can escape proteolytic digestion in newborn infants and is found in immunologically intact form in the stool of breast-fed infants (Davidson & Lönnerdal, 1987), it is known that some digestion of lactoferrin will occur (Goldman et al., 1990). Lactoferrin is known to be relatively resistant against attack by proteolytic enzymes (Brock et al., 1976); however, the two similar lobes of lactoferrin are connected via an extended helical loop which can be cleaved by proteases (Hutchens et al., 1991). We prepared and isolated a larger fragment of lactoferrin and explored the binding of this fragment to rhesus brush border membranes (Davidson & Lönnerdal, 1989). It was found that the fragment bound to the lactoferrin receptor in a saturable manner and with an affinity similar to that of intact lactoferrin. Competitive binding experiments, however, showed that intact lactoferrin more effectively competed for binding than the lactoferrin fragment, suggesting some preference for the intact form. The effect of degree of iron saturation of lactoferrin on receptor binding was also studied. Since lactoferrin in human milk is present in a molar excess to iron, only 3–5% of human milk lactoferrin is iron-saturated (Fransson & Lönnerdal, 1980). It was found that apo-lactoferrin, partially iron-saturated lactoferrin and holo-lactoferrin all bound to the lactoferrin receptor, but that half- or fully saturated lactoferrin were more effective to displace labelled lactoferrin in competitive binding experiments, suggesting a somewhat higher affinity for the receptor. Although lactoferrin in human milk primarily binds iron, a small fraction of it binds manganese (Lönnerdal et al., 1985). We therefore tested whether manganese-lactoferrin bound to the receptor. As some reports have indicated lactoferrin also can bind zinc, we tried to label lactoferrin with zinc under physiological conditions; however, these attempts were unsuccessful. This is possibly explained by the observation that zinc only binds to lactoferrin when the ionic strength is extremely low and that the binding in this case is very weak (Blakeborough and Salter, 1987). Thus, it is highly unlikely that any zinc is bound to lactoferrin under physiological conditions; this is corroborated by the finding that lactoferrin isolated from human milk under by mild conditions does not contain any zinc, only iron and manganese (Lönnerdal et al., 1985). Manganese-lactoferrin was found to bind to the intestinal lactoferrin receptor with an affinity similar to that of iron-lactoferrin.

Intestinal lactoferrin receptors have also been described in the mouse (Hu et al., 1988, 1990), which may explain earlier findings of high iron absorption from lactoferrin in mice (Fransson et al., 1983). This receptor appears to be less specific than the other receptors reported; it bound both bovine and human lactoferrin, although the affinity to the receptor was lower than that for mouse lactoferrin. Competitive binding experiments with the highly cationic protein lysozyme showed no effect, suggesting that receptor-binding is not due to non-specific ionic binding. Excess of fucosylated bovine serum albumin (BSA) did not inhibit binding of lactoferrin, indicating that the lactoferrin does not bind to the membranes via a fucose receptor. Optimal binding of mouse lactoferrin was found at pH 5.5 and binding appeared to be calcium-dependent. Both apo- and holo-lactoferrin bound to the receptor to similar extent.

We have recently isolated the lactoferrin receptor from human fetal and infant small intestine (Kawakami & Lönnerdal, 1991). The $K_d$ was found to be ~1 µM and bovine lactoferrin or human transferrin did not compete with human lactoferrin for binding. Recent kinetic studies in piglets have also demonstrated the presence of lactoferrin in this species (Gislason et al., 1993). The receptor was found to be specific; human lactoferrin, bovine lactoferrin and pig transferrin did not bind to the receptor. This was somewhat surprising as pig milk is known to contain transferrin as well as lactoferrin. The binding constant for pig lactoferrin was ~3 µM and it was found to be present in all segments of the small intestine. The number of binding sites and the affinity to lactoferrin were similar in all parts of the small intestine. During the neonatal period (0–21 days), no effect of age on receptor affinity or number of binding sites was found. However, intestines from older pigs were not studied and it is possible that the number of lactoferrin receptors is high during infancy, but that there is no pronounced change within this period.

## BIOCHEMICAL PROPERTIES OF THE INTESTINAL LACTOFERRIN
## RECEPTOR

It has been difficult to isolate significant quantities of the brush border membrane lactoferrin receptor which would allow detailed biochemical and structural analyses. However, some progress has been made by the use of affinity chromatography, utilizing immobilized human lactoferrin (Hu et al., 1990; Kawakami & Lönnerdal, 1991). By this method, detergent-solubilized brush border membrane proteins are passed through the affinity column and the receptor is eluted at low pH. However, it should be noted that when the mouse lactoferrin receptor was purified, intestines from 1, 000 mice yielded 110 μg of receptor (Hu et al., 1990)! The mouse lactoferrin receptor was described to have a molecular weight of about 130 kD and to consist of a single polypeptide chain. The isoelectric point was found to be at pH 5.8. Ligand-trapping experiments indicated that each receptor binds one molecule of lactoferrin. These data show that the lactoferrin receptor is structurally quite different from the transferrin receptor, which has a molecular weight of 180 kD and consists of two subunits of equal size (90 kD). The receptor was found to be a glycoprotein and that the glycans bind to concanavalin A and phytohemagglutinin A. Digestion by N-glucanase and N-acetyl-β-D-glucosaminidase B caused a reduction of the molecular weight by 25 kD, suggesting that this is the size of the glycan(s). Since endo-N-acetyl-β-D-glucosaminidase H was ineffective, it appears likely that the lactoferrin is glycosylated mainly by bi- and triantennary glycans. With an average molecular weight of 2.2 kD for a biantennary glycan, these investigators estimated the number of glycans per receptor to about 12.

The human lactoferrin receptor was found to have a molecular weight of 110 kD (Kawakami & Lönnerdal, 1991) which is somewhat similar to that reported for the mouse lactoferrin receptor. The human form, however, was found to consist of subunits with a molecular weight of 37 kD. The pH optimum for binding of human lactoferrin to its receptor was found to be 6.5–7.5, which would be consistent with the pH of the small intestine in infants. "Half-lactoferrin", prepared by digestion with pepsin at pH 3.0, bound to the receptor with an affinity similar to that of intact lactoferrin. Digestion of lactoferrin with peptide-N-glycanase F resulted in a deglycosylated form of human lactoferrin; binding affinity to the receptor was similar to that of native lactoferrin. These results indicate that the glycan(s) consists of N-linked oligosaccharides. Since the molecular weight of the lactoferrin receptor subunit was reduced by 4 kD in apparent molecular mass, it appears likely that each subunit contains two glycans, and that the intact receptor may contain six glycans. This is similar to what has been found for the human transferrin receptor (Huebers & Finch, 1987), but lower than the number estimated for the mouse lactoferrin receptor (Hu et al., 1990). Ligand-blotting of gels after SDS electrophoresis with alkaline phosphatase labelled lactoferrin showed that the solubilized receptor retains it binding capacity for lactoferrin. The amino acid composition of the human lactoferrin receptor was determined and indicated a subunit molecular weight of about 33–35 kD, which is consistent with the results from SDS gel electrophoresis.

We have subsequently used a human infant intestine cDNA library to clone the lactoferrin receptor cDNA (unpublished data). The nucleotide sequence has been determined and the amino acid sequence deduced from these data. Expression of the recombinant receptor will allow further studies on its structure and, also, a better understanding of its interaction with the lactoferrin molecule.

## REFERENCES

Blakeborough, P. and Salter, D.N. (1987). The intestinal transport of zinc studied using brush border membrane vesicles from the piglet. Br. J. Nutr. 57, 45–55.

Brines, R.D. and Brock, J.H. (1983). The effect of trypsin and chymotrypsin on the in vitro antimicrobial and iron-binding properties of lactoferrin in human milk and bovine colostrum: unusual resistance of human lactoferrin to proteolytic digestion. Biochim. Biophys. Acta 759, 229–235.

Brock, J.H., Arzabe, F., Lampreave, F. and Pineira, A. (1976). The effect of trypsin on bovine transferrin and lactoferrin. Biochim. Biophys. Acta 446, 214–225.

Cox, T.M., Mazurier, J., Spik, G., Montreuil, J. and Peters, T.J. (1979). Iron binding proteins and influx of iron across the duodenal brush-border. Evidence for specific lactotransferrin receptors in the human intestine. Biochim. Biophys. Acta 558, 129–141.

Davidson, L.A. and Lönnerdal, B. (1986). Isolation and characterization of Rhesus monkey milk lactoferrin. Pediatr. Res. 20, 197–201.

Davidson, L.A. and Lönnerdal, B. (1987). Lactoferrin and secretory IgA in the feces of exclusively breast-fed infants. Am. J. Clin. Nutr. 41, 852–861.

Davidson, L.A. and Lönnerdal, B. (1988). Specific binding of lactoferrin to brush-border membrane: ontogeny and effect of glycan chain. Am. J. Physiol. 254, G580–G585.

Davidson, L.A. and Lönnerdal, B. (1989). Fe-saturation and proteolysis of human lactoferrin: effect on brush-border receptor-mediated uptake of Fe and Mn. Am. J. Physiol. 257, G930–G934.

Fransson, G.B. and Lönnerdal, B. (1980). Iron in human milk. J. Pediatr. 2, 693–701.

Fransson, G.B., Keen, C.L. and Lönnerdal, B. (1983). Supplementation of milk with iron bound to lactoferrin using weanling mice. I. Effects on hematology and tissue iron. J. Pediatr. Gastroenterol. Nutr. 2, 693–700.

Gislason, J., Iyer, S., Hutchens, T.W. and Lönnerdal, B. (1993). Lactoferrin receptors in piglet intestine: lactoferrin binding properties, ontogeny and regional distribution in the gastrointestinal tract. J. Nutr. Biochem. 4, 528–533.

Goldman, A.S., Garza, C., Schanler, R.J. and Goldblum, R.M. (1990). Molecular forms of lactoferrin in the stool and urine from infants fed human milk. Pediatr. Res. 27, 252–255.

Hu, W.L., Mazurier, J., Montreuil, J. and Spik, G. (1990). Isolation and partial characterization of a lactotransferrin receptor from mouse intestinal brush border. Biochemistry 29, 535–541.

Hu, W.L., Mazurier, J., Sawatzki, G., Montreuil, J. and Spik, G. (1988). Lactotransferrin receptor of mouse small intestinal brush border. Biochem. J. 248, 435–441.

Huebers, H.A. and Finch, C.A. (1987). The physiology of transferrin and transferrin receptors. Physiol. Rev. 67, 520–582.

Hutchens, T.W., Henry, J.F. and Yip, T.-T. (1991). Structurally intact (78 kDa) forms of maternal lactoferrin purified from urine of preterm infants fed human milk. Identification of a trypsin-like proteolytic cleavage event in vivo that does not result in fragment dissociation. Proc. Natl. Acad. Sci. USA 88, 2994–2998.

Kawakami, H. and Lönnerdal, B. (1991). Isolation and function of a receptor for human lactoferrin in human fetal intestinal brush-border membranes. Am. J. Physiol. 261, G841–G846.

Kawakami, H., Dosako, S. and Lönnerdal, B. (1990). Iron uptake from transferrin and lactoferrin by rat intestinal brush-border membrane vesicles. Am. J. Physiol. 258, G535–G541.

Kawakami, H., Hiratsuka, M. and Dosako, S. (1988). Effects of iron-saturated lactoferrin on iron absorption. Agric. Biol. Chem. 52, 903–908.

Lönnerdal, B., Keen, C.L. and Hurley, L.S. (1985). Manganese binding proteins in human and cow's milk. Am. J. Clin. Nutr. 41, 550–559.

Masson, P.L. and Heremans, J.F. (1971). Lactoferrin in milk from different species. Comp. Biochem. Physiol. 39B, 119–129.

Mazurier, J., Montreuil, J. and Spik, G. (1985). Visualization of lactotransferrin brush-border receptors by ligand blotting. Biochim. Biophys. Acta 821, 453–460.

Sánchez, L., Calvo, M. and Brock, J.H. (1992) Biological role of lactoferrin. Am. J. Dis. Child. 67, 657–661.

# MATERNAL LACTOFERRIN IN THE URINE OF PRETERM INFANTS[*]

## Evidence for Retention of Structure and Function

Roger D. Knapp[1] and T. William Hutchens[2][†]

[1]USDA/ARS Children's Nutrition Research Center and
Department of Medicine
Baylor College of Medicine and Texas Children's Hospital
Houston, Texas 77030

[2]Department of Food Science and Technology and Department of Pediatrics
University of California, Davis, California, 95616

## ABSTRACT

Intact (i.e., 78-kDa) lactoferrin has been purified from the urine of preterm infants fed human milk. The maternal origin of this lactoferrin, and the integrity of its primary structure have been documented. Computer analyses of the circular dichroism spectra revealed a composite secondary structure for the urinary lactoferrin that was indistinguishable from that of purified human milk lactoferrin and similar to that observed in the crystal structure. Intact function was suggested by iron binding; an approximate 2:1 molar ratio of iron to lactoferrin was confirmed. Thus, maternal lactoferrin is absorbed intact by the preterm infant and appears to remain structurally and functionally intact within the circulatory system and during urinary excretion. It is possible, therefore, that maternal lactoferrin has an immunoregulatory influence in newborn infants fed human milk.

## INTRODUCTION

Lactoferrin is a bilobate, 78-kDa, metal-binding, secretory glycoprotein, perhaps best characterized for its iron-binding properties (1). It is the major protein in human colostral whey, where it exists predominantly (80–90%) in the apo (metal-free) form (2). The function and biochemical fate of lactoferrin in preterm and term infants fed human milk (3, 4) are not well known. We have recently shown that intact (i.e., 78-kDa) lactoferrin can be absorbed by the

---

[*] Address correspondence to: T. William Hutchens, Department of Food Science and Technology, University of California, Davis, California, 95616.

[†] Portions of this work were performed while TWH was in the Department of Pediatrics, Baylor College of Medicine and Texas Children's Hospital, Houston, Texas 77030.

gastrointestinal tract of preterm infants fed human milk and isolated from their urine (4, 5); incorporation of stable isotopes into human milk lactoferrin enabled us to demonstrate that the 78-kDa urinary lactoferrin had an exclusively maternal origin (5). In addition to the presence of intact urinary lactoferrin (minus the first two amino acid residues), however, we discovered a nicked (at Lys residue 283), but stable (78-kDa), form of urinary lactoferrin also of maternal origin (6). To assess the biological significance and potential regulatory implications of these findings, it is necessary to know whether the absorbed maternal lactoferrin retains any activity (i.e., structure and function) after absorption and processing *in vivo*. Because the high-resolution (2.8 Å) X-ray structure of human lactoferrin in both its iron-saturated or "closed" form and its metal-free or apo form (at least one conformer) is now known (7,8), we have investigated whether either (or both) of the two isolated forms of urinary lactoferrin retains any of its structure. Our investigations were based upon composite secondary structure assignments from circular dichroism (CD) spectra. These data, and our discovery of bound iron in the isolated urinary lactoferrin, suggest that the two 78-kDa forms of lactoferrin recovered from the urine of human milk-fed preterm infants are 1) functional, 2) structurally indistinguishable from one another, and 3) similar in overall secondary structure to that observed in the crystal structure of human milk lactoferrin.

## RESULTS

Urine was collected from very low birth weight, preterm infants who had been fed human milk proteins intrinsically labeled with stable isotopes of leucine and lysine (5). Urinary lactoferrin was purified to homogeneity (recovery > 85%) by affinity chromatography on immobilized single-stranded DNA and characterized as described previously (4–6, 9) using purified human milk lactoferrin as a reference standard (10). The isotopic enrichment and ratio of labeled amino acids in the isolated urinary lactoferrin was evaluated by GC/MS to verify maternal origin (5).

The iron-saturated status of the purified urinary lactoferrin was demonstrated by a visible absorbance spectra with a strong absorption maximum at 465 nm. The 465/280 absorption ratio for urinary lactoferrin was routinely 0.046–0.047 (0.044 ± 0.005; n = 13), identical to that obtained for purified human milk lactoferrin after saturation with iron in vitro. The purified, undenatured urinary lactoferrin was evaluated by X-ray fluorescence energy dispersion analy-

**Table 1.** Iron Content of Purified Urinary Lactoferrin Determined by X-ray Fluorescence Energy Dispersion

| | Source of purified lactoferrin | |
| --- | --- | --- |
| | **Pooled infant urine** | **Urine obtained from a single infant** |
| Concentration of purified urinary lactoferrin | 0.93 mg/mL (12 µM) | 5.4 mg/mL (69 µM) |
| 465/280 absorbance ratio | 0.042 | 0.047 |
| Protein-bound iron | 1.42 µg/mL (25.5 µM) | 8.74 µg/mL (156 µM) |
| Iron/lactoferrin ratio | 2.1/1 | 2.3/1 |

Urinary lactoferrin concentrations were determined by quantitative amino acid composition analyses and by the Pierce BCA micro assay procedure.
Protein-bound iron determined by X-ray fluorescence spectroscopy on a Kevex 0600 XRF spectrometer.

sis and found to contain two bound iron atoms per lactoferrin molecule (Table 1). Levels of Zn(II) or Cu(II) were insignificant (data not shown).

Because the purified, undenatured urinary lactoferrin contains a mixture of both an intact and a nicked (at residue 283) but stable (78-kDa) form of lactoferrin (6), the exact ratio of intact to nicked lactoferrin was determined before evaluation of secondary structure. Separation and quantitation of the intact and nicked urinary lactoferrin forms was accomplished by high-performance reverse-phase chromatography. The two resolved lactoferrin fragments, normally associated in a 78-kDa form *in vivo* and *in vitro* under nondenaturing conditions (6), appeared in constant relative ratios. Nearly equal amounts of intact urinary lactoferrin and nicked urinary lactoferrin were present in the preparations analyzed for secondary structure.

The CD spectra of the urinary lactoferrin and the human milk lactoferrin (that had not been fed to infants) were consistently indistinguishable. The variable selection method outlined by Manavalan and Johnson (11) and the simple matrix multiplication method described earlier by Compton and Johnson (12) were used to analyze the CD spectra (averaged from 9 separate experiments) for helix, parallel and anti-parallel b-structures, b-turns, and other elements of secondary structure. By either method of calculation, the sum of the five components of estimated secondary structure was between 0.96 and 1.05, within the range required for the successful interpretation of deconvoluted spectra (Table 2). To estimate the relative accuracy of these predictions, the calculated secondary structure predictions were compared to the secondary structure of iron-saturated lactoferrin derived from the X-ray crystal structure by observation (7); these values generally agreed with those calculated according to the algorithm DEFINE_Structure (13). The CD spectra of hen egg white lysozyme and horse heart myoglobin, recorded and evaluated (Table 2) to monitor the accuracy of our predictions, were within 2% of the values reported previously (12).

Several separate, but indirect, indicators of intact urinary lactoferrin surface structure have been evaluated (6); the findings support the idea that the maternal lactoferrin structure is

**Table 2.** Predicted Secondary Structure Calculated

| | | b-Structure | | | | |
|---|---|---|---|---|---|---|
| | Helix | Anti-parallel | Parallel | b-Turn | Other | Sum |
| CD: milk LF | | | | | | |
| Control 1 | 0.407 | 0.149 | 0.030 | 0.136 | 0.282 | 1.003 |
| Control 2 | 0.414 | 0.157 | 0.023 | 0.129 | 0.279 | 1.003 |
| CD: urine LF | 0.397 | 0.149 | 0.042 | 0.132 | 0.284 | 1.004 |
| Lysozyme (control) | | | | | | |
| present study | 0.347 | 0.195 | 0.028 | 0.219 | 0.261 | 1.050 |
| reported | 0.340 | 0.274 | 0.013 | 0.258 | 0.280 | 1.165 |
| | 0.41 | 0.24 | | | | |
| Myoglobin (control) | | | | | | |
| present study | 0.759 | −0.041 | 0.017 | 0.153 | 0.153 | 1.04 |
| reported | 0.76 | 0.015 | 0 | 0.157 | 0.164 | 1.08 |

Secondary structure calculated from averaged CD spectra (9 experiments) according to the variable selection algorithm of Manavalan and Johnson (12) with a 22-protein reference data set. Selection criteria included 1) no fraction < −0.05, 2) a sum between 0.9 and 1.1, and 3) an RMS of errors of less than or equal to 0.2 from circular dichroism spectra of human milk lactoferrin and maternal lactoferrin isolated from the urine of human milk-fed preterm infants.
LF structure observed from the crystal structure as reported by Anderson et al. (7).
Parallel and anti-parallel b-strand structures not delineated by Anderson et al. (7)

preserved during passage through the infant. The secondary structure estimates by CD, however, provide the first direct evidence of preserved structure. As we had expected from our evaluation of the X-ray crystal structure, the CD spectra revealed no change in any component of secondary structure from the loss of residues 1 and 2 from the amino terminus (both the intact and nicked forms). More surprisingly, although 50% of the urinary lactoferrin was shown to be nicked at residue 283 in the N2 domain, no apparent change was detected by CD in any component of secondary structure. Thus, the secondary structure of human milk lactoferrin was apparently not altered during its ingestion, absorption, circulation, and urinary excretion.

The urinary lactoferrin (both 78-kDa forms) was saturated with iron. This finding, together with the preserved far UV CD spectra, suggests conserved structure and function. The overall effects of absorbed lactoferrin on iron and trace metal metabolism of preterm infants cannot be extrapolated from this investigation. Lactoferrin in human milk exists primarily (80-90%) in the apo or metal-free form. We do not know whether the maternal lactoferrin present in the urine of these infants (up to 250 μg/ml) became saturated with iron in the gastrointestinal system, the circulatory system, or during filtration and urinary excretion.

If preterm infants can absorb intact lactoferrin from ingested human milk (5,6 ), it seems likely that lactoferrin is neither simply a source of dietary amino acids, nor merely a transient immunoprotective reagent localized to the gastrointestinal tract of these infants. The unexpectedly minimal degree of proteolytic damage sustained by the maternal lactoferrin in transit from the gut to urine (5, 6) illustrates the need for a closer examination of possible biological effects.

Fecal losses reportedly account for less than 5%-6% of the lactoferrin ingested by human milk-fed infants, either term or preterm (3). Given the unusual resistance of lactoferrin to proteolytic digestion, both *in vitro* and *in vivo*, it should not be assumed that the ingested lactoferrin thus far unaccounted for is simply degraded. Absorbed lactoferrin (with preserved structure) may influence a number of cellular events, including regulation of cell growth and differentiation. Evidence for the existence of lactoferrin receptors on intestinal brush border membranes has been reported (14). The growth of intestinal crypt cells, at least in vitro, is reportedly stimulated by incubation with lactoferrin (15). Lymphocytes with specific, high affinity lactoferrin receptors have also been reported (16). Lactoferrin has also been reported to alter expression of Fc receptors on developing lymphocytes (17). Numerous other investigations suggest both direct and indirect mechanisms for lactoferrin regulation of cells involved in inflammatory cell responses (e.g., 18). Further investigations are needed to discover how this molecule affects the growth and development of preterm infants who are fed human milk.

## REFERENCES AND NOTES

1. J. H. Brock, in Metalloproteins. Part 2: Metal Proteins with Non-Redox Roles. Topics in Molecular and Structural Biology, P. Harrison, Ed. (Verlag Chemie, Basel, 1985), 7, 183–262.

2. L. Woodhouse and B. Lonnerdal, Nutr. Res. 8, 853 (1988).

3. R. Schanler, R. Goldblum, C. Garza, A.S. Goldman, Pediatr. Res. 20, 711 (1986); L.A. Davidson and B. Lonnerdal, Acta Paediatr. Scand. 76, 733 (1987); A. Prentice, et al., Acta Paediatr. Scand. 76, 592 (1987).

4. T.W. Hutchens, J.F. Henry, T.-T. Yip, Clin. Chem. 35, 1928 (1989).

5. T.W. Hutchens, et al., Pediatr. Res. 29, 243 (1991).

6. T.W. Hutchens, J.F. Henry, T.-T. Yip, Proc. Natl. Acad. Sci. U.S.A. 88, 2994 (1991).

7. B.F. Anderson, H.M. Baker, D.W. Rice, E.N. Baker, J. Mol. Biol. 209, 711 (1989).

8. B.F. Anderson, H.M. Baker, G.E. Norris, S.V. Rumball, E.N. Baker, Nature 344, 784 (1990).

9. T.W. Hutchens, T.-T. Yip, J. Chromatogr. 536, 1 (1990).

10. T.W. Hutchens, J.S. Magnuson, T.-T. Yip, Pediatr. Res. 26, 618 (1989).

11. L.A. Compton and W. C. Johnson, Jr., Anal. Biochem. 155, 155 (1986).

12. P. Manavalan and W. C. Johnson, Jr., Anal. Biochem. 167, 76 (1987).

13. F.M. Richards and C.E. Kundrot, Proteins 3, 71 (1988).

14. T.M. Cox, J. Mazurier, G. Spik, J. Montreuil, T.J. Peters, Biochim. Biophys. Acta 588, 120 (1979); H.S. Birgens, N.E. Hansen, H. Karle, L.O. Kristensen, Br. J. Haematol. 54, 383 (1983); J. Mazurier, J. Montreuil, G. Spik, Biochim. Biophys. Acta 821, 453 (1985); L.A. Davidson and B. Lonnerdal, Am. J. Physiol. 254, G580 (1988); L.A. Davidson and B. Lonnerdal, Am. J. Physiol. 257, G930 (1989); D. Legrand, M. Metz-Boutigue, J. Jolles, P. Jolles, J. Montreuil, G. Spik, Biochim. Biophys. Acta 787, 90 (1984); D. Legrand, et al., Biochem. J. 236, 839 (1986).

15. B.L. Nichols, K.S. McKee, J.F. Henry, M. Putman, Pediatr. Res. 21, 563 (1987).

16. E. Rochard, D. Legrand, J. Mazurier, J. Montreuil, G. Spik, FEBS Lett. 255, 201 (1989).

17. L.V. Beletskaya and E.V. Gnezditskaya, Thymus 7, 377 (1985); E.V. Gnezditskaya, V.P. Bukhova, N.A. Zakharova, L.A. Malkina, Byull. Eksp. Biol. Med. 103, 447 (1987).

18. R.M. Bennett and T. Kokocinski, Br. J. Heamatol. 39, 509 (1978); G.C. Bagby, Jr., V.D. Rigas, R.M. Bennett, A.A. Vandenbark, J. Clin. Invest. 68, 56 (1981).

19. This project has been funded, in part, with federal funds from the U.S. Department of Agriculture, Agricultural Research Service under Cooperative Agreement number 58-6250-1-003 and a capital equipment grant from the National Science Foundation (PCM-8413751). The contents of this publication do not necessarily reflect the views or policies of the U.S. Department of Agriculture, nor does mention of trade names, commercial products, or organizations imply endorsement by the U.S. Government. The refined coordinates for human lactoferrin were kindly provided by Dr. Edward N. Baker and his colleagues at Massey University, New Zealand. We thank Dr. Richard J. Schanler and Pam Burns for their help with the collection of urine from preterm infants fed human milk. We thank Dr. Tai-Tung Yip and Mr. Joseph F. Henry for their help with the isolation and characterization of lactoferrin from milk and from infant urine samples.

# LACTOFERRIN GENE PROMOTER IN HUMAN AND MOUSE

## Analogous and Dissimilar Characteristics

Christina T. Teng

Laboratory of Reproductive and Developmental Toxicology
National Institute of Environmental Health Sciences, NIH
Research Triangle Park, North Carolina 27709

## ABSTRACT

Lactoferrin promoter and the 5′-flanking region of both human and mouse were isolated from a genomic library constructed with lambda phage. A 2.0 kbp Sac I fragment of the human clone (HLF031a.30) and a 3.0 kbp Eco R I/Hinc II fragment of the mouse clone (mL14p9E) containing the lactoferrin promoter, 5′-flanking region, first exon and partial intron were sequenced completely. There were many sequence homologies between human and mouse at the promoter/enhancer (1 to −363) region, yet substantial divergence was observed beyond this region. To determine the promoter activity, 5′-deletion mutants of the mouse lactoferrin gene were linked to a CAT-reporter plasmid and transfected into the human endometrium carcinoma cell line, RL95-2. We identified a number of positive and negative regulatory sequences as well as the estrogen-response element in the 5′-flanking region of the lactoferrin gene. The imperfect estrogen response elements of both human and mouse are functional as demonstrated by transfection experiments, band-shift assay and DNase I footprint analysis. The molecular mechanism that governs the estrogen-stimulated response, however, differs between human and mouse.

## INTRODUCTION

Expression of the lactoferrin gene in a variety of tissues is regulated differentially (Masson and Heremans, 1971; Green and Pastewka, 1978; Rado, et al, 1984; Pentecost and Teng, 1987; Teng et al 1989). In the reproductive tract and the mammary gland, hormones play a major role, while in neutrophil development is the key. In most wet surface mucosa, however, lactoferrin is expressed constitutively. To understand the expression and regulation of the lactoferrin gene in a tissue-specific manner, it is necessary to characterize the regulatory elements (both positive and negative) and unravel the complicated interaction with the various transcription factors that control the initiation of lactoferrin transcription.

We have previously demonstrated that lactoferrin is a major protein of uterine epithelial cells and the uterine secretory fluid in adult pseudopregnancy, early pregnancy, estrogen-treated immature mice and the estrous stage of the cycling animal (Teng, et al, 1986; Pentecost and Teng, 1987; Teng, et al, 1989; McMaster, et al 1991; Newbold et al, 1992). A patient with adenomatous hyperplasia of the uterus (chronically exposed to estrogen) expressed lactoferrin in the endometrium epithelium (Teng, et al 1992). The level of lactoferrin secretion in the endometrium and the vaginal mucus of a normal cycling woman varied at different stages (Tourville et al, 1970; Cohen, et al, 1987). These observations also implied that the hormone plays some roles in human endometrium. The presence of lactoferrin in the human reproductive tract was not limited to females; it has also been localized in the prostate and seminal vesicle of the male (Wichmann, et al, 1989). The role lactoferrin plays in the uterus under the influence of estrogen is not well understood, and whether or not androgen plays a role in lactoferrin expression in the male needs to be examined. In rodents, the secretion of lactoferrin in the uterus during the normal estrous cycle and early pregnancy correlates well with the secretion of immunoglobulins (Rachman et al, 1983, McMaster et al, 1991; Newbold et al, 1992). Thus, lactoferrin not only could be an antibacteria agent (Arnold et al, 1976) but could also play a role in immuno/inflammatory response by interacting with the macrophages and monocytes present in the uterine stroma at these stages (Tachi et al, 1981; Birgens, 1991; Broxmeyer, 1992). It has been reported that lactoferrin is the major coating antigen of the human sperm (Hekman and Rumke, 1969; Goodman and Young, 1981), therefore, lactoferrin might have additional functions in fertilization. Nevertheless, identification of a hormone-response element in the lactoferrin gene is the first step towards understanding how hormone regulates this gene in reproductive organs. In this study, we isolated the lactoferrin promoter and its 5′-flanking sequence from both human and mouse and compared the analogous and dissimilar characteristics between them. We identified a functional estrogen-response element in both human and mouse lactoferrin genes that confers estrogen action; however, the molecular mechanism that governs this action differs in these two species.

## EXPERIMENTAL PROCEDURE

### Materials

Enzymes were purchased from BRL. Radiolabeled compounds were purchased from New England Nuclear Corp. (Boston, MA). Antibody to COUP transcription factor was a gift from Dr. Tsai's laboratory (Cell Biology Department, Baylor College of Medicine, Houston, Texas). Estrogen- receptor expression plasmid HEO was a gift from Dr. P. Chambon (Laboratoire de Genetique Moleculaire des Eucaryotes du CNRS, France). Antibody to estrogen receptor was obtained from Abbott Laboratories (Abbott ER-ICA monoclonal H222 kit, Chicago, IL). Diethylstilbestrol (DES) was obtained from Sigma (St. Louis, MO). The CAT reporter plasmid was purchased from Promega (Madison, WI). Tissue culture components were obtained from Gibco BRL (Grand Island, NY). Oligodeoxyribonucleotides (oligo) were obtained from Research Genetics (Huntsville, Ala).

### Isolation of the Genomic Fragment

A human placental phage library (Clonetech HL 1067 J) was probed with an 1.3 kb 5′-fragment of human lactoferrin cDNA (p1212) (Panella et al, 1991). Positive clones were plaque-purified. Clone HLF031a contained a 16 kb insert and produced 7 fragments when digested with Sac I. These 7 fragments were subcloned into the Sac I site of pBluescript II SK+ (Stratagene) and screened with a 21 mer oligonucleotide probe (TP0, 5′-AGCAGGACGAG-GAAGACAAGT-3′) synthesized according to the 5′ end of the first exon of human lactoferrin cDNA. A 2.0 kbp fragment (HLF031a.30) hybridized to the labeled probe was purified and then sequenced. We screened the genomic library of mouse 129/J liver DNA clone in lambda Dash (Stratagene) for the mouse lactoferrin gene. The hybridization probes were a nick-translated full-length nucleotide cDNA (Pentecost and Teng, 1987), and end-labeled oligonucleotides specific to the 5'-regions of the cDNA. A 14 kbp genomic lambda clone (lambda J 14) containing the 7.6 kbp 5'-flanking and 6.5 kbp of lactoferrin gene sequences was isolated and plaque-purified. A DNA insert from the lambda J14 clone cut by EcoR1 at a single location resulted in 9 kbp and 5 kbp fragments. These fragments were subcloned in pGem 3Z. Further analysis of the subcloned plasmids by restriction mapping, Southern blotting and hybridization to specific oligonucleotides revealed that the 9 kbp fragment contained 2.6 kbp of 5'-flanking sequences and the first eight exons of the lactoferrin gene. A 3.0 kbp fragment containing the first exon, part of the first intron and 5'-flanking sequences was sequenced. Construction of plasmids for transfection experiments were described in previous publications (Panella et al, 1991; Liu and Teng, 1991; Liu and Teng, 1992a).

## Expression Assay

Two complementary oligonucleotides were synthesized according to the COUP/ERE sequence found in the promoter region of the human and mouse. These nucleotides were phosphorylated (T4 Kinase), annealed and ligated to either the Bgl II or Bam HI site of the SV40-CAT expression vector (Promega) as previously described. The presence and direction of inserts were confirmed by dideoxy sequencing of the double-stranded plasmids. Twice CsCl-purified plasmids were transfected into the human endometrial carcinoma cell line RL95-2 (ATCC #CRL 1617) at 50% confluency using the DNA calcium phosphate method according to the manufacturer (Pharmacia). In each experiment $2 \times 10^5$ cells were transfected with 5 µg of the plasmid and 0.25 µg of the β-galactosidase reference plasmid pCH 110 (Pharmacia) as an internal standard for transfection efficiency. Cell extracts were prepared 24–28 hours later and CAT enzyme activity was performed. The reaction products were analyzed with an ascending thin layer chromatography followed by X-ray autoradiography and liquid scintillation counting. To test estrogen responsiveness, cells were cotransfected with the HEO (estrogen receptor, 1 µg/well) expression plasmid. Incubations were carried out for 24 hours in the presence or absence of $10^{-8}$ M diethylstilbestrol (DES) (Sigma) before harvest. All experiments were repeated at least three times (duplicate dishes/experiment). Relative CAT activity was reported after normalization for β-galactosidase activity and values were expressed as the mean ± S.E.

## Oligonucleotide

The oligonucleotides and their complements used in the band shift assay were synthesized by Research Genetics and purified by gel electrophoresis:

mERM 5'-GATCGCATGCAAGTGTCACAGGTCAAGGTAACCCACAAATGCATGC-3';
hERM, 5'-GATCGTCTCACAGGTCAAGGCGATCTTCAA-3';
mut hERM 5'-GATCGTGTCACAGGTCAAGGCGATCTTCAA-3';
Ov coup, 5'-TTTCTATGGTGTCAAAGGTCAAACT-3'

## Preparation of Nuclear Extract from Human RL95-2 and Mouse Uterine Cells

Uterine nuclear extract was prepared from three-month old female CD-1 mice (Charles River, Wilmington MA) after being ovariectomized for two weeks and receiving (10 µg/kg body weight) diethylstilbestrol (DES) one hour before killing. The uterine nuclear protein extract used in band-shift assay was prepared as previously described (Liu and Teng, 1992).

RL95-2 cells were grown and transfected with 5 µg of HEO as described above. The cells were treated with $10^{-8}$ M DES one hour before harvest. Nuclear extract was prepared essentially the same as the mouse uterine nuclear extract described above except Dounce homogenizer was used to break up the cells.

## Band-Shift Assay

Complementary strands of oligonucleotides were annealed. The double-stranded oligonucleotides were labeled with [α-$^{32}$P] dGTP (3, 000 Ci/mmol; NEN ) by fill-in, and blunt-ended with Klenow enzyme (Boehringer Mannheim Biochemicals) and nucleotide triphosphates. A blunt-ended doubled-stranded oligonucleotide (ov coup) was end-labeled with [γ-$^{32}$P]ATP (6, 000 Ci/mmol; NEN) and T4 DNA kinase to the specific activity of $0.4$–$2.0 \times 10^8$ cpm/µg. The labeled probe was then gel-purified and used in band-shift assay. Four µg of poly (dI-dC); poly (dI-dC) (Pharmacia) was used as the nonspecific competitor in 10 µl of reaction mixture. The binding reactions were carried out at room temperature for 10 minutes before loading on the 5% nondenaturing polyacrylamide (19: 1, acrylamide: bisacrylamide) gels. Both nuclear extracts used in each reaction were 4 µg and the labeled probed was 0.3–0.4 ng. Gels were run in $0.5 \times$ TBE buffer at a constant voltage of 160 V, dried and autoradiographed with intensifying screens.

## DNase I Footprinting Protection Assay

DNase I protection of the lactoferrin 5′-flanking sequence with proteins from the Baculovirus-expressed mouse estrogen receptor, nuclear extract of RL-95-2 cells and DES-treated mouse uterus were performed according to the instructions specified by the manufacturer (Hot DNase I footprinting kit, Stratagene, La Jolla, CA). The Pvu II/Hinf I DNA fragment (−500 to −258, 242 bp) from the human and a Hinc II/Xba I DNA fragment (−589 to −291, 298 bp) from mouse lactoferrin 5′-flanking region were labeled and used for footprinting analyses. The interactions of labeled DNA and proteins were similar to the described for band-shift assay. At the end of the reaction, similar counts (in the range of $1 \times 10^4$ cpm per reaction) from various reactions were loaded onto a 6% sequencing gel; the DNA were separated by electrophoresis. Chemical reactions for the G and G/A on the same DNA fragment were conducted as described (Maniatis et al. 1982).

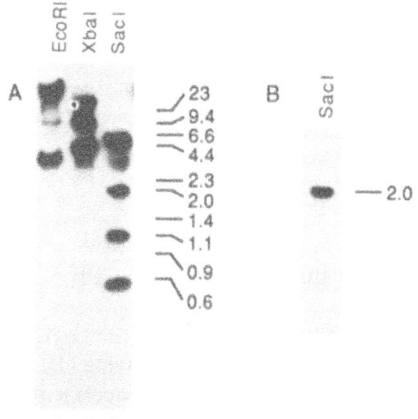

**Figure 1.** Isolation of human lactoferrin promoter and 5′-flanking sequence. Genomic clones containing the lactoferrin sequence were isolated and analyzed as described in Experimental Procedures. A. The HLF 031A.30 clone was restriction enzyme digested and probed with 5′-half of the lactoferrin cDNA. B. Sac I restriction fragments of HLF 031A.30 were subcloned into sk-Blue Script and probed with oligonucleotide to the 5′-end of the lactoferrin message.

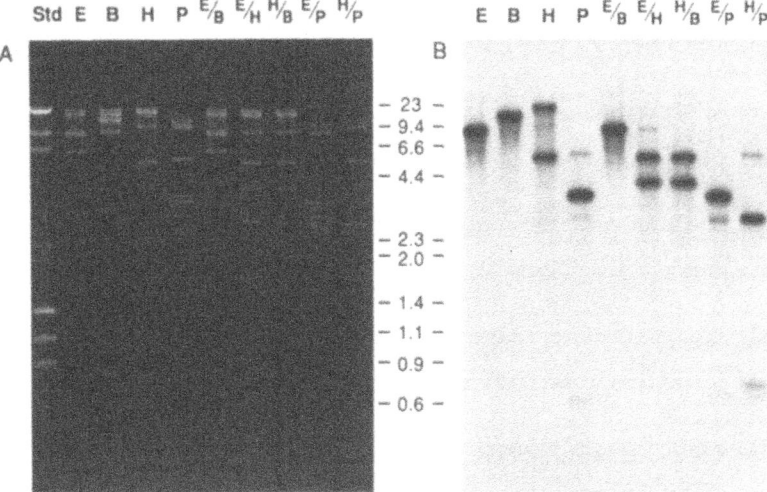

**Figure 2.** Isolation of mouse lactoferrin gene. Genomic clones containing the lactoferrin sequence were isolated and analyzed as described in Experimental Procedures. A. Stained-gel of the restriction digestion of lambda J 14 clone. B. Autoradiogram of the restriction digestion with total cDNA (T267)

## RESULTS

### Isolation of Human and Mouse Lactoferrin Promoter

A complete cDNA, encoding human (Panella et al, 1991) and mouse (Pentecost and Teng, 1987) lactoferrin was used initially to isolate the lactoferrin gene from a human placental genomic library in phage EMBL 3 and 129/J mouse liver genomic library in lambda Dash respectively. To identify the promoter of the gene, a 5'-end of the lactoferrin cDNA and an oligonucleotide based on the 5'-end of the mRNA were

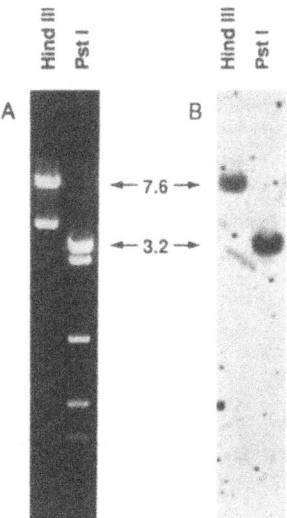

**Figure 3.** Identification of the mouse lactoferrin promoter and 5'-flanking sequence. The 9 kbp EcoRI fragment of the lambda J 14 clone was subcloned into pGem 3Z and digested with restriction enzymes. A. Stained-gel. B. Autoradiogram of the restriction fragments with oligonucleotide to the 5'-end of the message.

```
-1294 ...CGAGGATCATGGCTCACTGCCACCTTCATCTCCCAGGCTCAAATGGT   -1248  Human
           | ||    || |   |         |    || ||           ||
-1251 GACTTCAGCTCCCAGCACTGTGTTGCAAGGCTCATTCATAACTGCCTGTA   -1202  Mouse
               IAP

-1247 CCTCCCACTTTAG...CCTCCCAAGTAGCTGGGACCATAGGCATACACCA   -1201
      |||||  ||  ||    |  | || |  || |    |  |   |   |
-1201 CCTCCAGCTCCAGCTCCAGCTCCAGCTCCAGAGATGAGGTGTCTGAACCT   -1152

                             SAR
-1200 CCATGCTGGGCTAATTTTTGTATTTTTTGTAGAGATGGGGGTTTCCCTAT   -1151
      |   ||      |     | |||| |                       |
-1151 CTGGCCTCCATGGGTAGCTGCATTCATGTGCACATACCCCCACCATCACC   -1102

-1150 GAAGCCCAGGCTAGTCTTGAACTCCTGGGCTCAAGCGATCCTCCCATCTT   -1101
        || |  ||| |     |        | ||   ||| |        |
-1101 CCTAAACATACACGTATAAATAATTCCAAATTAATATATCATTAAAATGG   -1052

-1100 GGCCTCCCAAAGTGCTGGGATTACAGGCATGAGCCACTGTGCCCTGCCTA   -1051
      |       ||    || || |     || || |       |       |
-1051 CTTTTATTTTTTTGACAGGGTTTCTCTGTGTAACCCTGACTGTCCTGGAA   -1002
          SAR                              B₁

-1050 GTTACTCTTGGGCTAAGTTCACATCCATACACACAGGATATTCTTTCTGA   -1001
       | |||||  |   |        || ||| |       || |||
-1001 CTCACTCTGTAGACCAGGCTGGCCTGGAACTCAGAAATCTGCCTGTCTCT   -952

-1000 GGCCCCCAATGTGTCCCACAGGCACCATGCTGTATGTGACACTCCCCTAG   -951
      | | |||||| |        ||||| |
-951  GTCTCCCAAGTGCTGGGATTAAAGGCATGCGCCACCACCGCCGGGCTCAT   -902
                                           SP1

-950  AGATGGATGTTTAGTTTGCTTCCA.ACTGATTAATGGCATGCAGTGGTGC   -902
      |     || |  |  |  |   |   |   || |||||||  |  |   |
-901  AATATCATAATATGACTCTTAAGACATTCATTAAAAAGAAGTTCTCATTT   -852

-901  CTGGAAACATTTGTACCTGGGGTGCTGTGTGTCATGGGAATGTATTTACG   -852
      ||  |        |        |         |||  |        | |
-851  CTATACCACACATGTACACACACACACAATC.TCTCTCTCTCACACAC    -803

-851  AGATGTATTCTTAGAAGCAGTATTCTAGCTTTTGAATTTTAAAATCTGAC   -802
      | ||  |  |  |  |    ||          |   |  |  |   |  |
-802  ACATACACACACACACACACACACACACACAGACTCACACACACAA      -753
                            Z-DNA

-801  ATTTATGGCGATTGTTAAAATGAGGTTACCATTTCCTACTGAATACTATC   -752
      ||  |  |  |     ||  |  | | ||  | ||     |  | ||   |
-752  ATACACAGACACAAAAATAGACACGGTCACAGATAGACATAGACACAGAC   -703

-751  AACACCAAAAAAGAAGAAGGAGGAGATGGAGAAAAAAAAAGACAAAAAAAA   -702
      |   |||| | |||||L| | | |   |||   |     |   || |||
-702  AGACACACAGAGGAAGAGGAA..AAAAGATATGGAGGAAGGAATGAAAG    -655

-701  AAAAAGTGGTAGGGCATCTTAGCCATAGGGCATCTTTCTCATTGGCAAAT   -652
      |||| | | | |  | |      |||| |||| ||| |  | ||||||
-654  AAAAGGAGCAAAAGAAATCAAATGTCAGGGTCTCTTCCTCGCTAGCAAAT   -605

-651  AAGAACATGGAACCAGCCTTGGGTGGTGGCCATTCCCCTCTGAGGTCCCT   -602
      |    | |  || |  |     ||| | |||||| |||||||||||
-604  ...GAAGGGACACAGGTCAACCCTGGGTGGCATTCCTTTCTGAGGTCCTA   -558
```

**Fig 4a.** (legend - see page 190)

```
-601  GTCTGTTTTCTGGGAGCTGTATTGTGGGTCTCAGCAGGGCAGGGAGATAC  -552
      |  |  |||||  |||  |||||||| |  ||   |||       |   ||    ||
-557  GGTTATTTTC.GGGGGCTGTATGGCGGGTTTCAAGGCAGTGTGG....AC  -513
                           SP1

-551  CCCATGGGCAGCTTGCCTGAGACTCTGGGCAGCCTCTCTTTTCTCTGTCA  -502
      ||||  ||  |  |  ||  ||     ||| ||| |  ||| |  || ||| |
-512  CCCACAGGAACCCTGTGTGCAAGTCTAGGCTGACTCCGCTCTCCCTG.CG  -464

-501  GCTGTCCCTAGGCTGCTGCTGGGG..........GTGGTCGGGTCATCTT  -462
      |||||| |  |||||||||| || ||         |  |  |  || ||||
-463  GCTGTCACCGGGCTGCTGTTGTGGCCAGGCCTGAGCAGCTGCCTCTTCTT  -414

-461  TTCAACTCTCAGCTCACTGCTGAGCCAAGGTGAAAGCAAACCCACCTGCC  -412
      |  ||  |  |  |||  ||  |||||||||| | |||   |   | |||
-413  TAGAATCCACCACTCTTTGTCTAGCCAAGGAGGAAGGGGATTTGCTTGCT  -364

                                             GATA
-411  CTAACTGGCTCCTAGGCACCTTCAAGGTCATCTGCTGAAGAAGATAGCAG  -362
      | |   |||                                       ||
-363  CCATGCAGCT..................................TAAG  -350

         COUP/ERE
-361  TCTCACAGGTCAAGGCGATCTTCAAGTAAAGACCCTCTGCTCTGTGTCCT  -312
      | |||||||||||||| | |   ||| || |||||||| || | |  |||||
-349  TGTCACAGGTCAAGGTAACCCACAAATATAGACCCCCTACCCCATGTCC.  -300

                                  IAP
-311  GCCCTCTAGAAGGCACTGAGACCAGAGCTGGGACAGGGCTCAGGGGGCTG  -262
      |||||||||| | |||| | ||||| ||  ||| | | | ||||| |||
-299  CACCTCTAGAAAGTACTGGAAACAGAGAAAGGAGAAGACT.TGGGGACTG  -251

-261  CGACTCCTAGGGGCTTGCAGACCTAGTGGGAGAGAAAGAACATCGCAGCA  -212
      |||||   |  ||    | ||| |  ||||| ||| | ||  |
-250  TGACTC..TGATCCTGCAGAAGCTGGGTGGAGATTAAGGAAAT..CACTC  -205

-211  GCCAGGCAGAACCAGGACAGGTGAGGTGCAGGCTGGCTTTCCTCTCGCAG  -162
      |   |  | |||||| |  |||  | |       |||| | |||
-204  GGTTTCCTGTACCAGCGCCTGTGTAGGGGGTACTGGAGTCCCT.......  -162

-161  CGCGGTGTGGAGTCCTGTCCTGCCTCAGGGCTTTTCGGAGCCTGGATCCT  -112
            | |||| | ||| ||||||    | ||| |    |||
-161  ............GTTTCCTCCTTCTGGGCTCCAGGAAGCTGG...CCT    -129

-111  CAAGGAACAAGTAGACCTGGCCGCGGGGAGTGGGGAGGGAAGGGGTGTCT  -62
      |  | |||| || | ||||| | |   ||| |  ||| ||
-128  CTAAGAACTAGCACACCTGGTTGAGGGCAATGGGGCTGGAAGGCAGGCCT  -79

         CAAT·                                    TATA·
-61   ATTGGGCAACAGGGCGGGGCAAAGC................CCTGAATAA  -28
      ||||||||||  |||| |||||| |  |              |    ||||
-78   ATTGGGCAATAGGGTGGGGCCAGCCCGGTGAGGTCACCCAGCACAGATAA  -29

-27   AGGGGCGCAGGGCAGGCGCAAGTGGCAGAGCCTTCGTTTGCGAAGTCGCC  23
      |||| | | |||   | ||| |||  || |     | |  ||  |
-28   AGGGCCCCGGGGGAGAGGGCAGAAGCCAGGCTTGTCCT....CTAGGTCTC  18

                        M  K  L  V  F  L  V  L  L  F  L
24    TCCAGACCGCAGACATGAAACTTGTCTTCCTCGTCCTGCTGTTCCTCG   71
      | |||||| |||||||||   || || |  ||  || || || ||| |
19    CCAAGACCCACAGACATGAGGCTGCTCATCCCTTCCTTGATATTTCTTG   66
                        M  R  L  L  I  P  S  L  I  F  L
```

Fig 4a. (cont'd)

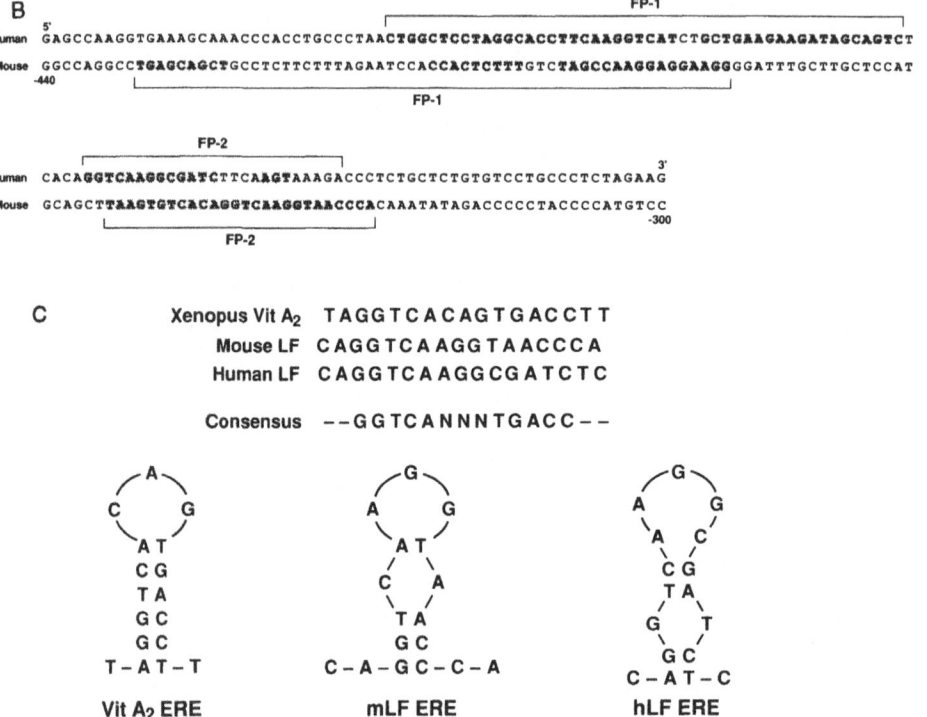

**Figure 4.** Comparison between human and mouse lactoferrin promoter/enhancer regions. A. The 1.2 kbp 5′-flanking sequence upstream from the transcription initiation site, 5′ untranslated sequence and the first eleven amino acids of the lactoferrin gene are compared between human and mouse. The GCG gap program is used for this analysis. Unusual sequences are underlined. Transcription factor binding sites are boxed. Potential transcription binding sites are dotted. B. Footprint region of the human and mouse lactoferrin gene. DNA fragment of the human (242 bp of Pvu II/Hinf I fragment, −500 to −258) and mouse (298 bp of Hinc II/Xba I fragment, −589 to −291) were end-labeled and interacted with nuclear extract of the RL95-2 cells and mouse uterine cells, respectively. Individual protected sequences are boxed. The regions being protected are denoted as FP-1 and FP-2. C. Compare the lactoferrin imperfect ERE to the vitellogenin perfect ERE.

also used (Fig. 1, 2 and 3). The human clone HLF 031a.30 contained a 16 kbp insert and produced 7 fragments when digested with Sac I. Four of the fragments were hybridized to the 5′-half of the cDNA (p1212) probe (Fig. 1 A lane 3). The human lactoferrin promoter was found in the 2.0 kbp Sac I fragment by hybridization to the synthetic oligonucleotide made against the 5′-end of the message (Fig. 1B). Mouse lactoferrin promoter was identified in a similar way as shown in Fig. 2 and 3. The 9 kbp fragment of the EcoR I cut, hybridized to the mouse cDNA (T267), was subcloned into pGem 3Z (Fig. 2B). Fragments hybridized to the 5′ oligonucleotide of the mRNA were identified (Fig. 3B). A 2.0 kbp Sac I fragment of the human and a 3.0 kbp Eco R I/Hinc II fragment of the mouse lactoferrin gene containing the lactoferrin promoter, 5′-flanking region, first exon and partial intron were sequenced completely.

## Promoter/Enhancer Region of the Human and Mouse Lactoferrin Genes

There were many sequence homologies between human and mouse at the promoter/enhancer (1 to −363) region, yet substantial divergence was observed beyond this region (Fig.4A). The promoter from both species contained noncanonical TATA and CAAT-like boxes in addition to a GC-rich sequence in this region (1 to −363). They shared 70% homology; their structural organization was compatible and the number of elements were conserved. The lactoferrin message from both species was initiated at G, 26 bp and 27 bp downstream from the TATA box of the human and mouse respectively. The elements conserved at the corresponding regions of the human and mouse lactoferrin gene were: noncanonical TATA, CAAT boxes, high GC content at the promoter region, the potential Pu.1/Spi.1 binding element (PU box) and the overlapping COUP-TF binding element and estrogen response element (ERM). These features suggest the existence of a similar manner of regulation by the binding factors in both species.

There were a number of DNA sequences in the 5′-flanking region of both human and mouse lactoferrin genes protected by nuclear extract from human endometrium adenocarcinoma cell and mouse uterus, respectively (Fig. 4B). The ERM sequence was among the protected regions (FP-2 region). The COUP-TF binding sequence was identical in human and mouse with the exception of one nucleotide difference, while the 3′ half of the ERE sequences of these two species were diverse. The structures of lactoferrin ERE and the Xenopus vitellogenin A2 ERE were shown in Fig. 4C.

### Identification of Estrogen-Response Element of the Lactoferrin Gene

The functional aspect of the ERM from both human and mouse has been characterized in detail (Liu and Teng, 1992; Teng et al, 1992). We found that the ERM confers estrogen-stimulated transcription to both homologous and heterologous promoters. Fig. 5 showed that the ERM of both human and mouse acted as an enhancer to confer estrogen-stimulated activity of a heterogous promoter. The enhancer activity, however, was much stronger in both orientations when the ERM was placed closer to the promoter. The level of stimulation depended on the level of estrogen receptors in the cell and the number of ERMs present in the construct. Under the same assay conditions, human and mouse ERMs responded to estrogen stimulation similarly.

### Differential Molecular Mechanism that Regulates Estrogen-Stimulated Transcription

Although the ERMs from both human and mouse responded similarly to estrogen stimulation in transient transfection experiments, the molecular mechanism that governs

| Reporter Constructs | Human ERM | | | Mouse ERM | | |
| --- | --- | --- | --- | --- | --- | --- |
| | CAT Activity (% Conversion) | | Fold Stimulation | CAT Activity (% Conversion) | | Fold Stimulation |
| | Control | DES | | Control | DES | |
| SV CAT | 2.4 ± 0.5 | 4.0 ± 0.9 | 1.6 | 2.7 ± 0.4 | 3.8 ± 1.0 | 1.4 |
| ■ SV CAT | 2.4 ± 0.5 | 46.0 ± 2.2 | 19.0 | 2.6 ± 0.3 | 51.6 ± 1.4 | 19.8 |
| ■ SV CAT | 2.0 ± 0.4 | 47.7 ± 1.2 | 23.8 | 2.5 ± 0.4 | 49.7 ± 1.1 | 20.0 |
| ■ 2.6 kb SV CAT | 2.3 ± 0.2 | 16.2 ± 2.1 | 7.0 | 2.7 ± 0.4 | 28.2 ± 1.2 | 10.4 |
| ■ SV CAT | 2.3 ± 0.3 | 15.0 ± 2.5 | 6.6 | 2.9 ± 0.2 | 26.8 ± 1.9 | 9.2 |

**Figure 5.** Lactoferrin ERM acted as an enhancer to heterologous promoter in response to DES stimulation. Left, schematic description of the ERM-CAT chimeric plasmids with SV-40 promoter. The orientation and position in the ERM insert are indicated by the arrow. Center, estrogen-responsiveness of the human ERM. Right, estrogen-responsiveness of the mouse ERM.

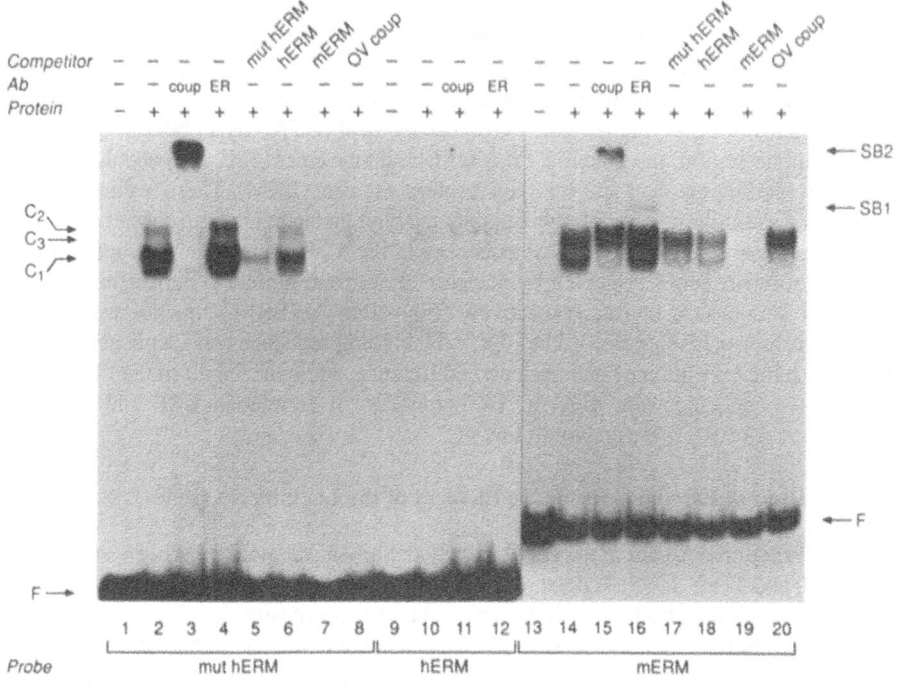

**Figure 6.** COUP-TF from RL95-2 nuclear extract interacted with mouse ERM and mutated human ERM specifically but not with wild-type human ERE. Oligonucleotides and the complements of human ERM (lanes 9–12), mutated human ERM (lanes 1–8) and mouse ERM (lane 13–20) were labeled and interacted with the nuclear extract of the HEO-transfected RL95-2 cells. The protein-DNA complexes (C1, C2 and C3) were analyzed by band-shift assay as described in the Experimental Procedures. Four µg of nuclear protein was used in each lane and a 50-fold excess of unlabeled oligonucleotide was used in the competing reactions (lanes 5–8 and 17–20). To detect the presence of estrogen receptor (ER) and COUP-TF (coup) in the protein-DNA complexes, nuclear extractions were preincubated with estrogen receptor antibody (lane 4, 12 and 16) or COUP-TF antibody (lanes 3, 11 and 15) for 10 minutes before the addition of reaction mixture. The supershifted protein-DNA complex with ER antibody (Ab) was designated SB 1; those with COUP-TF antibody were designated SB 2, respectively. F. Free probes from the double-stranded oligonucleotides.

estrogen action is different between human and mouse. From both band-shift and the DNase I footprinting protection assays, we found that the COUP-TF and estrogen receptor could bind to the mouse ERM individually but not simultaneously, whereas the human ERM binds estrogen receptor only. The reason that human ERM did not bind COUP-TF is due to a single mutation at the COUP-TF binding element. Fig. 6 shows the double-stranded wild-type human ERM (hERM 5'-GTCTCACAGGTCAAGGCGATC-3') has very little interaction with the RL95-2 nuclear proteins (lanes 9) in a band shift assay, whereas the double-stranded wild type mouse ERM (mERM 5'-GTGTCACAGGTCAAGGTAACC-3') binds nuclear protein efficiently (C1 and C2, lane14). As we mutated the third nucleotide at the 5'-end of the human ERM from C to G (mut hERM 5'-GTGTCACAGGTCAAGGCGATC-3'), this double-stranded oligonucleotide bound nuclear protein readily (lane 2). Using antibody to the COUP-TF, we demonstrated that COUP-TF is the major component of the protein-DNA complexes (lanes 3 and15). Therefore, the lack of interaction between COUP-TF and the human ERM was due to a single mutation at the COUP-TF binding element. The COUP element from chicken ovablumin gene (5'-TTTCTATGGTGTCAAAGGTCAAACT-3') competed most effectively

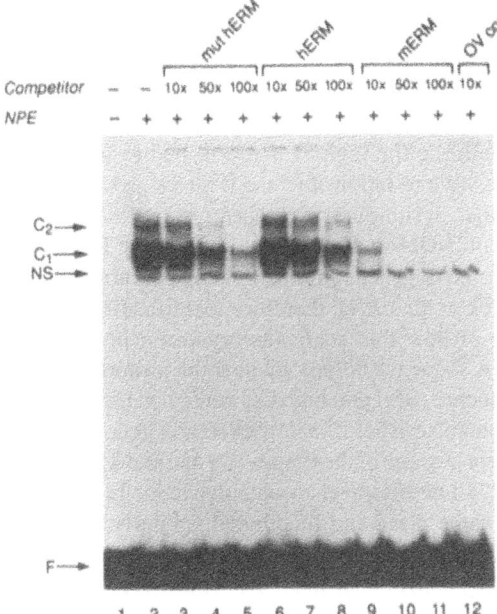

**Figure 7.** Competition efficiency of various lactoferrin ERM and ovalbumin COUP. Oligonucleotides of the human mutated ERM were used as the probe. The experimental conditions were identical as described in Fig. 6 except that various concentrations of unlabeled oligonucleotides were used as the competitor.

with the labeled mERM and mut hERM (lanes 8 and 20) for the COUP-TF in the nuclear extract of RL95-2 cells. Mouse ERM was a better competitor than the human mut ERM for COUP-TF binding (compare lanes 5 and 7; lanes 17 and 19), whereas at least ten-fold more human wild-type ERM was required to show any competition (lanes 6 and 18). Similar observations were demonstrated in Fig. 7. The order of COUP-TF binding to the lactoferrin ERM and ovalbumin COUP is as follows: OV COUP > mERM > mut hERM > hERM. The estrogen receptor-ERM complex (C3) was much easier to detect with the mouse (lane 13–20) than with the human (lane 1–12), although we were able to demonstrate such interaction in DNase I footprinting protection assay of the human (Teng et al, 1992). A weak interaction of bacculovirus-expressed estrogen receptor and human ERM (both wild-type and mutated) was observed by band-shift assay (Teng, unpublished data). From the DNase I footprinting and methylation interference studies, we found that COUP-TF interferes with estrogen receptor binding to the mouse ERM by direct competition for the overlapping site (Liu et al, 1992). In human, it is possible that COUP-TF does not play any direct role in estrogen-stimulated transcription due to the weak interaction between the protein and the binding element. The expression of the lactoferrin gene in mouse uterus is very sensitive to the level of circulating estrogen. The lactoferrin message and protein fluctuate with the estrogen level in uterus during the estrous cycle (Newbold et al, 1992; Walmer et al, 1992), whereas in human, the fluctuation of lactoferrin expression is not as obvious as in mouse during the menstrual cycle. Nevertheless, clinical studies showed that estrogen does effect lactoferrin expression in human endometrium. The differential molecular mechanism that regulates estrogen responses in human and mouse could explain these in vivo observations.

## DISCUSSION

This study compares the lactoferrin gene structurally and the estrogen-regulated expression in human and mouse. The experimental results of the present study reveal several unique features of the lactoferrin promoter in both species. One of these is the unusual DNA sequence present in both human and mouse. A purine-rich sequence was found at −751 to −696 (51/56 A or G) in human, whereas RY repeats (at −809 to −753) (Spitzner et al. 1990) and four TAGACA repeats were found immediately preceeding the purine-rich sequences (at −694 to −638; 53/57 A or G) in mouse. The functional significance of these DNA sequences in the lactoferrin gene is unclear; however, they could be the regions that attached to the nuclear scaffold (Gasser and Laemmli, 1986) and sensitive to topoisomerase II cleavage (Spitzner et al, 1990). The RY repeats (alternating purine: pyrimidine) have been identified in many naturally occuring DNA and exist in the form of Z-DNA (Wells et al, 1988). Large RY repeats (40–60 bp) such as the one in the mouse lactoferrin gene, could be the hotspots for recombination and gene conversion in vivo (Kilpatrick et al, 1984); therefore identification of these structures might facilitate the understanding of how this gene was organized in chromatin structure. Human and mouse shared high DNA sequence homology near the promoter region. The noncanonical TATA box, CAAT-like sequence, SP 1 and high GC content were conserved in human and mouse. These features are the characteristics of a housekeeping gene that could explain the constitutive expression of lactoferrin in some of the tissues (Dynan and Tjian, 1983; McKnight and Tjian, 1986; Maniatis et al, 1987). Lactoferrin also contained inducible elements such as COUP/ERE which is located in the same place and functioned similarly in the two species. The other concensus transcription factor binding elements, such as the PU.1 box and the acute-phase reaction element, were present in both human and mouse, however, in different copy number and at different locations on the gene. It was interesting to note that the erythroid growth factor GATA-1 binding site (Tsai et al, 1989) is present not only in the human lactoferrin gene, but also interacts with the GATA-1 protein from the MEL cells (Johnston et al, 1992). This GATA-1 binding site, however, was absent from the present sequence data of the mouse gene.

Our current finding that the imperfect ERE of both species confers estrogen-stimulated transcription in a receptor and hormone-dependent manner is the first step towards understanding the estrogen regulation of the lactoferrin gene in uterine tissue. The unique character of this ERE is its composite nature. Both the COUP-TF binding element and the estrogen-receptor binding element of the mouse interacted with their respective transcription factors; therefore, competition exists between the two transcription factors for the overlapping sequence of the COUP/ERE element. In fact, we have demonstrated that COUP-TF and the estrogen- receptor did interfer with each others' binding (Liu et al, 1992), and the level of either factor in the cell determined the estrogen-responsiveness. The human COUP-TF binding element has one nucleotide that differs from the mouse which did not bind COUP-TF in band-shift assay and DNase I footprint analysis (Teng et al, 1992). As compared to the concensus sequence of the COUP-TF binding element (Direct repeats, AGGTCANAGGTCA, Sagami et al, 1986; Kliewer et al, 1992), there were three mutations in human, versus two in mouse at the 5′-half. This could explain the difference in binding. It is important to note that the COUP-TF does not interact directly with the human COUP/ERE element, thus the molecular mechanism that governs the estrogen-stimulated transcription differs between these two species. To demonstrate this difference, the following model is proposed (Fig. 8): COUP-TF binds to the ERM in mouse uterine cells at an uninduced state; when estrogen is provided, the activated estrogen receptor replaces COUP-TF binding and initiates transcription. Because COUP-TF may not occupy the ERM constantly in human, it is possible that other protein factors can interact with COUP-TF and together play a positive or negative role in estrogen-regulated lactoferrin expression.

In conclusion, we have cloned and sequenced a 2.0 kbp of human and 3.0 kbp of mouse lactoferrin promoter and its 5′-flanking region. There are analogous and dissimilar characteristics between these two promoters. The complicated structural organization of the lactoferrin promoter could explain the differential regulation of this gene in various tissue and cell

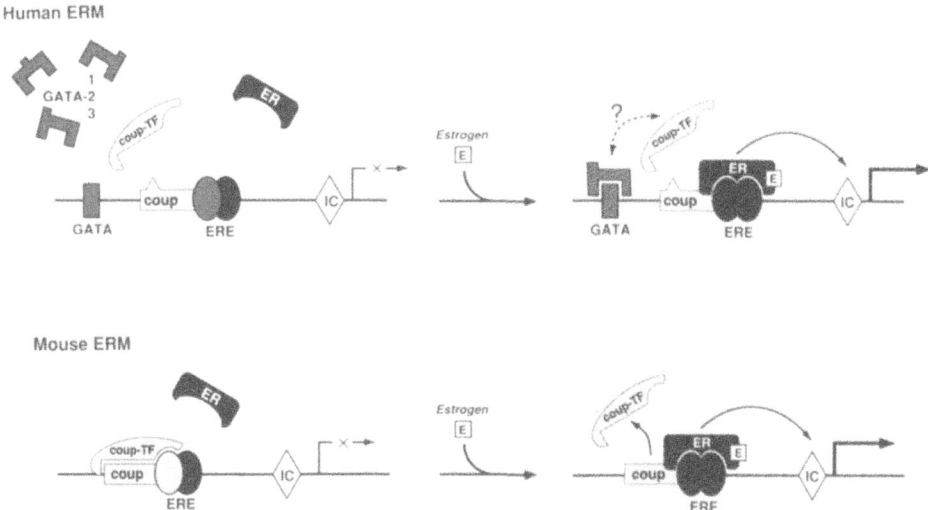

**Figure 8.** A proposed model on the differential molecular mechanisms that regulates the estrogen action on human and mouse lactoferrin ERM.

types. The most exciting finding is that the molecular mechanism that governs estrogen-stimulated transcription is different in human and mouse. In short, the same gene in two different species responds to the same stimulant through differential molecular means.

## REFERENCES

Arnold RR, Cole MF, McGhee JR. (1976) A bactericidal effect for human lactoferrin. Science 197: 263–265.

Birgens HS. (1991) The interaction of lactoferrin with human monocytes. Dan. Med. Bull. 38: 244–252.

Broxmeyer HE. (1992) Suppressor cytokines and regulation of myelopoiesis; biology and possible clinical uses. Am. J. Pediatr. Hemato. Oncol. 14: 22–30

Dynan WS, Tjian R. (1983) The promoter-specific transcription factor Sp1 binds to upstream sequences in the SV40 early promoter. Cell 35: 79–87.

Gasser SM, Laemmli UK. (1986) Cohabitation of scaffold binding regions with upstream/enhancer elements of three developmentally regulated genes of D. melanogaster. Cell 46: 521–530.

Goodman AS, Young LG. (1981) Immunological identification of lactoferrin as a shared antigen on radioiodinated human sperm surface and in radioiodinated human seminal plasma. J. Reprod. Immunol. 3: 99–108.

Green MR, Pastewka JV. (1978) Lactoferrin is a marker for prolactin response in mouse mammary explants. Endocrinology 103: 1510–1513.

Hekman A, Rumke P. (1969) The antigens of human seminal plasma (with special reference to lactoferrin as a spermatozoa-coating antigen). P549–552. In H. Peeters. (ed.) Protides Biological Fluids Colloquium. Vol.16. Pergamon Press Oxford.

Johnston JJ, Rintels P, Chung J, Sather J, Benz EJ, Berliner N. (1992) Lactoferrin gene promoter: Structural integrity and nonexpression in HL60 cells. Blood 79: 2998–3006.

Kilewer SA, Umesono K, Heyman RA, Mangelsdorf DJ, Dyck JA, Evans RM. (1992) Retinoid X receptor-COUP-TF interactions modulate retinoic acid signaling. Proc. Natl. Acad. Sci. USA 89: 1448–1452.

Kilpatrick MW, Klysik J, Singleton CK, Zarling DA, Jovin TM, Hanau LH, Erlanger BF, Wells RD. (1984) Intervening sequences in human fetal globin genes adopt left-handed Z helices. J. Biol. Chem. 259: 7268–7274.

Liu YH, Teng CT (1991) Characterization of estrogen responsive lactoferrin gene promoter. J. Biol. Chem. 266: 21880–21885

Liu YH, Teng CT. (1992) Estrogen response module of the mouse lactoferrin gene contains overlapping chicken ovalbumin upstream promoter transcription factor and estrogen receptor-binding elements. Mol. Endocrinol. 6: 355–364.

Liu YH, Yang N, Teng CT. (1993) COUP-TF acts as competitive repressor for estrogen-receptor mediated activation of the mouse lactoferrin gene. Mol. Cell. Biol. 13: 1836–1846.

Maniatis T, Goodbourn S, Fischer JA. (1987) Regulation of inducible and tissue-specific gene expression. Science 236: 1237–1245.

Maniatis T, Fritsch EF, Sarubrook J. (1982) Maxam Gilbert chemical degredation of DNA method. p. 13.11–13.13. In Molecular Cloning: A Laboratory Manual. Cold Spring Harbor Laboratory, Cold Spring Harbor N.Y.

Masson PL, Heremans JF. (1971) Lactoferrin in milk from different species. Comp. Biochem. Physiol.-B. 39: 119–129.

McKnight S, Tjian R. (1986) Transcriptional selectivity of viral genes in mammalian cells. Cell 46: 795–805.

McMaster MT, Teng CT, Dey SK, Andrews GK. (1992) Lactoferrin in the mouse uterus: analyses of the preimplantation period and regulation by ovarian steroids. Mol. Endocrinol. 6: 101–111.

Newbold RR, Teng CT, Beckman WC Jr., Jefferson WN, Hanson RB, Miller JV, McLachlan JA. (1992) Fluctuations of lactoferrin protein and mRNA in the reproductive tract of the mouse during the estrous cycle. Biol. Reprod. 47: 903–915.

Panella TJ, Liu Yh, Huang AT, Teng CT. (1991) Polymorphism and altered methylation of the lactoferrin gene in normal leukocytes, leukemic cells and breast cancer. Cancer Res. 51: 3037–3043.

Pentecost BT, Teng CT. (1987) Lactotransferrin is the major estrogen inducible protein of mouse uterine secretions. J. Biol. Chem. 262: 10134–10139.

Rachman F, Casimiri V, Psychoyos A, Bernard O. (1983) Immunoglobulins in the mouse uterus during the oestrous cycle. J. Reprod. Fert. 69: 17–21.

Rado TA, Bollekens J, St Laurent G, Parker L, Benz EJ. (1984) Lactoferrin biosynthesis during granulocytopoiesis. Blood 64: 1103–1109.

Sagami I, Tsai SY, Wang H, Tsai MJ, O'Malley BW. (1986) Identification of two factors required for transcription of the ovalbumin gene. Mol. Cell. Biol. 6: 4259–4267.

Spitzner JR, Chung IK, Muller MT. (1990) Eukaryotic toposiomerase II preferentially cleaves alternating purine-pyrimidine repeats. Nucleic Acids Res. 18: 1–11.

Teng CT, Liu YH, Yang NY, Walmer D, Panella T. (1992) Differential molecular mechanism of the estrogen action that regulates lactoferrin gene in human and mouse. Mol. Endocrinol. 6: 1969–1981.

Teng CT, Pentecost BT, Chen YH, Newbold RR, Eddy EM, McLachlan JA. (1989) Lactoferrin gene expression in the mouse uterus and mammary gland. Endocrinology 124: 992–999.

Teng CT, walker MP, Bhattacharyya SN, Klapper DG, DiAugustine RP, McLachlan JA. (1986) Purification and properties of an oestrogen-stimulated mouse uterine glycoprotein (approx. 70kDa). Biochem. J. 240: 413–422.

Tachi C, Tachi S, Knyszynski A, Linder HR. (1981) Possible involvement of macrophages in embryo-maternal relationships during embryo implantation in the rat. J. Exp. Zool. 217: 81–92.

Tourville DR, Ogra SS, Lippes J, Tomasi Jr. TB. (1970) The human female reproductive tract: Immunohistological localization of γA, γG, γM secretory "piece, " and lactoferrin. Am. J. Obstet. Gynecol. 108: 1102–1108.

Tsai SF, Martin D, Zon LI, D'Andrea A, Wong G, Orkin S. (1989) Cloning of cDNA for the major DNA-binding protein of the erythroid lineage through expression in mammalian cells. Nature 339: 446–451.

Walmer DK, Wrona MA, Hughes CL, Nelson KG. (1992) Lactoferrin expression in the mouse reproductive tract during the natural estrous cycle: correlation with circulating estradiol and progesterone. Endocrinology 131: 1458–1466.

Wells RD, Collier DA, Hanvey JC, Shimizu M, Wohlrab F. (1988) The chemistry and biology of unusual DNA structures adopted by oligopurine.oligopyrimidine sequences. FASEB J. 2: 2939–2949.

Wichmann L, Vaalasti A. Vaalasti T, Tuohimaa P. (1989) Localization of lactoferrin in the male reproductive tract. Internat. J. Andrology 12: 179–186.

# LACTOFERRIN cDNA

**Expression and In Vitro Mutagenesis**

John W. Tweedie,[*] Heather B. Bain, Catherine L. Day, H. Hale Nicholson,
Paul E. Mead, Bhavwanti Sheth, and Kathryn M. Stowell

Department of Chemistry and Biochemistry
Massey University
Palmerston North, New Zealand

## SUMMARY

The full length copy DNA (cDNA) for human lactoferrin has been synthesised by the polymerase chain reaction (PCR) using sequence specific primers. The template was first strand cDNA, synthesised from human bone marrow RNA using oligo(dT) to prime DNA synthesis by MMLV reverse transcriptase. The full-length human lactoferrin cDNA has been expressed in baby hamster kidney (BHK) cells using the expression vector pNUT. The protein expressed from the cloned cDNA is secreted into the culture medium and yields of up to 40 mg per litre have been obtained.

A mutant protein corresponding to the N-lobe of human lactoferrin ($Lf_N$) has also been expressed in BHK cells. The cDNA coding for this protein was produced by the introduction of stop codons into the region of the cDNA corresponding to the helix linking the N- and C-lobes of the native protein. $Lf_N$ is also expressed as a secreted protein and has been obtained in high yield. $Lf_N$ binds iron and has UV/Vis and ESR spectra which are virtually identical to the native protein. However, the pH at which iron is released from $Lf_N$ is quite different to the pH of iron release from the native and the full-length recombinant protein. A number of mutations have been introduced into $Lf_N$ by site-directed mutagenesis and the mutant proteins expressed in BHK cells. These mutations involve the iron binding ligands and have been designed to introduce some of the changes found in the C-lobe of melanotransferrin into $Lf_N$.

An attempt has been made to express a protein corresponding to the C-lobe of human lactoferrin ($Lf_C$) by attaching the sequence for the signal peptide of lactoferrin to the cDNA sequences coding for the C-lobe.

---

[*] To whom correspondence should be addressed

## INTRODUCTION

Lactoferrin is of great interest across a wide variety of disciplines because of its occurrence in many physiological secretions and in the secondary granules of neutrophils. The iron-binding properties of the protein have been well documented (Brock, 1985). As well as binding iron, lactoferrin has distinctive antibacterial properties which have been ascribed to the high affinity of the protein for iron (Dalmastri et al, 1988) and also to the presence of a specific antibacterial domain on the protein surface (Bellamy et al, 1992). A number of other physiological roles have been postulated for lactoferrin in the modulation of the immune and inflammatory responses and as a growth factor (Mazurier et al, 1989).

The properties of lactoferrin which would appear to be of prime importance to the biological role(s) of the protein are its ability to tightly but reversibly bind iron and its capacity to interact with many different types of cell (Birgens et al, 1983; Kawakami and Lonnerdal, 1991; Ziere et al, 1992).

The tertiary structure of human lactoferrin has been solved at high resolution for both the iron-bound ($Fe_2Lf$) and iron-free forms (apoLf) (Anderson et al,1989, 1990). We have chosen to extend these studies on the three-dimensional structure of lactoferrin by expression of the cDNA for the protein in a cell-free system (Stowell et al, 1991) and to carry out a program of site-directed mutagenesis of the recombinant lactoferrin produced by this system. Our initial objectives have been to express the recombinant N-terminal (Day et al, 1992) and C-terminal lobes of human lactoferrin and to introduce changes into the iron-binding sites of the recombinant protein. In our mutagenesis studies we have sequentially introduced, into the recombinant N-terminal protein, some of the changes to the iron-binding ligands which occur in the C-terminal lobe of melanotransferrin. It was initially postulated, on the basis of sequence homology and structural inference, that this lobe of melanotransferrin would not contain a functional iron-binding site (Baker et al, 1987). This hypothesis has recently been confirmed experimentally (Baker et al, 1992).

We report here some of the properties of the recombinant full-length lactoferrin, the recombinant N-terminal half-molecule and of some of the mutants produced by the introduction of changes to the iron-binding ligands in the N-terminal half-lactoferrin. Our results have established the validity of this approach to the study of the structure-function relationships of lactoferrin and have revealed that interactions between the two lobes of lactoferrin may be important in modulating the release of iron from the protein at low pH.

## EXPERIMENTAL

### Synthesis of cDNA for Human Lactoferrin

Total RNA isolated from human bone marrow was used as the template for first strand cDNA synthesis using an oligo(dT) primer and AMV reverse transcriptase. The first strand cDNA was used directly as a template for the polymerase chain reaction (PCR) using oligonucleotide primers specific for the 5'- and 3'-regions of human lactoferrin mRNA (Rado et al. 1987, Rado et al, unpublished). After preliminary identification of the 2.3 kb PCR product by restriction endonuclease mapping the hLf cDNA was cloned into the plasmid vector pGEM-I to produce pGEM:hLf. A single pGEM:hLf clone was selected and the hLf cDNA insert was transferred into the single-strand viral vector m13 for sequencing to verify the identity of the cDNA, and to check for the introduction of errors during PCR. The cDNA insert from the same pGEM:hLf clone was also introduced into the vector pNUT for expression in cells in tissue culture.

## Expression of pNUT:hLf in BHK Cells

Baby hamster kidney (BHK) cells were grown at 37°C under an atmosphere of 95% air:5% $CO_2$ in a medium consisting of a 1:1 mixture of Dulbeccos's modified Eagle's medium (DMEM) and F-12 nutrient mixture supplemented with foetal calf serum (10%) and the antibiotics streptomycin (100 µg/ml) and penicillin (100 units/ml). The expression plasmid pNUT:hLf was introduced into BHK cells by co-precipitation of the plasmid DNA with calcium phosphate on to a layer of cells at 70% confluence. After 24 hrs the medium was supplemented with methotrexate (0.5 mM) to select transformants. The successful expression of recombinant lactoferrin was verified by immunoprecipitation and/or western blotting using affinity purified antibodies to human lactoferrin. When the synthesis of lactoferrin by transfected cells was confirmed the culture was grown on a large scale and transferred to roller bottles for recombinant protein production. Once the cultures were established in roller bottles the medium (70–100 ml), containing the secreted recombinant protein, was replaced daily and retained for subsequent isolation of lactoferrin. Roller bottle cultures could be maintained for up to 4–5 weeks before extensive cell death was evident.

## Cloning and Expression of the N-terminal Lobe of Human Lactoferrin

A cDNA corresponding to the N-terminal lobe of human lactoferrin was obtained by PCR using the pGEM:hLf plasmid as template. The 5'-primer used was the same as that used above to obtain the full-length cDNA. However the 3'-primer corresponded to part of the cDNA sequence linking the two lobes of the protein. This primer was mutagenic, containing two in-frame translational stop codons inserted into the sequence following the codon for lysine 333, the last residue in the N-lobe before the linking helix. The PCR product (1.1 kb) was introduced into pGEM-3Zf(+) and the identity of a single clone was verified by restriction analysis and sequencing of the cDNA insert. The $Lf_N$ cDNA was then cloned into pNUT for expression as pNUT:$Lf_N$. Transfection and expression of pNUT:$Lf_N$ in BHK cells was performed as described above for pNUT:hLf.

## Synthesis of a cDNA Construct for Expression of the C-terminal Lobe of Human Lactoferrin

A cDNA designed for synthesis and secretion of the C-terminal lobe of human lactoferrin was constructed by PCR of a 1.5 kb, 5'-truncated fragment of hLf cDNA. The 5'-primer used contained a sequence corresponding to the last 27 nucleotides of the signal peptide of hLf, followed by the first 24 nucleotides coding for the C-terminal lobe of hLf, beginning with the sequence coding for $Ala_{340}$. The 3'-primer was the same as that used previously to obtain the full-length hLf cDNA from first-strand cDNA. The PCR product (~1.2 kb), containing the sequence coding for the C-terminal most 9 amino acids of the signal peptide, was then digested with ApaI, which has a unique restriction site within the hLf signal peptide, and BgIII, which cleaves at a site within the C-terminal sequence. The 0.5 kb ApaI-BgIII fragment produced was cloned into the large fragment of pGEM:hLf produced by digestion with ApaI and BgIII. After verification by DNA sequencing, the $hLf_C$ cDNA insert was cloned into pNUT for expression as pNUT:$hLf_C$. Transfection and expression of pNUT:$hLf_C$ in BHK cells was as described above. A second $hLf_C$ construct was made using a different 5'-primer for PCR. As before this primer contained sequences corresponding to the hLf signal peptide, including the ApaI restriction site. However the first 15 nucleotides following the putative signal cleavage site were designed to produce a C-terminal lobe which had 5 amino acids at the amino terminus identical to those of the N-terminus of hLf and $Lf_N$. This strategy was designed to produce a recombinant C-terminal pre-protein which had a signal peptidase cleavage site with the same amino acid sequence as in hLf and Lf$_N$. All subsequent steps were as described above for the first construct.

### In vitro Mutagenesis

Mutagenesis of cloned lactoferrin cDNA was carried out by annealing appropriate mutagenic oligonucleotides to circular, single-stranded, uracil-containing template DNA produced from pGEM3Zf(+) into which the appropriate cDNA had been inserted (Kunkel et al., 1987). Second strand DNA was produced by extension of the mutagenic oligonucleotides with the Klenow fragment of *E. coli* DNA polymerase I, followed by ligation with T$_4$ DNA ligase. Plasmid-containing clones were isolated after transformation of *E. coli* and analysed by digestion with diagnostic restriction endonucleases. The presence of the desired mutation was verified by DNA sequencing. The mutant cDNA was then cloned into pNUT for expression in BHK cells as described above.

### Purification of Recombinant Lactoferrins

Lactoferrins were isolated by ion exchange chromatography from the medium harvested from BHK cells. Approximately 5 g (wet weight) of CM-Sephadex in 0.01 M HEPES, pH 7.8, 0.2 M NaCl was added to each litre of medium and stirred gently overnight at 4°C. The medium was removed by filtration on a sintered glass funnel and the resin washed on the filter with 0.025 M Tris, pH 7.8, 0.2 M NaCl to remove non-absorbing proteins. Lactoferrins were eluted from the resin with 0.025 M Tris, pH 7.8, 0.8 M NaCl. Minor impurities remaining at this stage were removed, after dialysis against 0.01 M HEPES, pH 7.8, 0.2 M NaCl, by chromatography on a column of CM-Sephadex equilibrated in the same buffer. The column was washed with HEPES buffer containing 0.35 M NaCl and lactoferrin was then eluted with HEPES buffer containing 0.4 M NaCl.

### Characterisation of Recombinant Lactoferrins

Purified recombinant lactoferrins were characterised by SDS-polyacrylamide gel electrophoresis, protein blotting using affinity purified antibodies to human lactoferrin, UV/visible and ESR spectroscopy, N-terminal amino acid sequence analysis and by determination of the of the pH at which iron was released from the proteins.

## RESULTS

### Expression and Initial Characterisation of Recombinant Human Lactoferrin (hLf)

The recombinant human lactoferrin expressed from the cloned cDNA in BHK cells was virtually identical to that isolated from human milk. The purified recombinant protein had a mobility on SDS-polyacrylamide gel electrophoresis which corresponded to that of lactoferrin from human milk (Figure 1). The sequence of the first 21 amino acids of the protein was identical to that of milk lactoferrin (data not shown), showing that the recombinant pre-protein was correctly processed during synthesis and secretion by BHK cells. When the recombinant lactoferrin was deglycosylated by PNGase most of the protein showed a decrease in apparent molecular mass to a size corresponding to that of human milk lactoferrin which had been treated in the same way (Figure 2). However a small fraction of the recombinant protein was resistant to deglycosylation by PNGase, suggesting that there was some heterogeneity in the carbohydrate side chains of the recombinant lactoferrin and that at least some of these chains were different to those in the native protein. The recombinant lactoferrin was isolated from the cell

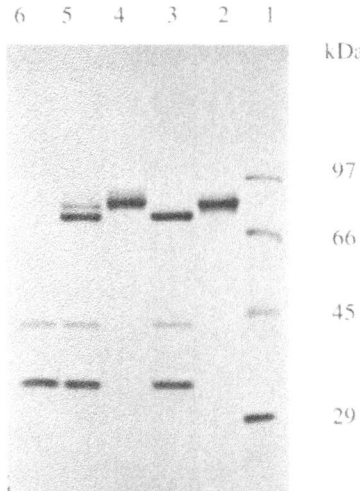

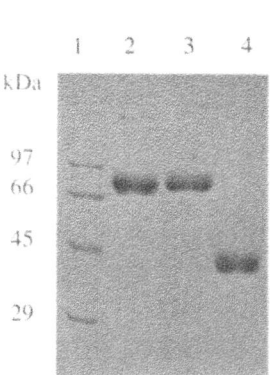

**Figure 1.** SDS-polyacrylamide gel electrophoresis of native and recombinant lactoferrins. Proteins (5 μg) were denatured by boiling in SDS-urea and separated on an 8% SDS polyacrylamide gel. Visualisation of protein bands was by staining with Coomassie Brilliant blue. Lane 1: Molecular weight protein standards. Lane 2: Native human milk lactoferrin. Lane 3: Recombinant full-length human lactoferrin. Lane 4: Recombinant N-terminal half-molecule of human lactoferrin.

**Figure 2.** Deglycosylation of native and recombinant lactoferrins by PNGase. Proteins were separated by gel electrophoresis as described in Figure 1. Lane 1: Molecular weight protein standards. lane 2: Human milk lactoferrin. Lane 3: PNGase treated human milk lactoferrin. Lane 4: Recombinant full-length human lactoferrin. Lane 5: PNGase treated recombinant human lactoferrin. Lane 6: PNGase.

culture medium in a fully iron-saturated form with a UV/visible spectrum identical to that of fully iron-saturated native lactoferrin with a $\lambda_{max}$ in the visible region at 465 nm (data not shown). This initial characterisation suggests that the recombinant lactoferrin was virtually identical to the native protein found in human milk.

Recombinant lactoferrin was virtually the sole protein secreted by BHK cells which had been transfected with pNUT:hLf. Lactoferrin was immunoprecipitated from the medium of BHK cells grown in the presence of $^{35}$S-methionine and the amount of label in the lactoferrin immunoprecipitate was compared to that in a TCA precipitate of total protein from the same volume of medium (Table 1). The results show that, within the limits of precision of the

**Table 1.** Lactoferrin Synthesis and Secretion by Baby Hamster Kidney (BHK) Cells

| BHK cells | Radioactivity in culture medium | | |
| --- | --- | --- | --- |
| | Total protein (CPM) | Lactoferrin immunoprecipitate (CPM) | Lactoferrin synthesis (%) |
| Not transfected | 7561 | 49 | 0.5 |
| Transfected with pNUT:hLf | 12285 | 12560 | 102 |

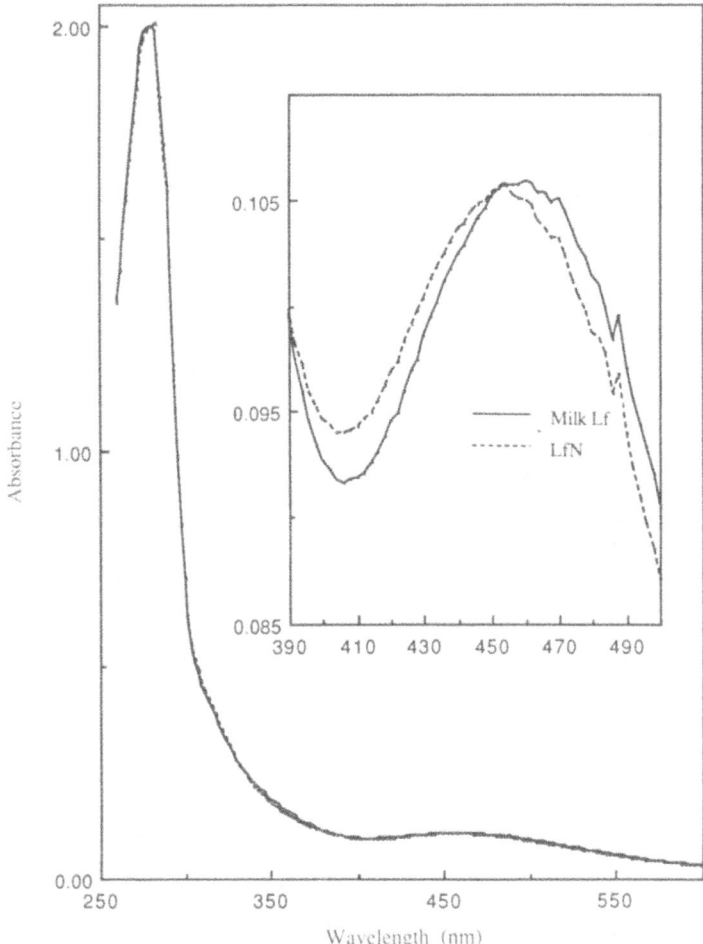

**Figure 3.** Spectra of human milk lactoferrin and recombinant Lf$_N$. UV/Visible spectra were obtained using a Hewlett-Packard diode array spectrophotometer.

experiment, all of the radiolabeled protein secreted by the transfected cells was human lactoferrin.

The amount of recombinant lactoferrin in the medium of cultures of transfected BHK cells was estimated from the yield of the protein after isolation from the medium using CM-Sephadex. Routinely concentrations of recombinant lactoferrin of between 20 to 30 mg/litre were maintained throughout the life of the roller bottle cultures. There was a tendency for the yield to decrease towards the lower value in the later stages of the culture. The total yield of recombinant lactoferrin from a single roller bottle culture maintained for 4–5 weeks was 40–60 mg.

## Expression of the N-terminal Lobe of Lactoferrin (Lf$_N$)

The recombinant N-terminal lobe of human lactoferrin was produced from the cDNA which encodes amino acids 1–333 of the mature sequence of hLf. This cDNA was introduced into pNUT and the resulting plasmid (pNUT:Lf$_N$) transfected into BHK cells The transfected cells secreted a protein which had a mobility corresponding to a $M_r$ of ~40 kDa on SDS-polyacrylamide gel electrophoresis (Figure 1). This is the approximate $M_r$ expected for a protein comprising the first 333 amino acids of human lactoferrin and containing a single N-linked glycosyl group. The protein could be completely deglycosylated by PNGase, as judged by the decrease in $M_r$ following treatment with this enzyme (data not shown). The absorption spectrum of iron-saturated $Lf_N$ was slightly different to that of hLf and recombinant hLf with the $\lambda_{max}$ shifted from 466 nm for native and recombinant full-length lactoferrin to 454 nm for $Lf_N$ (Figure 3).

## Iron Release from Lactoferrin with Changes in pH

The most pronounced difference we have found between full-length lactoferrin and the N- terminal half molecule is the shift in the pH dependence of iron release for $Lf_N$. This result is shown in Figure 4. Both the native and recombinant full-length proteins showed virtually identical pH dependence curves for iron release. However $Lf_N$ released iron at higher pH values than did either the native or recombinant full-length lactoferrin. Both native and recombinant full-length proteins were half-saturated with iron at pH 3 while half of the iron had been lost by the N-terminal half molecule by pH 4.8. The curve for the release of iron with pH from $Lf_N$ is similar to the curve obtained with human serum transferrin when determined under the same conditions (Figure 4).

## Mutants of the N-terminal Lobe of Lactoferrin

Two mutants involving iron-binding ligands have been introduced singly and in combination into the cDNA for $Lf_N$. These mutations are $Lf_N$:D60S and $Lf_N$:R121S, which correspond to two of the changes in the iron binding ligands which occur in the C-terminal lobe of human melanotransferrin (D402S, R466S) (Garratt and Jhoti 1992). The cDNA's for all three of these mutants ($Lf_N$:D60S, $Lf_N$:R121S and $Lf_N$:D60S; R121S) were successfully expressed in BHK cells, although the amounts of protein isolated from the medium were somewhat lower than for recombinant Lf and $Lf_N$. The change of aspartate to serine in the $Lf_N$:D60S single mutant shifted the $\lambda_{max}$ in the visible region to ~434 nm, a decrease of 20 nm from the $\lambda_{max}$ of 454 nm for $Lf_N$. However the substitution of serine for arginine in the $Lf_N$:R121S mutant produced a protein with a spectrum identical to that of $Lf_N$. The double mutant ($Lf_N$:D60S; R121S) had a visible absorption maximum at 474 nm, a shift of 20 nm to longer wavelengths compared to $Lf_N$. The $\lambda_{max}$ values in the visible region for these proteins are shown in Table 2.

The effect of pH on iron release from these mutant N-lobe proteins was also examined (data not shown). The $Lf_N$:D60S protein released iron at a higher pH than did $Lf_N$, whereas $Lf_N$:R121S behaved in essentially the same fashion as $Lf_N$ in this experiment, with 50% of the iron being released at pH 4.8. The pH dependence of iron release by the double mutant ($Lf_N$:D60S; RI21S) followed the same curve as that for $Lf_N$:D60S.

**Table 2.** Wavelength of Maximum Absorption for Mutants of Recombinant Lactoferrin.

| Lactoferrin | $\gamma_{max}$ (nm) |
|---|---|
| Lf (native and recombinant) | 465 |
| $Lf_N$ | 454 |
| $Lf_{N:D60S}$ | 434 |
| $Lf_{N:R121S}$ | 454 |
| $Lf_{N:D60S:R121S}$ | 474 |

UV/Visible spectra were obtained using a Hewlett-Packard diode array spectrophotometer.

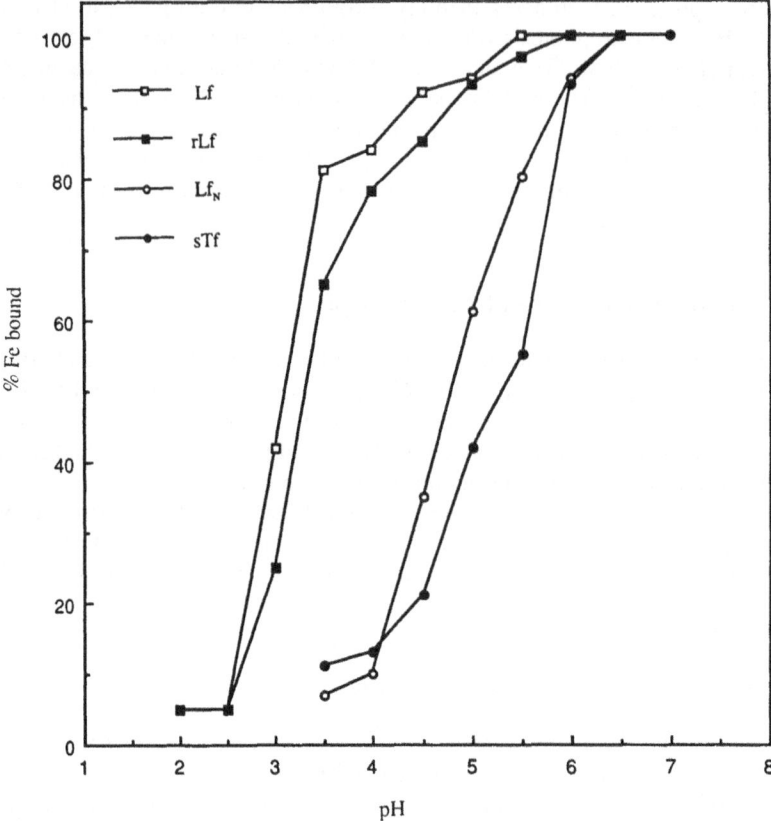

**Figure 4.** Release of iron from transferrins with change in pH. Samples of human milk lactoferrin (Lf), recombinant full-length lactoferrin (rLf), the recombinant N-lobe of human lactoferrin ($Lf_N$) and human serum transferrin (sTf) were dialysed for 48 hours against buffers of the appropriate pH. The spectra of the dialysed protein samples were determined and the percentage of iron bound calculated from the ratio of the absorbance at $\lambda_{max}$ in the dialysed sample to the absorbance of the fully iron-saturated protein.

## Construction of a cDNA for Expression of the C-terminal Lobe of Human Lactoferrin

The cDNA for expression and secretion of the C-lobe of human lactoferrin was constructed as described in the experimental section. Two constructs were prepared and transfected into BHK cells. The basic strategy was to attach sequences coding for the signal peptide of the pre-protein of human lactoferrin to the cDNA for the C-lobe of the protein. In the first construct the signal peptide sequences were attached directly to the cDNA adjacent to the codon for the alanine residue at position 340 in human lactoferrin. The second construct contained the same pre-sequence but a glycine was substituted for the alanine at position 340 and an arginine for the alanine at position 343. This construct reproduces the signal peptide plus the sequence of the first five amino acids of the mature full-length and N-lobe proteins and should provide an adequate signal peptidase site. The N-terminal pre-protein sequences coded for by these cDNA constructs are shown in Table 3, together with the N-terminal pre-protein sequences coded for by the cDNA's for recombinant Lf and $Lf_N$.

These cDNA constructs were introduced into BHK cells and the presence of the cDNA in the transfected cells was demonstrated by PCR using specific primers. BHK cells transfected

**Table 3.** Sequences at the N-terminal End of Recombinant Lactoferrin Pre-proteins.

| | |
|---|---|
| Sequence at the N-terminus of the pre-protein for recombinant Lf and Lf$_N$ | ↓<br>MKLVFLVLLFLGALGLCLAGRRRRS |
| Predicted sequence at the N-terminus for the pre-protein for the recombinant C-lobe (Construct 1) | ↓<br>MKLVFLVLLFLGALGLCLAARRARV |
| Predicted sequence at the N-terminus of the pre-protein for the recombinant C-lobe (Construct 2) | ↓<br>MKLVFLVLLFLGALGLCLAGRRRRV |

with the C-lobe cDNA constructs produced a diagnostic PCR product which was not evident when the template DNA for the PCR was from non-transfected BHK cells (Figure 5). RNA was isolated from transfected cells, separated by electrophoresis on agarose gels containing formaldehyde and transferred to nitrocellulose membrane by blotting. The transfected BHK cells produced a mRNA of 1.2 kb which hybridised specifically to a human lactoferrin cDNA probe (Figure 6). This shows that the transfected cDNA was successfully transcribed by the BHK cells.

We have attempted to demonstrate the presence of C-lobe protein in the medium from transfected cells by immunoprecipitation using affinity purified polyclonal antibodies to human lactoferrin and also by absorption of the protein on to CM-Sephadex. In an attempt to determine whether the C-lobe protein was synthesised but not secreted from the cells we prepared immunoprecipitates from [35]S-methionine labeled intracellular proteins of BHK cells transfected by the C-lobe cDNA construct. All of these procedures were shown to be capable of demonstrating synthesis of recombinant lactoferrin from cells containing other human lactoferrin cDNA constructs but did not show expression from either of the two C-lobe constructs.

## CONCLUSIONS

The cloning and expression of the cDNA for lactoferrin and its variants reported here provides an opportunity to probe the structure-function relationships of the protein by site-di-

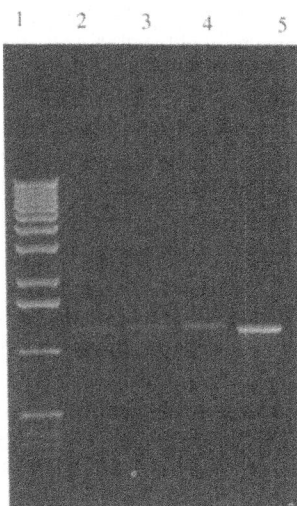

**Figure 5.** PCR analysis of DNA extracted from BHK cells. DNA was isolated from BHK cells and was used as a template for PCR with 5'- and 3'- primers specific for the coding sequences of the C-lobe of human lactoferrin. Lane 1: BRL 1 kb DNA ladder. Lanes 2 &3: DNA isolated from BHK cells transfected with pNUT:hLf$_C$ (forward). Lane 4: DNA isolated from BHK cells transfected with pNUT:hLf$_C$ (reverse). Lane 5: Human lactoferrin cDNA (positive control). Analysis of PCR reactions without template DNA or without primers showed no amplified products.

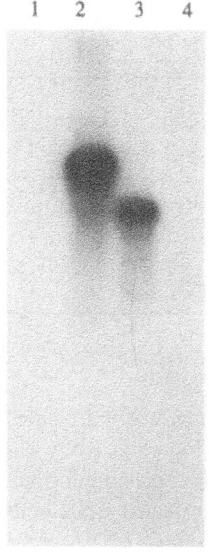

**Figure 6.** Hybridisation of a labeled human lactoferrin cDNA probe to RNA isolated from BHK cells. Total RNA was isolated from BHK cells and separated by electro-phoresis on agarose gels containing formaldehyde. The RNA was transferred to nitrocellulose and hybridised to a probe of $^{32}$P-labeled human lactoferrin cDNA. After washing in $0.1^x$ SSC at 68°C the nitrocellulose membrane was exposed to X-ray film. Lane 1: RNA from non-transfected BI-IK cells. Lane 2: RNA from cells transfected with pNUT:hLf. Lane 3: RNA from cells transfected with pNUT:hLf$_C$(forward). Lane 4: RNA from cells transfected with pNUT:hLf$_C$(reverse).

rected mutagenesis. Amounts of recombinant lactoferrins sufficient for crystallographic analysis can be obtained readily using the BHK cell expression system. Analysis of the full-length recombinant lactoferrin suggests that the protein is identical to the native milk protein, apart from some differences in the nature of the N-linked carbohydrate chains. The fact that a small proportion of the protein is resistant to deglycosylation by PNGase suggests that some of the carbohydrate chains have a triantennary structure. Similar differences between the carbohydrate groups of recombinant proteins produced in cells in tissue culture and those of the native protein have been reported by others (Tsuda et al, 1988). These differences in the nature of the carbohydrate do not affect any of the properties of the protein which we have examined. The metal binding characteristics of the recombinant N-lobe at neutral pH appear to be very similar to those of native lactoferrin. The electronic and ESR spectra of the two proteins are similar, indicating that the ligands involved in iron binding have virtually identical geometries. The significance of the small shift in $\lambda_{max}$ (465 to 454) shown by Lf$_N$ is not apparent at present but may be able to be interpreted once the tertiary structure has been determined. However the recombinant N-lobe demonstrates a greatly reduced stability at low pH values, as shown by the displacement of the iron release curve to about two pH units higher than that for native lactoferrin. The pH for 50% release of iron (pH$_{0.5}$) is increased from ~3.0 for Lf to ~4.8 for Lf$_N$. This change suggests that interactions between the two lobes contribute to the stability of lactoferrin at low pH values. The effect of the removal of the interlobe interactions demonstrates clearly that structural changes occurring at a distance from the iron binding site can markedly influence the stability of iron binding at low pH. This stabilisation by interlobe contacts does not appear to occur with serum transferrin, which shows the same dependence of iron release on pH as does Lf$_N$. Site specific mutagenesis of the residues involved in the interlobe contact of the full-length lactoferrin should help in determining the interactions responsible for the stability of iron binding by lactoferrin at low pH.

The influence of interactions between the lobes on the release of iron at low pH has physiological implications in terms of the possible role of lactoferrin in iron delivery into cells. The marked stability of iron binding by lactoferrin at low pH has raised some doubts about the release of iron from lactoferrin in acidified endocytic vesicles as has been shown to occur with serum transferrin. In this context it has recently been shown (Bali and Aisen, 1992) that interaction with the transferrin receptor accelerates release of iron from the C-lobe of serum transferrin at low pH values. The influence of the transferrin receptor on iron release from transferrin also shows that interactions at the surface of the protein can influence the release of iron from the iron-binding site.

The role of some of the ligands involved in iron binding has been examined by the production of $Lf_N$:D60S, $Lf_N$:R121S and $Lf_N$:D60S;R121S. All of these mutants bound iron at neutral pH and their spectra in the visible range showed only slight changes in the position of the $\lambda_{max}$ (Table 2). The significance of these changes with respect to the geometry of the iron binding site may become clearer when the tertiary structure of these mutant proteins has been determined. It is clear however that the changes we have made to the iron binding site by the D60S and R121 S mutations do not abolish the capacity of the protein to bind iron. These R121S mutations were chosen because they occur in the corresponding positions in the C-lobe of melanotransferrin. This protein has recently been shown to bind only a single iron atom (Baker et al. 1992), presumably in the N-lobe since the C-lobe contains a number of mutations in residues which could influence iron binding. The carboxylate oxygen from Asp60 in $Lf_N$ is bound directly to iron and a change to serine at this position clearly influences the iron binding site as shown by the shift in the $\lambda_{max}$ in the visible region. However this change alone is not sufficient to abolish iron binding by $Lf_N$. The change of Arg121 to serine does not appear to influence the iron-binding capacity of the N-lobe protein. The visible region spectrum of $Lf_N$:R121S is identical to that of $Lf_N$ and the pH dependence curve for iron release by this mutant is marginally shifted to lower pH values (data not shown). The arginine at position 121 of $Lf_N$ forms a hydrogen bond to the carboxylate anion. It is possible that a serine at this position can form an equivalent hydrogen bond, which may explain why this mutation does not appear to significantly affect iron binding.

The double mutant, incorporating both of these changes, shows changes in the visible spectrum ($\lambda_{max}$, 474 nm) which are different to those occurring in the single mutants. This suggests that the iron binding site is influenced by the combined effect of these two changes but the protein is still clearly capable of binding iron.

These results show that the loss of the capacity for iron-binding by the C-lobe of melanotransferrin is due to more than the effect of the two mutations corresponding to D60S and R121S that we have introduced into $Lf_N$. There are several additional changes in melanotransferrin at positions close to the iron binding site of the C-lobe which are likely to interact with the D60S and R121S changes to abolish the iron binding capacity of this lobe.

We are unable to provide an explanation for the failure of the cDNA constructs for the production of the C-lobe of lactoferrin to be expressed by the BHK cells. The pNUT:hLf$_C$ plasmid DNA is incorporated into the BHK cell DNA following transfection (Figure 5) and, on induction of the metallothionine promotor with zinc, the cDNA is transcribed to produce a mRNA of the size predicted from the cDNA sequence (Figure 6). Attempts to identify the recombinant C-lobe using affinity purified polyclonal antibodies to human lactoferrin (Stowell at al. 1991) have shown that there was no lactoferrin secreted into the medium from transfected BHK cells Immunoprecipitates prepared against the intracellular proteins of cells transfected with pNUT:hLf$_C$ showed that there was no lactoferrin within the cells. These results suggest that the mRNA transcript is not translated by the BHK cell ribosomes or that if translation is occurring the resulting protein is not processed correctly and is rapidly degraded within the cells.

## ACKNOWLEDGEMENTS

This work was supported by the Health Research Council of New Zealand, the Palmerston. North Medical Research Foundation, the New Zealand University Grants Committee,

AgResearch NZ and the Massey University Research Fund. CLD was supported by a New Zealand Universities Vice-Chancellors Committee Scholarship.

# REFERENCES

Anderson, B.F., Baker, H.M., Norris, G.E., Rice, D.W. and Baker, E.N. (1989). Structure of human lactoferrin: Structure analysis and refinement at 2.8 Å resolution. J. Mol. Biol. 209: 711–734.

Anderson, B.F., Baker, H.M., Norris, G.E., Rumball, S.V. and Baker, E.N. (1990). Apolactoferrin structure demonstrates ligand-induced conformational change in transferrin. Nature (London) 334: 785–787.

Baker, E.N., Baker, H.M., Smith, C.A., Stebbins, M.R., Kahn, M., Hellström, K.E. and Hellström, I. (1992). Human melanotransferrin (p97) has only one functional iron binding site. FEBS Letters 298: 215–218.

Baker, E.N., Rumball, S.V. and Anderson, B.M. (1997). Transferrins: Insights into structure and function from studies on lactoferrin. Trends in Biochemical Science 12: 350–353.

Bali, P.K. and Aisen, P. (1992) Receptor-modulated iron release from transferrin: Different effects on N- and C-terminal sites. Biochemistry 30: 9947–9952.

Bellamy, W., Takase, M., Yamauchi, K., Wakabayashi, H., Kawase, K. and Tomira, M. (1992). Identification of the bactericidal domain of lactoferrin. Biochirm Biophys Acta 1121: 130–136.

Birgens, H.S., Hansen, N.E., Karle, H. and Kristensen, L.O. (1983). Receptor binding of lactoferrin by human monocytes. Br. J. Haematol. 54: 383–391.

Brock, J.H. (1985). Transferrins. In Metalloproteins (Harrison, P.M., ed.), Part 2. pp 183–262, Macmillan, London.

Dalmastn, C., Valenti, P., Visca, P., Vittorioso, P and Orsi, N. (1988). Enhanced anti-microbial activity of lactoferrin by binding to the bacterial surface. Microbiologica 11, 225–230.

Day, C.L., Stowell, K.M., Baker, E.N. and Tweedie, J.W. (1992). Studies of the N-terminal half of human lactoferrin produced from the cloned cDNA demonstrate that interlobe interactions modulate iron release. J. Biol. Chem. 267: 13857–13862.

Garratt, R.C. and Jhoti, H. (1992) A molecular model for the rumour-associated antigen, p 97 suggests a Zn-binding function. FEBS Letters 305: 55–61.

Kawakami, H. and Lonnerdal, B. ( 1991 ). Isolation and function of a receptor for human lactoferrin in human fetal brush-border membranes. Am. J. Physiol. 261 (Gastrointest. Liver Physiol. 24): G841–G846.

Kunkel, T.A., Roberts, J.D. and Zakour, R.A. (1987) Rapid and efficient site-specific mutagenesis without phenotypic selection. Methods in Enzymol. ??? 154: 367–382.

Mazurier, J., Legrand, D., Hu, W-L., Montreuil, J. and Spik, G. (1989). Expression of human lactotransferrin receptors in phytohemaglutinin-stimulated human peripheral blood lymphocytes. Isolation of the receptors by antiligand affinity chromatography. Eur. J. Biochem. 179: 481–487.

Rado, T.A., Wei, X. and Benz, E.J., Jr. (1987). Isolation of lactoferrin cDNA from a human myeloid library and expression of mRNA during normal and leukemic myelopoiesis. Blood 70: 989–993.

Stowell, K.M., Rado, T.A., Funk, W.D. and Tweedie, J.W. (1991). Expression of cloned human lactoferrin in baby-hamster kidney cells. Biochem. J. 276: 349–355.

Tsuda, E., Goto, M., Murakami, A., Akai, K., Ueda, M., Kawanishi, G., Takahashi, N., Sasaki, R., Chiba, H., Ishihara, H., Mori, M., Tejima, S., Endo, S. and Arata, Y. (1988) Comparative study of N-linked oligosaccharide of urinary and recombinant erythropoietins. Biochemistry 27: 5646–5654

Ziere, G.J., van Dilk, M.C.M., Bijsterbosch, M.K. and van Berkel, T.J.C. (1992) Lactoferrin uptake by the rat liver. J. Biol. Chem. 267: 11229–11235.

# ANTIMICROBIAL PEPTIDES OF LACTOFERRIN

Mamoru Tomita, Mitsunori Takase, Hiroyuki Wakabayashi, and
Wayne Bellamy

Nutritional Science Laboratory
Morinaga Milk Industry Co. Ltd.
1-83-5 Higashihara, Zama City
Kanagawa 228, Japan

## SUMMARY

Lactoferrin was found to contain an antimicrobial sequence near its N-terminus which appears to function by a mechanism distinct from iron chelation. Antimicrobial peptides representing this domain were isolated following pepsin cleavage of human lactoferrin and bovine lactoferrin. The antimicrobial sequence was found to consist mainly of a loop of 18 amino acid residues formed by a disulfide bond between cysteine residues 20 and 37 of human lactoferrin, or 19 and 36 of bovine lactoferrin. The identified domain contains a high proportion of basic residues, like various other antimicrobial peptides known to target microbial membranes and it appears to be located on the surface of the folded protein allowing its interaction with surface components of microbial cells. The isolated domain, "lactoferricin", was shown to have potent broad spectrum antimicrobial properties and its effect was lethal causing a rapid loss of colony-forming capability. Such evidence points to the conclusion that this domain is the structural region responsible for the microbicidal properties of lactoferrin. The evidence also suggests the possibility that active peptides produced by enzymatic digestion of lactoferrin may contribute to the host defense against microbial disease.

## INTRODUCTION

Lactoferrin has broad-spectrum antimicrobial properties and appears to function as an important component of the host defense system active at mucosal surfaces, and in colostrum and milk. It has been suggested that the antimicrobial activity of this protein results from its ability to chelate iron and deprive microorganisms of this essential nutrient (Reiter, 1983; Finkelstein et al., 1983; Bullen et al., 1987). Evidence from several studies, however, suggests the existence of other mechanisms of action. Immunofluorescence studies indicate that lactoferrin binds directly to the surface of various microorganisms (Arnold et al., 1977; Dalmastri et al., 1988). Apo-lactoferrin causes a rapid loss of microbial viability (Arnold et al., 1977; Arnold et al., 1980; Bortner et al., 1986; Kalmar et al., 1988) that does not occur in iron-deficient medium alone (Arnold et al., 1982). It induces the release of lipopolysaccharide from the surface of Gram-negative bacteria and alters the permeability properties of the outer membrane (Ellison et al., 1988; Ellison et al., 1990). The structural features of the lactoferrin molecule responsible for these effects are unknown.

In the present study, to determine whether the entire lactoferrin molecule is required for activity, we have examined the antimicrobial properties of hydrolysates prepared by enzymatic digestion of lactoferrin. A potently active peptide representing a previously unknown antimicrobial domain within the lactoferrin molecule was isolated and characterized, providing new insight into the structure-function relationship responsible for the antimicrobial properties of lactoferrin (Some aspects of this work have been published elsewhere (Tomita et al., 1991; Bellamy et al., 1992).

## EXPERIMENTAL

### Enzymatic Hydrolysis of Lactoferrin

Human lactoferrin (Sigma) or bovine lactoferrin (Morinaga) was dissolved in distilled water at a concentration of 5% (w/v) and the pH was adjusted to 2.5 (aspartic proteinase) or 7.0 (neutral proteinase). Hydrolysis was performed using a proteinase concentration of 3% (w/w of substrate) at 37°C for 4 h. The reaction was terminated by heating at 80°C for 15 min. Hydrolysates were adjusted to pH 7.0, any precipitate formed was removed by centrifugation or filtration, and the soluble material was freeze-dried.

### Antimicrobial Assays

For assay of minimal inhibitory concentration (MIC) strains were cultured for 16 to 20 h in a basal medium of 1% peptone or PYG medium (containing 1% peptone, 0.05% yeast extract, and 1% glucose) supplemented with defined concentrations of each test agent. For assays of killing activity, strains were incubated in PYG medium with or without lactoferricin B. Initially, and after 60 min of treatment, serial 10-fold dilutions were prepared in basal medium and plated onto suitable agar media for enumeration. The limit of detection was 100 cfu/ml. Obligate anaerobes were incubated in an anaerobic glove box under an atmosphere of 85% $N_2$, 10% $CO_2$, and 5% $H_2$. The Campypak gas system (BBL) was used for cultivation of *Campylobacter jejuni*.

### Reverse-Phase HPLC

Lactoferrin hydrolysate was fractionated on a column of TSK-GEL 120T ($6.0 \times 150$ mm; Tosoh) using a mixture 80:20 of eluents A (0.05% TFA) and B (90% acetonitrile in 0.05% TFA) for 10 min followed by a linear gradient of A-B from 80:20 to 40:60 for 30 min at 0.8 ml/min. The active peptides obtained were further purified on a column of COSMOSIL 5CN-R ($4.6 \times 150$ min; Nacalai Tesque) eluted with eluent A for 5 min, followed by a linear gradient of eluents A:B from 100:0 to 55:45 for 40 min at 0.8 ml/min. After each step, fractions were collected, dried under vacuum in a centrifugal evaporator, and assayed for antibacterial activity. The purity of the isolated peptides was estimated to be at least 95%.

### Peptide Sequence Analysis

Peptide sequence analysis was performed using an Applied Biosystems Model 470A gas-phase protein sequencer.

## Animal Studies

Female Wistar rats (140 g) were fed a diet containing 40% bovine lactoferrin with spaced feeding for 4 days. Gastrointestinal contents (stomach, small intestine, caecum, and colon) were sampled at 2, 4, and 8 h after feeding. Samples were heated at 100°C for 15 min, homogenized, centrifuged at 18,000 rpm at 4°C for 15 min and the supernatants were filtered. The samples were then fractionated by HPLC as indicated above.

## RESULTS

To test whether active peptides are produced upon enzymatic digestion of lactoferrin, bovine lactoferrin was treated with various commercial proteases and the hydrolysates were assayed for activity against *E. coli* 0111 (Table 1). Hydrolysates produced at pH 2.5 using pig pepsin, cod pepsin, or acid protease from *Penicillium duponti* showed very strong antibacterial activity. In contrast, hydrolysates produced at pH 7.0 using trypsin, papain or other neutral proteases showed little or no activity. In the case of the aspartic proteases, the minimal concentration of hydrolysate required for complete inhibition of bacterial growth was about 100 µg/ml. With undigested lactoferrin, about 2000 µg/ml was required for the same inhibitory effect. These findings indicate that active peptides more potent than lactoferrin are produced by various enzymes similar to gastric pepsin. This enzyme is known to cleave at the carboxyl terminus of phenylalanine and leucine residues in the substrate (Webb, 1984).

In an effort to isolate the active peptides, the pepsin hydrolysate of bovine lactoferrin was fractionated by reverse- phase HPLC (data not presented). Each of the peptide peaks was collected and tested for activity against *E. coli* 0111. However, only one antimicrobial peptide was detected. We named this active peptide "lactoferricin". The letter B was appended to denote the bovine species. The isolated peptide was sequenced by automated Edman degradation and found to consist of 25 amino acid residues (Fig. 1A) corresponding to the sequence of residues 17 to 41 near the N-terminus of bovine lactoferrin (Pierce et al., 1991). It has a somewhat

**Table 1.** Activity of Lactoferrin Hydrolysates against *E. coli*

| Enzyme for Hydrolysis | Rel. Activity | (MIC) |
|---|---|---|
| Pepsin, pig | +++ | (100 µg/ml) |
| Pepsin, cod | +++ | (100 µg/ml) |
| Protease, *Penicillium duponti* | +++ | (100 µg/ml) |
| Trypsin, pig | – | |
| Papain, papaya | – | |
| Protease, *Streptomyces griseus* | – | |
| Protease, *Aspergillus oryzae* | – | |
| Bioprase, *Bacillus subtilis* | – | |
| CONTROL, undigested lactoferrin | + | (2000 µg/ml) |

*E. coli* 0111 was cultured in 1% peptone, pH 6.8.
Cell viability was assayed after 14 h at 37°C.
MIC, minimal inhibitory concentration.

A

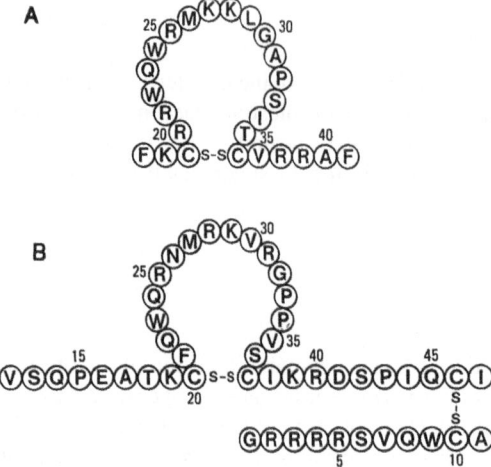

B

**Figure 1.** Primary Structure of Lactoferricin B (A) and Lactoferricin H (B). Single-letter code is used to indicate the amino acid sequence of each peptide. Numbers indicate sequence positions in bovine and human lactoferrins.

circular structure due to the presence of a disulfide bond between its two cysteine residues. The molecular weight of this peptide is 3,126 as estimated from the amino acid sequence. We have isolated the same active peptide from the gastric contents of rats fed bovine lactoferrin (Fig. 2). This peptide was collected and sequenced, and its primary structure was confirmed to be identical to lactoferricin B. This is the first evidence to demonstrate that active peptides of lactoferrin can be generated by gastric pepsin digestion *in vivo*.

To establish whether similar active peptides may be produced from lactoferrins of other mammalian species, the antimicrobial properties of hydrolysates of human lactoferrin were investigated. In the case of human lactoferrin, as well, a single antimicrobial peptide generated by pepsin cleavage was isolated and sequenced (Fig. 1B). This active peptide was named "lactoferricin H", where H denotes human. Its sequence consists of 47 amino acid residues, corresponding to the sequence of residues 1 to 47 at the N-terminal end of human lactoferrin

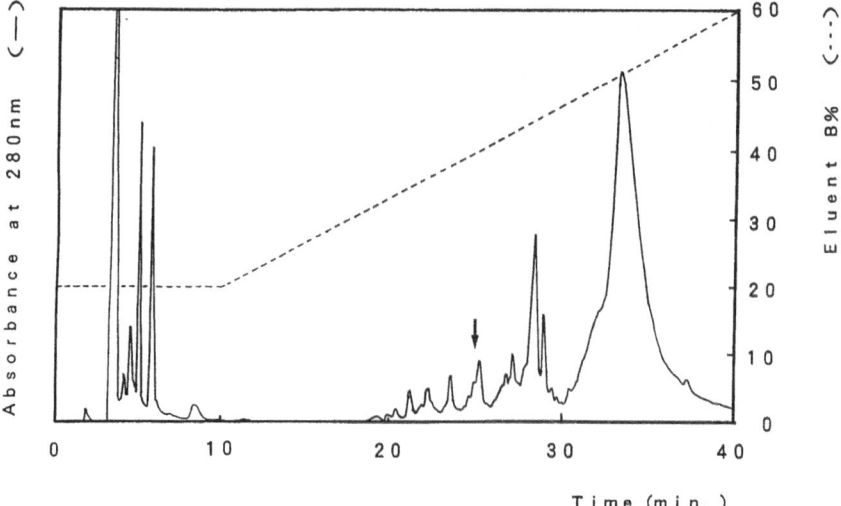

**Figure 2.** Reverse-phase HPLC profile of the gastric contents of rats fed bovine lactoferrin. Arrow indicates the position of the active peptide recovered for sequence analysis.

**Table 2.** Activity of Lactoferrin and its Peptides against *E. coli* 0111

| Lactoferrin or derivative | MIC | |
|---|---|---|
| | µg/ml | (µM) |
| Human lactoferrin | 3000 | (37) |
| Human lactoferrin, pepsin hydrolysed | 500 | |
| Lactoferricin H | 100 | (18) |
| Bovine lactoferrin | 2000 | (25) |
| Bovine lactoferrin, pepsin hydrolysed | 100 | |
| Lactoferricin B | 6 | (2) |

MIC, minimal inhibitory concentration.

(Metz-Boutigue et al., 1984). It includes two subfragments connected by a disulfide bond. The total molecular weight is 5,558 as estimated from the amino acid sequence. Notably, the sequence of lactoferricin H includes the same loop region present in lactoferricin B.

The relative effectiveness of human lactoferrin, bovine lactoferrin, and the active peptides was examined against *E. coli* 0111 (Table 2). In the case of human lactoferrin, as well as bovine lactoferrin, the hydrolysate produced by pepsin cleavage showed stronger activity than the undigested protein. By comparing MIC values on a molar basis, it could be estimated that lactoferricin H is about 2-fold more effective than human lactoferrin and lactoferricin B is about 12-fold more effective than bovine lactoferrin. Lactoferricin B appears to be about 9-fold more effective than lactoferricin H.

In an attempt to define the range of susceptible organisms, the effectiveness of lactoferricin B was tested against a variety of microbial species (Table 3). Concentrations required to cause complete inhibition of growth varied within the range of 0.3 to 45 µg/ml depending on the strain and the medium used. The susceptible organisms represent a physiologically diverse range of types, including Gram-negative and Gram-positive bacteria (rods, cocci, aerobes and anaerobes), yeasts and molds. Interestingly, certain microbial species, including *Pseudomonas fluorescens*, *Enterococcus faecalis*, and *Aspergillus fumigatus*, were not susceptible to inhibition. We intend to investigate the physiological features responsible for their unusual resistance to this peptide.

To determine whether the effect of lactoferricin B against susceptible organisms is lethal the number of viable cells was monitored following exposure to the active peptide. Using *E. coli* 0111 (Fig. 2A) or *Campylobacter jejuni* JCM-2013 (Fig. 2B) as the test organism, no loss of viability was observed with the untreated control; however, in the presence of the active peptide, the number of viable cells decreased rapidly in a concentration-dependent manner. In a similar experiment, other bacterial strains were treated with lactoferricin B for 60 minutes and the number of viable cells remaining was compared with untreated controls (Table 4). Lactoferricin B caused a profound loss of cell viability with many of the strains examined. *E. coli*, *Klebsiella*, *Bacteroides*, *Staphylococcus*, *Streptococcus*, *Corynebacterium*, and *Listeria* strains were highly susceptible to its killing effect. *Bifidobacterium* species are known to colonize the intestinal tract of healthy breast-fed infants (Benno et al., 1984). Bifidobacteria showed some loss of cell viability in the presence of lactoferricin B, but these bacteria appear to be less susceptible to its killing effect than many opportunistic pathogens. It is tempting to speculate that active peptides of lactoferrin might help to promote the establishment of a bifidobacterial flora in nursing infants. Of course, further studies are required to test this hypothesis.

**Table 3.** Susceptibility of Various Microorganisms to Inhibition by Lactoferricin B

| | MIC (µg/ml) | |
|---|---|---|
| Strain | Peptone | PYG |
| **Gram-negative bacteria** | | |
| *Escherichia coli* IID-861 | 6 | 9 |
| *Salmonella enteritidis* IID-604 | 12 | 18 |
| *Klebsiella pneumoniae* JCM-i662T | 9 | 12 |
| *Proteus vulgaris* JCM-i668T | 12 | 45 |
| *Yersinia enterocolitica* IID-981 | 6 | 24 |
| *Pseudomonas aeruginosa* IF0-3446 | 9 | 24 |
| *Pseudomonas fluorescens* IF0-14160 | > 60 | > 60 |
| **Gram-positive bacteria** | | |
| *Staphylococcus aureus* JCM-2179 | 3 | 6 |
| *Streptococcus mutans* JCM-5705T | 2 | 6 |
| *Listeria monocytogenes* JCM-7673 | 0.3 | 1 |
| *Corynebacterium renale* JCM-1322 | 0.6 | 1 |
| *Bacillus subtilis* ATCC-6633 | 0.6 | 2 |
| *Clostridium perfringens* ATCC-6013 | 12 | 24 |
| *Enterococcus faecalis* ATCC-E19433 | > 60 | > 60 |
| **Yeasts** | | |
| *Candida albicans* JCM-i542T | 18 | 24 |
| *Candida albicans* JCM-2900 | 18 | 24 |
| *Cryptococcus uniguttulatus* JCM-3685 | 1 | 6 |
| **Molds** | | |
| *Nannizia incurvata* JCM-1906 | 3 | 9 |
| *Sporothrix cyanescens* JCM-2114 | 9 | 18 |
| *Penicillium vermiculatum* JCM-5595 | 6 | 45 |
| *Aspergillus fumigatus* JCM-1739 | > 60 | > 60 |

MIC, minimal inhibitory concentration.

The active peptide, "lactoferricin", appears to represent a lethal domain released from the lactoferrin molecule by enzymatic cleavage. The antimicrobial sequence consists mainly of a loop of 18 amino acid residues formed by a disulfide bond between cysteine residues 20 and 37 of human lactoferrin, or 19 and 36 of bovine lactoferrin (Fig. 1), located near the N-terminus of the lactoferrin molecule. The sites of glycosylation, and the metal- binding residues of lactoferrin are known to be located in other regions of the molecule (Metz-Boutigue et al., 1984; Pierce et al., 1991). It seems likely that the identified domain contributes to the antimicrobial activity of the intact protein. Moreover, it would appear that the mechanism of action of this domain does not require the metal-chelating functions of lactoferrin.

A striking feature of the identified domain is the relatively high proportion of basic amino acid residues. In the case of bovine lactoferrin 8 of the 25 residues in this region are basic residues. The corresponding sequence of human lactoferrin contains 7 basic residues. As shown in Table 4, a similar feature is observed in various other antimicrobial

**Table 4.** Lethal Effect of Lactoferricin B against Various Bacteria.

| Microorganism | Mean CFU/ml | | Survival (%) |
|---|---|---|---|
| | Control | Treated | |
| *Escherichia coli* IID-861 | $1.2 \times 10^5$ | < 100 | < 0.08 |
| *Klebsiella pneumoniae* JCM-i662T | $3.2 \times 10^6$ | < 100 | < 0.01 |
| *Bacteroides vulgatus* MMI-S601 | $6.0 \times 10^5$ | 500 | 0.08 |
| *Staphylococcus aureus* JCM-2413 | $2.4 \times 10^6$ | 400 | 0.02 |
| *Streptococcus mutans* JCM-5175 | $3.0 \times 10^4$ | < 100 | < 0.33 |
| *Corynebacterium renale* JCM-1322 | $6.4 \times 10^5$ | < 100 | < 0.02 |
| *Listeria monocytogenes* JCM-7673 | $5.1 \times 10^4$ | < 100 | < 0.20 |
| *Bifidobacterium bifidum* ATCC-15696 | $5.0 \times 10^4$ | $5.7 \times 10^4$ | 114.0 |
| *Bifidobacterium breve* ATCC-15700 | $4.0 \times 10^5$ | $4.6 \times 10^4$ | 12.0 |
| *Bifidobacterium longum* ATCC-15707 | $4.8 \times 10^5$ | $3.1 \times 10^4$ | 7.0 |

Lactoferricin B, none (control) or 31 µg/ml (treated). Viable cells were assayed after 60 min incubation in PYG medium.

peptides, for example, magainins from frog skin (Bevins and Zasloff, 1990), cecropins from the hemolymph of insects (Boman and Haltmark, 1987), and defensins from mammalian neutrophils (Lehrer et al., 1991). These cationic peptides are known to display an affinity for biological membranes. They have the ability to alter the permeability properties of artificial membranes (Christenson et al., 1988; Westerhoff et al., 1989; Hill et al., 1991) and their lethal effect against various microorganisms is thought to result from the ability to disrupt essential functions of the cytoplasmic membrane. It seems likely that the lethal domain of lactoferrin functions by a similar mechanism of action.

**Table 5.** Comparison of Lactoferricin and Other Antimicrobial Peptides

| Peptide | Residues | Sequence |
|---|---|---|
| Lactoferricin B | 25 | F-**K**-C-**RR**-WQW-**R**-M-**KK**-LGAPSITCV-**RR**-AF |
| Lactoferricin H | 25 | T-**K**-CFQWQ-**R**-NM-**RK**-V-**R**-GPPVSCI-**KR**-DS |
| Magainin 1 | 23 | GIG-**K**-FL-**H**-SAG-**K**-FG-**K**-AFVGEIM-**K**-S |
| Magainin 2 | 23 | GIG-**K**-FL-**H**-SA-**KK**-FG-**K**-AFVGEIMNS |
| Cecropin A | 37 | **K**-W-**K**-LF-**KK**-IE-**K**-VGQNI-**R**-DGII-**K**-AGPAVAVVGQATQAI-**K*** |
| Cecropin B | 35 | **K**-W-**K**-VF-**KK**-IE-**K**-MG-**R**-NI-**R**-NGIV-**K**-AGPAIAVLGEA-**K**-AL* |
| Defensin NP-1 | 33 | VVCAC-**RR**-ALCLP-**R**-Q-**RR**-AGFC-**R**-I-**R**-G-**R**-I-H-PLCC-**RR** |
| Defensin NP-2 | 33 | VVCAC-**RR**-ALCLPLQ-**RR**-AGFC-**R**-I-**R**-G-**R**-I-H-PLCC-**RR** |

Single-letter code is used to indicate the amino acid sequence of each peptide.
Basic residues are separated from adjoining residues by hyphens and are shown in **boldface**.
An **asterisk** indicates that the C-terminal residue is amidated.

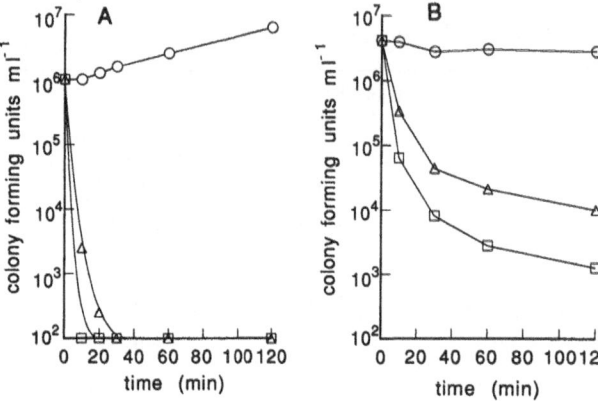

**Figure 3.** Lethal Effect of Lactoferricin B against *E. coli* 0111 (A) and *Campylobacter jejuni* JCM-2013 (B). Each strain was incubated at 30°C in PYG medium containing lactoferricin B at 15 µg/ml (△) or 31 µg/ml (□). Control (○) had no added peptide. The limit of detection was $10^2$ CFU/ml.

In apo-lactoferrin, this cationic domain is located on the outer surface of the folded protein (Fig. 4) where it seems to be available for interaction with target sites on the surface of microbial cells. Dr. Richard Ellison now working at the University of Massachusetts has reported that apo-lactoferrin directly damages the outer membrane of Gram-negative bacteria (Ellison et al., 1988; Ellison et al., 1990), increasing their sensitivity to killing by lysozyme (Ellison and Giehl, 1991). In collaboration with Dr. Ellison, it has been demonstrated that lactoferricin B displays similar activity. In fact, the active peptide is even more potent than apo-lactoferrin. Such evidence suggests the identified domain is the structural region responsible for the membrane-disruptive properties of apo-lactoferrin.

It has been reported that saturation of lactoferrin with iron abolishes its bactericidal properties (Arnold et al., 1977) and its ability to release LPS from the outer membrane of Gram-negative bacteria (Ellison et al., 1988). A substantial conformational change in the

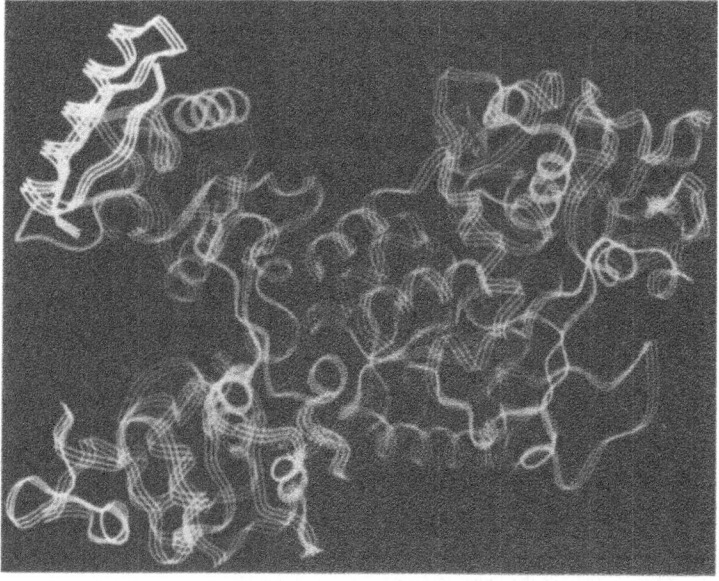

**Figure 4.** Three-Dimensional Structure of Apo-lactoferrin. The antimicrobial domain identified in the present study is highlighted. (Diagram kindly provided by Professor E.N. Baker, Massey University, New Zealand).

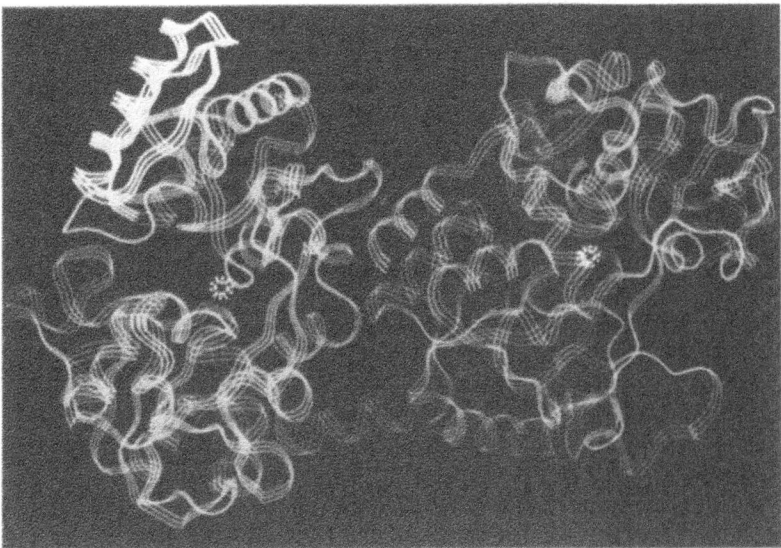

**Figure 5.** Three-Dimensional Structure of Holo-lactoferrin. The antimicrobial domain identified in the present study is highlighted. (Diagram kindly provided by Professor E.N. Baker, Massey University, New Zealand).

lactoferrin molecule occurs upon binding of iron, especially in the N-lobe containing the highlighted domain (compare Fig. 4 and 5). Nevertheless, the position of the antimicrobial sequence in holo-lactoferrin is very similar to apo-lactoferrin. The conformational change does not seem to limit its availability on the surface of the folded molecule. It remains to be determined why the activity of this domain is lost in the presence of iron.

## CONCLUSIONS

We have demonstrated that a potently active peptide is generated upon pepsin digestion of lactoferrin, both *in vitro* and *in vivo*. This peptide, named "lactoferricin", has potent antimicrobial properties. A wide variety of bacteria and fungi appear to be susceptible to its lethal effect, suggesting the possibility that active peptides of lactoferrin might contribute to the host defense against microbial disease. If such peptides are produced in substantial quantities *in vivo* they may have important physiological significance, especially in neonates. Our study has also led to the identification of a previously unknown antimicrobial sequence within the lactoferrin molecule. The evidence suggests this domain is the structural region responsible for the membrane-disruptive properties of lactoferrin and its lethal effect against various microorganisms. Hopefully, further studies will lead to a full understanding of the antimicrobial mechanism of lactoferrin.

## REFERENCES

Arnold, R.R., Cole, M.F., McGhee, J.R. (1977) Bactericidal activity of human lactoferrin. Science 127: 263–265.

Arnold, R.R., Brewer, M., Gauthier, J. (1980) Bactericidal activity of human lactoferrin: sensitivity of a variety of microorganisms. Infect Immun 28: 893–898.

Arnold, R.R., Russell, J.E., Champion, W.J., Brewer, M., Gauthier, J. (1982) Bactericidal activity of human lactoferrin: differentiation from the stasis of iron deprivation. Infect Immun 35: 792–797.

Bellamy, W., Takase, M., Yamauchi, K., Wakabayashi, H., Kawase, K., Tomita, M. (1992) Identification of the bactericidal domain of lactoferrin. Biochim et Biophys Acta 1121: 130–136.

Benno, Y., Sawada, K., Mitsuoka, T. (1984). The intestinal microflora of infants: composition of fecal flora in breast-fed and bottle-fed infants. Microbiol and Immunol 28: 975–986.

Bevins, C.L., Zasloff, M. (1990) Peptides from frog skin. Ann Rev Biochem 59: 395–414.

Boman, H.G., Hultmark, D. (1987) Cell-free immunity in insects. Ann Rev Microbiol 41: 103–126.

Bortner, C.A., Miller, R.D., Arnold, R.R. (1986) Bactericidal effect of lactoferrin on *Legionella pneumophila*. Infect and Immun 51: 373–377.

Bullen, J.J., Rogers, H.J., Griffiths, E. (1987) Role of iron in bacterial infection. Current Topics in Microbiology and Immunology, 80: 1–35.

Christenson, B., Fink, J., Merrifield, R.B., Mauzerall, D. (1988) Channel-forming properties of cecropins and related model compounds incorporated into planar lipid membranes. Proc Nat Acad Sci U.S.A. 85: 5072–5076.

Dalmastri, C., Valenti, P., Visca, P. Vittorioso, P., Orsi, N. (1988) Enhanced antibacterial activity of lactoferrin by binding to the bacterial surface. Microbiologica 11: 225–230.

Ellison, R.T., Giehl, T.J., Laforce, F.M. (1988) Damage of the outer membrane of enteric Gram-negative bacteria by lactoferrin and transferrin. Infect and Immun 56: 2774–2781.

Ellison, R.T., Laforce, F.M., Giehl, T.J., Boose, D.S., Dunn, B.E. (1990) Lactoferrin and transferrin damage of the Gram-negative outer membrane is modulated by $Ca^{2+}$ and $Mg^{2+}$. Gen Microbiol 136: 1437–1446.

Ellison, R.T., Giehl, T.J. (1991) Killing of Gram-negative bacteria by lactoferrin and lysozyme. Clin Invest 88: 1080–1091.

Finkelstein, R.A., Sciortino, C.V., McIntosh, M.A. (1983) Role of iron in microbe-host interactions Rev Infect Dis 5: 5759–5777.

Hill, C.P., Yee, J., Selsted, M.E., Eisenberg, D. (1991) Crystal structure of Defensin HNP-3, an amphiphilic dimer: mechanisms of membrane permeabilization. Science 251: 1481–1485.

Kalmar, J.R., Arnold, R.R. (1988) Killing of *Actinobacillus actinomycetemcomitans* by human lactoferrin. Infect and Immun 56: 2552–2557.

Lehrer, R.I., Ganz, T., Selsted, M.E. (1991) Defensins: endogenous antibiotic peptides of animal cells. Cell 64: 229–230.

Metz-Boutigue, M.M., Jolles, J., Mazurier, J., Schoentgen, F., Legrand, D., Spik, G., Montreuil, J., Jolles, P. (1984) Human lactotransferrin: amino acid sequence and structural comparisons with other transferrins. Eur J Biochem 145: 659–678.

Pierce, A., Colavizza, D., Bennaissa, M., Maes, P., Tartar, A., Montreuil, J., Spik, G. (1991) Molecular cloning and sequence analysis of bovine lactotransferrin. Eur J Biochem 196: 177–184.

Reiter, B. (1983) The biological significance of lactoferrin. Int J Tissue Reactions 5: 87–96.

Tomita, M., Bellamy, W., Takase, M., Yamauchi, K., Wakabayashi, H., Kawase, K. (1991) Potent antibacterial peptides generated by pepsin digestion of bovine lactoferrin. J Dairy Sci 74: 4137–4142.

Webb, E. C., Enzyme Nomenclature. Academic Press, San Diego, (1984).

Westerhoff, H.V., Juretic, D., Hendler, R.W., Zasloff, M. (1989) Magainins and the disruption of membrane-linked free-energy transduction. Proc Nat Acad Sci U.S.A. 86: 6597–6601.

# PHYSICOCHEMICAL AND ANTIBACTERIAL PROPERTIES OF LACTOFERRIN AND ITS HYDROLYSATE PRODUCED BY HEAT TREATMENT AT ACIDIC pH

Hitoshi Saito, Mitsunori Takase, Yoshitaka Tamura, Seiichi Shimamura, and Mamoru Tomita

Nutritional Science Laboratory
Morinaga Milk Industry Co., Ltd.
1-83-5 Higashihara, Zama City
Kanagawa 228, Japan

## ABSTRACT

In order to apply functionally active lactoferrin (Lf) to food products, the effect of pH on the heat stability of Lf was studied. Lf was easily denatured to an insoluble state by heat treatment under neutral or alkaline conditions, above pH 6. In contrast, it remained soluble after heat treatment under acidic conditions at pH 2 to 5, and the HPLC pattern of Lf heat-treated at pH 4 at 100°C for 5 min was the same as that of native Lf. Lf was found to be very thermostable at pH 4, and could be pasteurized or sterilized without any significant loss of its physicochemical properties. Lf was hydrolyzed by heat treatment at pH 2 to 3 at above 100°C, and its iron binding capacity and antigenicity were lost. But the antibacterial activity of the hydrolysate was found to be much stronger than that of native Lf. The antibacterial component of Lf hydrolysate produced by heat treatment at acidic pH was verified to be a peptide including the sequence of residues 1-54 from the N-terminal end of the bovine Lf molecule.

## INTRODUCTION

Lf is well known to be an iron binding glycoprotein in physiological fluids, such as milk. Many biological functions have been attributed to Lf, including antibacterial activity, growth promotion for *Bifdobacterium*, regulation of iron absorption, cell growth activity, and regulation of the immune system. Bovine Lf is now purified from skim milk or sweet cheese whey on an industrial scale by using an ion exchange resin, and Lf is expected to be useful as a functionally active component for use in foods, drugs, and cosmetics. Generally, heat treatment for pasteurization is necessary to produce these products. When Lf is applied to these products as a bioactive component, the heat stability of Lf is very important. It has been reported that Lf is easily inactivated in milk (5) or buffer (2) by heat treatment at near neutral pH. However, we have found that Lf is thermostable under acidic conditions, especially at pH 4 (1). While examining the heat stability of Lf, we found that the antibacterial activity of acid hydrolysate of Lf was much stronger than that of native Lf (6). In this paper, these phenomena are described further, and the amino acid sequences of the antibacterial peptides generated by acid hydrolysis were determined.

## EXPERIMENTAL

### Heat Treatment

Bovine Lf was purchased from Milei Produkte nach Maß (Stuttgart, Germany). The purity of the Lf obtained was 99%, and the amount of iron was 20.1 mg/100 g of Lf.

Lf was dissolved in distilled water (5%;wt/wt), and adjusted to pH 2 to 11 with HCl or NaOH. Solutions of Lf at various pH were heat-treated at 80 to 120°C for 5 min, and were cooled rapidly after heat treatment.

### High Performance Liquid Chromatography

Each heat-treated Lf solution was filtered through a 0.45 μm filter and 25 μL of the filtrate was applied to a reverse-phase HPLC column (Asahipac C4P-50, 4.6 × 150 mm, Asahi Chemical Industry Co., Ltd.). The following running conditions were used: eluent, 0.03% TFA/CH$_3$CN; gradient, 30–50% CH$_3$CN in 25 min; Flow rate: 0.8 ml/min; and detection, UV-280 nm.

### Iron Binding Capacity

Heat-treated and native Lf were dissolved in Tris buffer (pH 7.6) to give a final concentration of 1%, and the protein was saturated with iron by the addition of Fe-nitrilo-triacetic acid solution. The iron binding capacity was determined spectrophotometrically at 464 nm, the characteristic color of Fe-Lf.

### Antigenicity

Antigenicity of heat-treated Lf was determined by immunoelectrophoresis or an ELISA inhibition test using rabbit anti-bovine Lf antiserum. Remaining antigenicity was calculated comparatively with that of native Lf.

### Antibacterial Activity

**Growth Inhibitory Effect.** Heat-treated Lf or native Lf were added to 1% peptone medium to give a final concentration of 10 to 1000 μg/ml. Precultured *Escherichia coli* O-111 cells were added using an inoculum of 10$^6$ CFU/ml, and the cultures were incubated aerobically at 37°C for 10 h. Bacterial growth was determined spectrophotometrically at 660 nm, and the growth rate (%) was calculated. (Growth rate (%) = [Abs. 660 nm (experiment)/Abs. 660 nm (control)] × 100).

**Bactericidal Effect.** The acid hydrolysate of Lf or native Lf was added to 0.01 M phosphate buffer (pH 7.0) to give a final concentration of 100 μg/ml. The washed bacterial cells after centrifugation were suspended in 0.85% saline, and added to the buffer solutions using an inoculum of $10^7$ CFU/ml. The cell suspensions were incubated aerobically at 37°C for 5 h. After incubation, colony forming units (CFU) were enumerated.

### Fractionation of Acid Hydrolysate of Lf

The acid hydrolysate of Lf was fractionated by reverse-phase HPLC (column VYDAC-C4, 4.6 × 250 mm, The Separations Group, Hesperia, CA), and separated into twelve fractions. The following conditions were used: eluent, 0.05% TFA/CH$_3$CN; gradient, 10–40% CH$_3$CN in 5-65 min; flow rate, 1.0 ml/min; and detection, UV-220 nm. Each of these fractions was assayed for growth inhibitory effect, and bacterial growth rates (%) were calculated. Amino acid sequences of those isolated peptides which had strong antibacterial activity, were analysed using a peptide sequencer (477 A, Applied Biosystems Inc., San Jose, CA).

## RESULTS AND DISCUSSION

### Physicochemical Properties of Heat-Treated Lf

Solutions of Lf at various pH were heated, and the change of the solutions was observed visually. Solutions heat-treated at pH 2 to 5 remained soluble, but turbidity and gel formation occurred after heat treatment at pH 6 to 11. Upon heat treatment at pH above 11, Lf was hydrolysed by alkali and emitted a sulfurous smell. Figure 1 shows the HPLC patterns of Lf heat-treated at pH 2 to 6 at 90 to 120°C for 5 min. The HPLC patterns of Lf heat-treated at 90°C were similar to that of native Lf, but at pH 2, 3, and 6, the peak of Lf was slightly decreased. After heat treatment at 100°C at pH 2 or 3, minor peaks were observed, that represent degraded Lf fragments generated by acid hydrolysis. In contrast, a broad peak was observed after heat treatment at 100°C at pH 6, that probably represents a degraded or conformationally altered form of Lf. The HPLC patterns of Lf heat treated at 100°C at pH 4 and 5 were similar to that of native Lf. By heat treatment at 120°C for 5 min, Lf was degraded independent of pH.

Figure 2 shows the iron binding capacity of heat-treated Lf under various pH conditions. At pH 6, the iron binding capacity of Lf was diminished slightly by heat treatment at 80 or 90°C, but that of Lf heat-treated at pH 2 to 5 at 80 to 90°C was almost the same as that of native Lf. The iron binding capacity of Lf heat-treated at pH 4 and 100°C was about 90%, compared with that of native Lf. The iron binding capacity of Lf heat-treated at pH 2, 3, or 5 was decreased

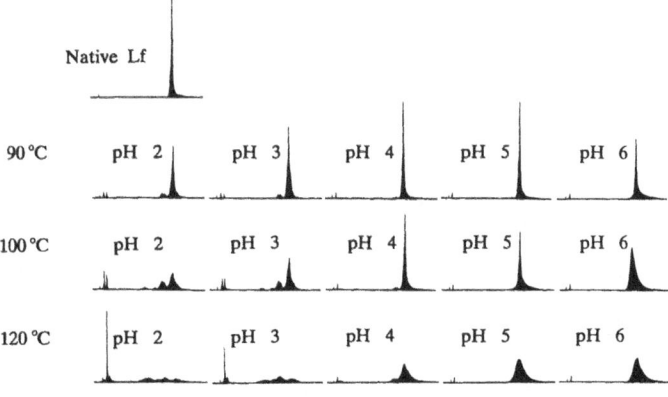

**Figure 1.** HPLC pattern of heat-treated Lf.

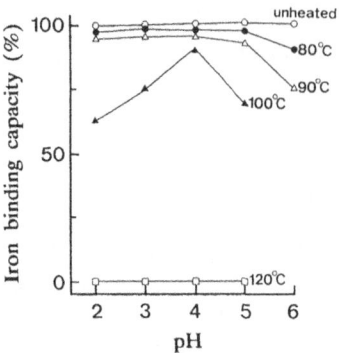

**Figure 2.** Iron binding capacity of heat-treated Lf.

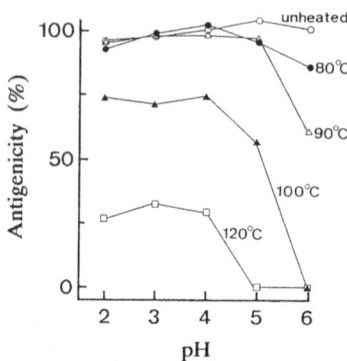

**Figure 3.** Antigenicity of heat-treated Lf.

to 60 to 75%, and at pH 6, the capacity was almost abolished. Upon heat treatment at 120°C, all samples lost iron binding capacity.

Antigenicity of Lf heat-treated at various pH conditions was analysed by immunoelectrophoresis (Figure 3). Lf was stable at acidic pH, especially at pH below 4, but was unstable near pH 6. Some antigenicity was retained after heating at 120°C at pH 2, 3, or 4. This might represent remaining antigenicity of Lf fragments generated by heat treatment at acidic pH.

The growth inhibitory effect of Lf heat-treated at 100°C at pH 5 or 6 for 5 min was almost abolished, but that of Lf heat-treated at pH 4 was similar to that of native Lf (Figure 4). After heat treatment at 120°C at pH 5 or 6, the growth inhibitory effect of heat-treated Lf was completely abolished. The effect of Lf heat-treated at 100°C or 120°C at pH 2 or 3 for 5 min was much stronger than that of native Lf, although the Lf was hydrolysed. This result suggests that some of the Lf fragments generated by acid hydrolysis had strong antibacterial activity. From the HPLC elution patterns and physicochemical properties, Lf appeared to be most thermostable at pH 4. These phenomena might be related to the isoelectric point of Lf (about 8) (4). Generally, proteins are most unstable at their isoelectric points, and proteins are hydrolysed by heat treatment at strong acid or alkali pH. For Lf, pH 4 is probably a balance point, that is far from its isoelectric point and not so strong acid conditions hydrolysed the protein.

In order to apply Lf to general food products, applicability of ultra high temperature (UHT) sterilization and stability in drinks were investigated. Lf was dissolved in distilled water

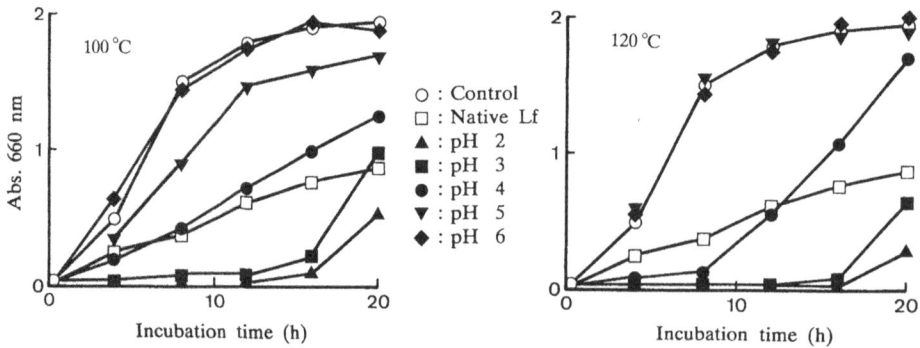

**Figure 4.** Antibacterial activity of heat-treated Lf.

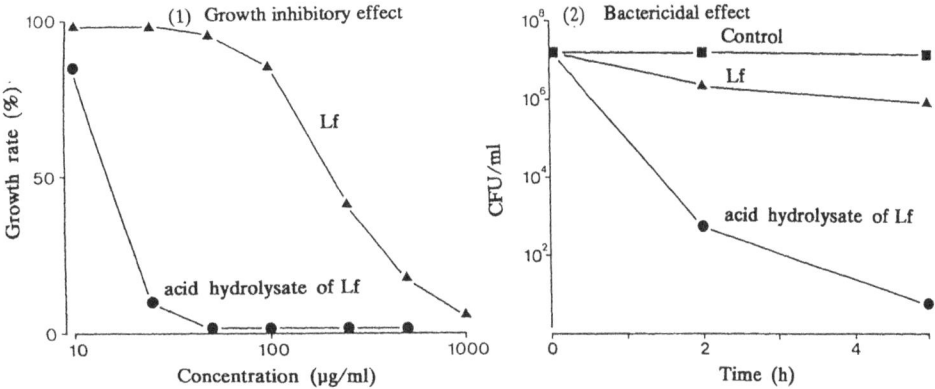

**Figure 5.** Antibacterial activity of acid hydrolysate of Lf. (1) Growth inhibitory effect. (2) Bactericidal effect.

to give a final concentration of 1%, and the pH was adjusted to 4 by HCl. The solution was sterilized (70°C for 3 min followed by 130°C for 2 sec) at our UHT test plant. After UHT sterilization, the iron binding capacity retained was 97% compared with native Lf. Stability in drinks was investigated as follows. Lf was dissolved in orange juice (pH 4.5, ingredients: sugar 5.0%, orange perfume 0.2%, citric acid 0.1%, sodium citrate 0.05%, carbonated water 94.65%), and was heated at 80°C for 15 min at our retort test plant. After heat treatment, the sample was cooled and stored at 5°C for 2 weeks. The remaining Lf was determined by HPLC, and was 90% of that in an unheated sample.

### Antibacterial Activity of Acid Hydrolysate of Lf

Antibacterial activity of Lf heat-treated at 100 or 120°C at pH 2 or 3 for 5 min was found to be stronger than that of native Lf, although it was hydrolysed to many fragments, so more suitable conditions for making antibacterial hydrolysate of Lf were investigated. Antibacterial activity of heat-treated Lf treated at 120°C at pH 2 for 15 min was verified to be very strong. This acid hydrolysate of Lf had no iron binding capacity, and had less antigenicity $(1/10^6)$ than that of native Lf as indicated by ELISA inhibition test. Figure 5-(1) shows the growth inhibitory effect of this acid hydrolysate of Lf and native Lf. Lf showed some growth inhibitory effect at concentrations above 100 µg/ml, but did not completely inhibit bacterial growth even at 1000 µg/ml. The acid hydrolysate of Lf showed growth inhibitory effect even at 10 µg/ml and completely inhibited bacterial growth at 50 µg/ml. Its activity was not affected by addition of iron (data not shown). This means that the antibacterial activity of the acid hydrolysate differs from that of native Lf. Figure 5-(2) shows the bactericidal effect of the acid hydrolysate of Lf and native Lf. Acid hydrolysate of Lf showed a very strong bactericidal effect, and the CFU decreased to about $1/10^6$ of inoculum after 5 h incubation while the activity of native Lf was not as strong.

Fractionation of the acid hydrolysate of Lf by reverse-phase HPLC column is shown in Figure 6, and the growth inhibitory effect of each fraction is shown in Table 1. The threshold concentration of the growth inhibitory effect of native Lf was about 100 µg/ml, and that of acid hydrolysate (pre-fractionation) was 10 to 50 µg/ml. Fractions 5, 6, 7, and 8, especially fraction 5, showed very strong activity, and the threshold concentration of the growth inhibitory effect of fraction 5 was less than 5 µg/ml. Fraction 5 inhibited bacterial growth completely at only 10 µg/ml. Antibacterial peptides in fractions 5, 6, 7, and 8 were analysed using a peptide sequencer, and the N-terminus of the peptides in these fractions was confirmed to be Ala-Pro-

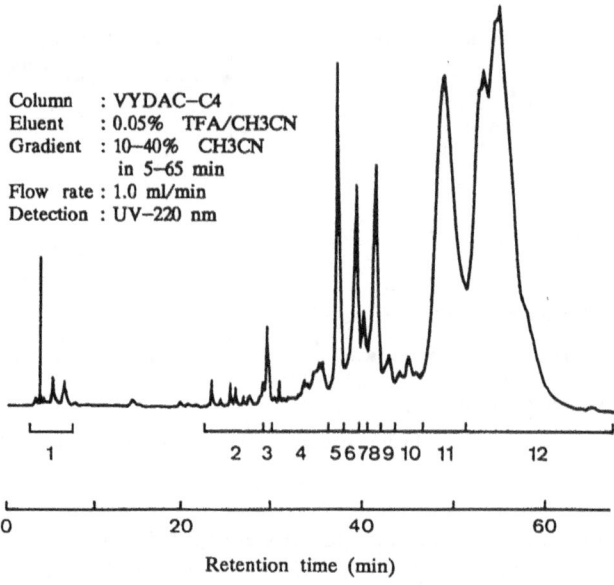

**Figure 6.** Fractionation of acid hydrolysate of Lf.

**Table 1.** Growth Inhibitory Effect of HPLC Fractions of Acid Hydrolysate of Lf.[a]

| Incubation time: | 5 h | | | | | 10 h | | | | |
|---|---|---|---|---|---|---|---|---|---|---|
| Concentration (μg/ml): | 5 | 10 | 50 | 100 | 250 | 5 | 10 | 50 | 100 | 250 |
| Native Lactoferrin (Lf) | 104 | 91 | 88 | 86 | 86 | 97 | 97 | 99 | 86 | 31 |
| Acid hydolysis of Lf | 102 | 75 | 9 | 4 | | 96 | 97 | 2 | 1 | |
| Fraction 1 | 101 | 96 | 144 | | | 99 | 101 | 117 | | |
| 2 | 107 | 109 | 108 | | | 102 | 105 | 110 | | |
| 3 | 90 | 90 | 99 | | | 96 | 95 | 101 | | |
| 4 | 100 | 106 | 36 | | | 100 | 101 | 69 | | |
| 5 | 42 | 3 | 2 | | | 84 | 1 | 0 | | |
| 6 | 93 | 2 | 4 | | | 92 | 13 | 2 | | |
| 7 | 105 | 9 | 6 | | | 94 | 18 | 1 | | |
| 8 | 108 | 10 | 10 | | | 90 | 21 | 1 | | |
| 9 | 99 | 98 | 9 | | | 95 | 85 | 2 | | |
| 10 | 99 | 70 | 10 | | | 92 | 88 | 3 | | |
| 11 | 109 | 104 | 103 | | | 96 | 95 | 101 | | |
| 12 | 106 | 105 | 109 | | | 95 | 96 | 86 | | |

[a]Growth rate (%)

```
  1  APRKNVRWCTISQPEWFKCRRWQWRMKKLGAPSITCVRRAFALECIRAIAEKKADAVTLD
 61  GGMVFEAGRDPYKLRPVAAEIYGTKESPQTHYYAVAVVKKGSNFQLDQLQGRKSCHTGLG
121  RSAGWVIPMGILRPYLSWTESLEPLQGAVAKFFSASCVPCIDRQAYPNLCQLCKGEGENQ
181  CACSSREPYFGYSGAFKCLQDGAGDVAFVKETTVFENLPEKADRDQYELLCLNNSRAPVD
241  AFKECHLAQVPSHAVVARSVDGKEDLIWKLLSKAQEKFGKNKSRSFQLFGSPPGQRDLLF
301  KDSALGFLRIPSKVDSALYLASRYLTTLKNLRETAEEVKARYTRVVWCAVGPEEQKKCQQ
361  WSQQSGQNVTCATASTTDDCIVLVLKGEADALNLDGGYIYTAGKCGLVPVLAENRKSSKY
421  SSLDCVLRPTEGYLAVAVVKKANEGLTWNSLKDKKSCHTAVDRTAGWNIPMGLIVNQTGS
481  CAFDEFFSQSCAPGRDPKSRLCALCAGDDQGLDKCVPNSKEKYYGYTGAFRCLAEDVGDV
541  AFVKNDTVWENTNGESTADWAKNLNREDFRLLCLDGTRKPVTEAQSCHLAVAPNHAVVSR
601  SDRAAHVKQVLLHQQALFGKNGKNCPDKFCLFKSETKNLLFNDNTECLAKLGGRPTYEEY
661  LGTEYVTAIANLKKCSTSPLLEACAFLTR
```

**Figure 7.** Antibacterial component of acid hydrolysate of Lf. The outside square shows the antibacterial peptide generated by acid hydrolysis, and the inside square shows lactoferricin B, the antibacterial peptide generated by peptic digestion.

Arg-Lys-Asn-, which is the same sequence as the N-terminal end of bovine Lf. The amino acid sequence of the peptide in fraction 5, having the strongest antibacterial activity, was analysed further, and was determined to include the sequence of residues 1–54 from N-terminal end of the Lf molecule. This peptide generated by acid hydrolysis contained the amino acid sequence of lactoferricin B, an antibacterial peptide generated by peptic digestion comprised of residues 17–41 of bovine Lf (3). Fraction 6, 7, and 8 are thought to share the same amino acid sequence, although the number of amino acid residues in these peptides differ. These antibacterial peptides probably act by the same mechanism.

## CONCLUSIONS

Lf was found to be very thermostable at near pH 4, and could be pasteurized or sterilized by UHT procedures without any significant loss of its physicochemical properties. By using this heat treatment method, industrially purified Lf is expected to be applied to food, drug, and cosmetic products as a bioactive component in the future. Lf was found to contain a bactericidal domain latently, that was released and activated by acid or enzymatic hydrolysis. These Lf hydrolysates are expected to be useful as natural antibacterial materials.

## REFERENCES

1. Abe, H., Saito, H., Miyakawa, H., Tamura, Y., Shimamura, S., Nagao, E., Tomita, M. (1991) Heat stability of bovine lactoferrin at acidic pH. J. Dairy Sci. 74:65.

2. Baer, A., Oroz, M., Blanc, B. (1979) Serological and fluorescence studies of heat stability of bovine lactoferrin. J. Dairy Res. 46:83.

3. Bellamy, W., Takase, M., Yamauchi, K., Wakabayashi, H., Kawase, K., Tomita, M. (1992) Identification of the bactericidal domain of lactoferrin.Biochim. Biophys. Acta 1121:130.

4. Groves, M.L. (1971) Milk proteins: chemistry and molecular biology. Vol. 3. H.A. McKenzie, ed. Academic Press, New York, NY, p. 367.

5. Ruegg, M., Moor, U., Blanc, B. (1977) A calorimetric study of the thermal denaturation of whey proteins in simulated milk ultrafiltrate. J. Dairy Res. 44:509.

6. Saito, H., Miyakawa, H., Tamura, Y., Shimamura, S., Tomita, M. (1991) Potent bactericidal activity of bovine lactoferrin hydrolysate produced by heat treatment at acidic pH. J. Dairy Sci. 74:3724.

# A COMPARISON OF THE THREE-DIMENSIONAL STRUCTURES OF HUMAN LACTOFERRIN IN ITS IRON FREE AND IRON SATURATED FORMS

B.F. Anderson, G.E. Norris, S.V. Rumball, D.H. Thomas, and E.N. Baker

Department of Chemistry
Biochemistry
Massey University
Palmerston North, New Zealand

The structures of human lactoferrin in its iron free[1] (apoLf) and iron saturated[2] ($Fe_2Lf$) forms have been determined crystallographically at high resolution (2.0 and 2.2 Å respectively). The root-mean-square errors in positions of the individual atoms are of the order of 0.2–0.3 Å, which means that it is possible to make meaningful comparisons between the two structures with a view to understanding how iron binding and release are accomplished.

The structure of $Fe_2Lf$ (Figure 1) shows that the single polypeptide chain is divided into two lobes, (N and C), that are connected by a short α helix. Each lobe is further divided into two domains (N1 and N2 in the N-lobe, C1 and C2 in the C-lobe), with the metal binding site of each lobe located deep in the cleft between the domains.

The two metal binding sites are essentially identical; ligands are provided by an aspartic acid, a histidine, and two tyrosine residues, as well as by an anion (usually carbonate) which is essential for iron binding. The two oligosaccharide attachment sites are found to be on the outside of each lobe on the N2 and C2 domains, while the antibacterial region identified by Bellamy et al[3] is located on the surface of the N1 domain.

In the structure of apoLf (Figure 2), the binding cleft of the N-lobe is wide open while that of the C-lobe remains closed. Binding of iron to the N-lobe of apoLf is accompanied by a 54° rotation of the N2 domain (and a physical movement of some parts of the structure of 30 Å) with the result that the cation is buried some 10–14Å within the protein molecule and so isolated completely from the surrounding medium. Nevertheless superposition of each of the individual $Fe_2Lf$ domains onto their apoLf equivalents indicate that this large conformational change occurs with little disturbance to the polypeptide chain folding and that the N2 domain moves as a rigid body (rms differences in position for the main chain atoms of domains N1 (residues 4–90, 252–332), N2 (91–251), C1 (345–433, 596–663), and C2 (434–595) are 0.55 Å, 0.61 Å, 0.45 Å and 0.41 Å respectively.)

This conformational change has two elements; a hinge between the two domains and a pivot where domain N2 moves against N1. The hinge is formed in the two anti-parallel strands of beta sheet which lie behind the iron site and here residues 90–92 and 250–252 flex in concert to move the N2 domain. Changes in the torsion angles are small (maximum 45°) and do not

*Lactoferrin: Structure and Function*
Edited by T.W. Hutchens *et al.*, Plenum Press, New York, 1994

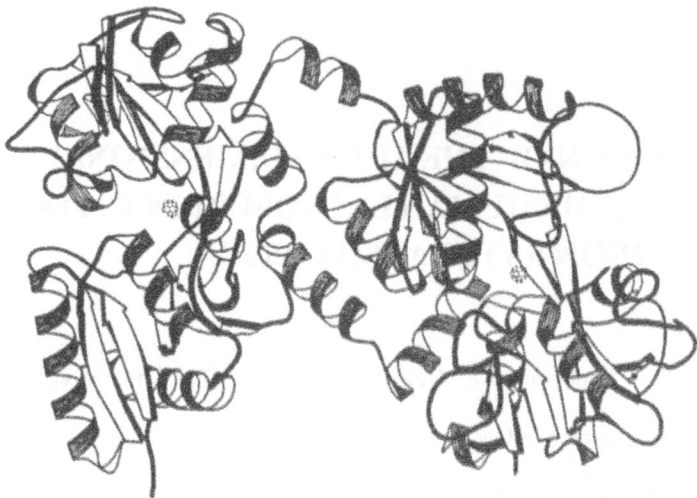

**Figure 1.** Cartoon representation of the Fe$_2$Lf structure. Domains are: $\frac{\text{N2 C1}}{\text{N1 C2}}$

move the structure through energetically unfavorable conformations and the interchain hydrogen bonds remain intact. Nevertheless the resulting movement is considerable and emphasises the flexibility inherent in beta sheet structures.

The pivot is found in a hydrophobic patch between helix 5 (residues 121–134) which lies along the length of the moving N2 domain and helix 11 (321–333) which extends like a post away from the N1 domain. This interface (residues 127, 129,130 and 327, 328, 330) is completely hydrophobic and consequently provides a non-interactive, 'greasy' surface for the two helices to move over while at the same time holding the two domains adjacent. The C-termini of the two helices are tethered together through hydrogen bonds between Arg 133 and the main chain oxygens of of residues 330 and 333. There are no other specific interactions in apoLf but upon iron binding the N-termini of helices 5 and 11 approach each other so that the OH of Tyr 324 can form bonds with OG1 of Thr 122 and ND2 of Asn 126. This has the appearance of a latch which may contribute to the stability of the holo-protein.

The movement of helix 5 is accompanied by a rotation of 24° in the same sense of helix 12 (residues 333–343) which is set at right angles to helix 11 and provides the connection between the two lobes. There is thus a tendency to reposition the C-lobe when iron binds to the N-lobe but this is counteracted by a set of hydrogen bonds between the C-terminus of the molecule (residues 689, 691) and residues 90, 243, 249 of the N-lobe. The final result is that the C-lobe undergoes a rotation of 6° and a shift of 1.9 Å. The terminating helix moves away from the N1 domain and what were direct hydrogen bonds are now mediated by water molecules. It is possible that these concerted movements enable cooperativity in iron binding at the two sites.

The significance of the closed C-lobe in the apoLf structure has been discussed elsewhere[4,5]. It is consistent with a lesser flexibility of the C-lobe, which makes the closed configuration a stable structure (though not necessarily the *only* stable structure) even without a metal atom bound. The lesser flexibility probably arises in part, from the presence of a disulphide bridge, 483–677, which links the two helices equivalent to helices 5 and 11, and which may thus constrain the movement of the C2 domain.

The conformational change seen here has its parallels in other binding proteins. Within the transferrin family, a recent structure analysis of human transferrin with one ferric ion bound

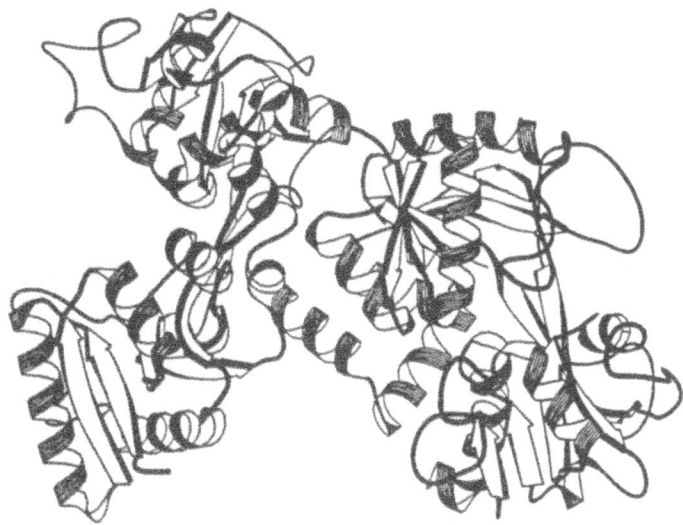

**Figure 2.** Cartoon representation of the apoLf structure. Domains are: $\begin{smallmatrix} N2 & C1 \\ N1 & C2 \end{smallmatrix}$

shows that the single iron atom is bound within the closed C-lobe, while the empty N-lobe binding cleft is wide open, as in apoLf, with a very similar domain movement. In the bacterial binding proteins specific for sugars, ions and amino acids, two-domain structures very similar in folding to each  lobe of Lf are found[6,7]. Both open and closed structures have been characterised for these proteins,  and energy calculations analysing the hinge-bending motion indicate that there is little energy  difference between the two states[8]. This is consistent with what we observe for Lf, and in particular with  the unexpected discovery of a closed, but iron-free, C-lobe in apoLf.

## CONCLUSIONS

It would appear from the comparison of lactoferrin, serum transferrin and the bacterial  binding proteins, that binding of substrates is accompanied by a large conformational change which has  very localised effects on the polypeptide chain. Although a number of specific differences between  the unliganded open and liganded closed forms of the N-lobe of lactoferrin have been pinpointed, further analysis will be required to fully understand the mechanism of the change.

## ACKNOWLEDGEMENTS

This work is supported by the U.S. National Institute of Child Health and Human Development, the Wellcome Trust, the New Zealand Health Research Council, the New Zealand Dairy Research Institute, and Massey University.

## REFERENCES

1. Anderson, B.F., Baker, H.M., Norris, G.E., Rumball, S,V., and Baker,  E.N.,  (1990). Nature (London) 344, 784–787.

2. Anderson, B.F., Baker, H.M., Norris, G.E., Rice, D.W., and Baker, E.N., (1989). J. Mol. Biol., 209, 711–734.

3. Bellamy, W., Takase, M., Yamauchi, K., Wakabayashi, H., Kawase, K., and Tomita, M., (1992). Biochimica et Biophysica Acta 1121, 130–136.

4. Baker, E.N., Anderson, B.F., Baker, H.M., Haridas, M., Norris, G.E., Rumball, S.V., and Smith, C.A., (1990). Pure Appl. Chem., 62, 1067–1070.

5. Baker, E.N., Anderson, B.F., Baker, H.M., Day, C.L., Haridas, M., Norris, G.E., Rumball, S.V., Smith, C.A. and Thomas, D.H., (1992). This Volume.

6. Quiocho, F.A., (1990). Phil. Trans. R. Soc. Lond., B326, 341–351.

7. Baker, E.N., Rumball, S.V., and Anderson, B.F., (1987). Trends in Biochem. Sci., 12, 350–353.

8. Mao, B., Pear, M.R., McCammon, J.A., and Quiocho, F.A., (1982). J. Biol. Chem., 257, 1131.

# BOVINE LACTOFERRIN

## Isolation and Characterisation of Genomic Regulatory Sequences

Heather Bain and John Tweedie

Department of Chemistry and Biochemistry
Massey University
Palmerston North
New Zealand

Lactoferrin is found in milk, polymorphonuclear leukocytes and a variety of vertebrate secretions (Masson *et al,* 1966; Baggiolini *et al,* 1970). The levels of lactoferrin differ considerably depending upon the species and the lactational status of the mammary gland being investigated. Lactoferrin concentrations up to 6 mg/ml have been measured within human colostrum, gradually declining to 1–2 mg/ml later in lactation (Lonnerdal *et al,* 1976). The concentration of lactoferrin in bovine colostrum however ranges between 1–5 mg/ml, falling to approximately 0.1 mg/ml during late lactation (Smith & Schanbacher, 1977). It has also been documented that during periods of mastitis and during involution, bovine lactoferrin lacteal secretion levels increase dramatically (Welty *et al,* 1976).

As yet the mechanism by which lactoferrin is regulated is unknown. Indirect evidence implies that lactoferrin may be regulated in a similar manner to serum transferrin, a member of the same family of iron binding proteins (Van Vugt *et al,* 1975). The human transferrin gene is expressed and regulated by a complex process involving multiple regulatory elements acting in an organ- and tissue-specific manner (McKnight *et al,* 1983; Bowman *et al,* 1988). Analysis of the human transferrin gene has revealed that these interactions involve the 5′ untranslated

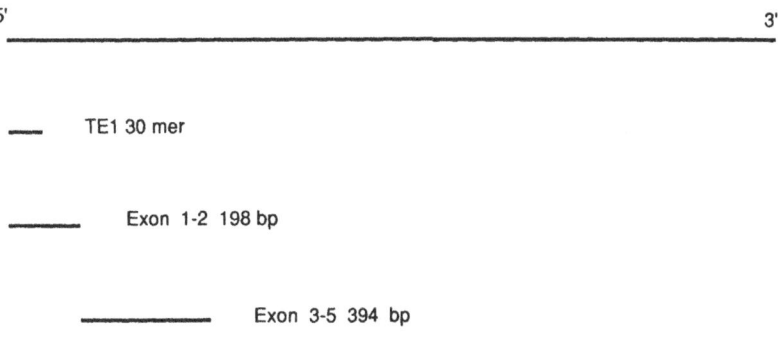

**Figure 1.** Probes from bovine lactoferrin cDNA (2364 bp).

*Lactoferrin: Structure and Function*
Edited by T.W. Hutchens *et al.*, Plenum Press, New York, 1994

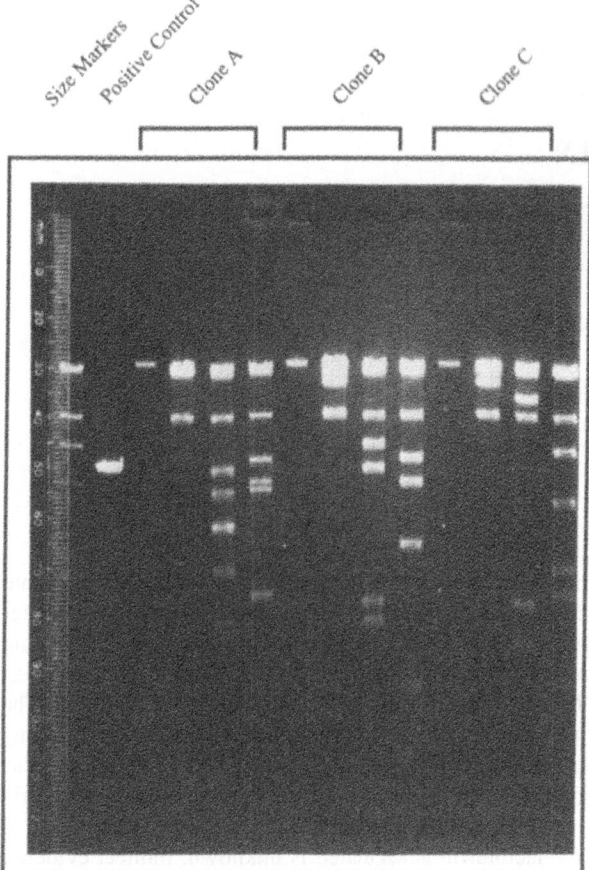

**Figure 2.** Restriction digests of lactoferrin genomic clones.

region of the gene which is responsive to both hormones and iron levels (Idzerda *et al*, 1986; McKnight *et al*, 1980). The study of the 5′ promoter regions of lactoferrin should reveal similar binding sites and lead to some understanding of the mechanism of expression. The principal aim of this investigation is to isolate and characterise genomic regulatory DNA sequences for bovine lactoferrin.

## EXPERIMENTAL

A commercial EMBL-3 bovine genomic library was screened twice. The probes utilised in the screening process were initially isolated from a partial bovine lactoferrin (bLf) cDNA clone as 5′ untranslated sequences were unavailable. A 394 base pair fragment covering exons III to V inclusively (see Figure 1) was used to screen the genomic library. This produced three putative bLf clones which when analysed by restriction endonuclease digestion were shown to be non-identical (Figure 2).

Following the isolation of the complete bLf cDNA including the 5′ most region, a 198 base pair fragment corresponding to exon I and II was used to identify 5′ clones (Figure 1). Clone A was the sole clone containing DNA fragments that hybridised to this probe. A 2 kb fragment was cloned and the DNA sequence determined. Part of this clone contained 96% homology with exon II of the bLf cDNA sequence (Figure 3). This clone was also shown to

```
Genomic   301 CTCTTTGGCCTCTTTCTCCCAGGACTGTGTCTGGCTGCCCCGAGGAAAAA 350
                  |||        ||  ||||||||||||||||||||||||||||||
cDNA       65 .....GTCCCTTGGAGCCCTTGGACTGTGTCTGGCTGCCCCGAGGAAAAA 109


Genomic   351 CGTTCGATGGTGTACCATCTCCCAACCTGAGTGGTTCAAATGCCGCCGAT 400
              ||||||||||||||||||||||||||||||| ||||||||||||||||||||
cDNA      110 CGTTCGATGGTGTACCATCTCCCAACCCGAGTGGTTCAAATGCCGCCGAT 159


Genomic   401 GGCAGTGGAGGATGAAGAAGCTGGGTGCTCCCTCTATCACCTGTGTGAGG 450
              |||||||||||||||||||||||||||||||||||||||||||||||||||
cDNA      160 GGCAGTGGAGGATGAAGAAGCTGGGTGCTCCCTCTATCACCTGTGTGAGG 209


Genomic   451 AGGGCCTTTGCCTTGGAATGTATCCGGGCCATCGCGGTGAGTCCAGCGCG 500
              ||+|||||||||||||||||||||||||||||||||||||| ||    |||
cDNA      210 AGGGCCTTTGCCTTGGAATGTATCCGGGCCATCGCGGAGAAAAAGGCGGA 259
```

**Figure 3.** Bovine lactoferrin genomic sequence compared to BLf cDNA sequence. Underlined sequences represent intron–extron splice junctions.

include part of the sequence of intron I. Sequences corresponding to exon I were not detected within this fragment.

Further southern blot analysis of the putative bLf clones with an oligonucleotide (TEl) synthesized to the 5′ most end of the bLf cDNA (Figure 1), demonstrated the absence of exon I within all the isolated clones. Subsequent screening of the same commercial library has produced no additional clones.

Recently another commercial lambda Dash II bovine genomic library has been purchased. Currently this is being screened with a bLf exon 1 cDNA probe to obtain 5′ untranslated regions of the bovine lactoferrin gene.

## REFERENCES

Baggiolini, M., deDuve, C., Masson, P.L. & Heremans, J.F. (1970) *J. Exp. Med.* 131, 559–570

Bowman, B.H., Yang, F. & Adrian, G.S. (1988) *Adv. Genet.* 25, 1–38

Idzerda, R.L., Huebers, H., Finch, C.A. & McKnight, G.S. (1986) *Proc. Natl. Acad. Sci. USA* 83, 3723–3727

Lonnerdal, B., Forsum, E. & Hambraeus, L. (1976) *Am. J. Clin. Nutr.* 29, 1127–1133

Masson, P.L., Heremans, J.F. & Dive, C.H. (1966) *Clin. Chim. Acta* 14, 735–739

McKnight, G.S., Lee, D.C. Hemmaplardh, D., Finch, C.A. & Palmiter, R.D. (1980a) J. *Biol. Chem.* 255, 144–147

McKnight, G.S., Hammer, R.E., Kuenzel, E.A. & Brinster, R.L. (1983) *Cell* 34, 335341

Schanbacher, F.L. & Smith, K.L. (1977) *J. Dai. Sci.* 58, 1048–1062

Smith, K.L. & Schanbacher, F.L. (1977) *J. Amer. Vet. Med. Assn.* 170, 1224–1227

Van Vugt, H., van Gool, J., Ladiges, N.C.J.J. & Boers, W. (1975) *Quart. J. Exp. Physiol.* 60, 79–88

Welty, F.K., Smith, K.L. & Schanbacher, F.L. (1976) *J. Dai. Sci.* 59, 224–231.

# X-RAY STRUCTURAL ANALYSIS OF BOVINE LACTOFERRIN AT 2.5 Å RESOLUTION

M. Haridas, Bryan F. Anderson, Heather M. Baker, Gillian E. Norris and
Edward N. Baker

Department of Chemistry and Biochemistry
Massey University
Palmerston North, New Zealand

Although the three-dimensional structure of human lactoferrin, in various functional states, has been determined by X-ray crystallographic analysis at high resolution (Baker *et al*, this volume) there are good reasons to investigate the structures of lactoferrins of other species. The sequence variations which occur between species result in subtle changes in properties and offer the opportunity to more closely analyse the relationships between structure and function. Moreover the structural work on human lactoferrin has identified various elements of flexibility (Baker *et al*, 1991), which could be expressed in species variations.

The amino acid sequence of bovine lactoferrin shows ~70% sequence identity with that of human lactoferrin (Mead and Tweedie, 1990; Pierce *et al*, 1991), but its pattern of glycosylation is different; the sequence contains 5 potential glycosylation sites, of which 4 are glycosylated (Pierce *et al*, 1991). Bovine lactoferrin also binds iron somewhat less strongly than human lactoferrin (Aisen and Leibman, 1972) and releases it more readily as the pH is lowered (Legrand *et al*, 1990). To investigate the structural basis of these differences an X-ray structure analysis of bovine lactoferrin was undertaken.

## EXPERIMENTAL

Bovine lactoferrin was purified from colostrum following the method of Norris *et al* (1986). The bovine protein is more difficult to crystallize than human lactoferrin. For some preparations the use of preparative isoelectric focusing appeared to aid crystallization, but by far the most important factor proved to be the use of high protein concentrations, greater than 100 mg/ml. Reproducible crystallization was achieved by microdialysis (50 µl cells from Cambridge Repetition Engineers, UK) of a 150 mg/ml diferric bovine lactoferrin solution against 0.05 M Tris-HCI, pH 7.8, containing 8% MPD (2-methyl-2,4-pentanediol) and 6.5% ethanol (v/v). Crystals up to $1.0 \times 0.5 \times 0.5$ mm grew in a period of 2–4 weeks. The crystals are orthorhombic, a = 138.4, b = 87.1, c = 73.6 Å, space group $P2_12_12_1$, as previously described (Norris *et al*, 1986). Attempts to fully deglycosylate bovine lactoferrin, using the methods successfully employed for human lactoferrin (Norris *et al*, 1989) resulted only in partially-deglycosylated material, which was not used in crystallization studies.

*Lactoferrin: Structure and Function*
Edited by T.W. Hutchens *et al.*, Plenum Press, New York, 1994

X-ray diffraction data to 2.5 Å resolution were collected by Weissenberg photography using imaging plates at the synchrotron radiation source of the Photon Factory (Tsukuba, Japan). This allowed a full set of X-ray data to be collected with less than 2 hours exposure. The final data set comprised some 27299 reflections to 2.5 Å resolution. The structure of bovine lactoferrin was solved by molecular replacement, using the human diferric lactoferrin structure as a search model (but with all sidechains, beyond $C_\beta$ ions and water molecules omitted). After finding the correct position and orientation of the bovine lactoferrin molecules in the unit cell, those sidechains which are common to the bovine and human proteins were re-included in the model, in their human orientations. This meant that 180 amino acids (26%) were still without sidechains. The model was refined by rigid body least squares methods (program CORELS), first as the whole molecule, then as two lobes, then as four domains, to give an R factor of 0.384 to 4 Å resolution. Further refinement was by restrained least squares methods (program PROLSQ) with manual rebuilding of the structure carried out on an interactive computer graphics system (program FRODO) and sidechains being added as they appeared in electron density maps. The present protein model is ~90% complete and gives good agreement with the observed X-ray data (R= 0.239 to 2.5 Å resolution). Refinement is continuing; some sidechains have yet to be included, while some density for three of the four glycan chains (at Asn 368, Asn 476 and Asn 545) is visible, and these will also have to be modelled.

## RESULTS AND DISCUSSION

The structure of bovine lactoferrin is very closely similar to that of human lactoferrin, even though the crystal packing is completely different. As for human lactoferrin, the polypeptide chain is folded into two globular lobes, the N-lobe and the C-lobe, representing the N-terminal and C-terminal halves of the molecule, and each lobe is further subdivided into two domains (N1 and N2, C1 and C2) with the specific iron binding sites in the interdomain cleft of each lobe (Figure 1). The connecting peptide between the lobes is helical, as in the human structure, and over most of the molecule atomic positions between the two species agree to within 1.0 Å. The extra disulphide bridge, 160–183, which is present in bovine lactoferrin but not in human lactoferrin, does not seem to cause any disturbance of the polypeptide chain

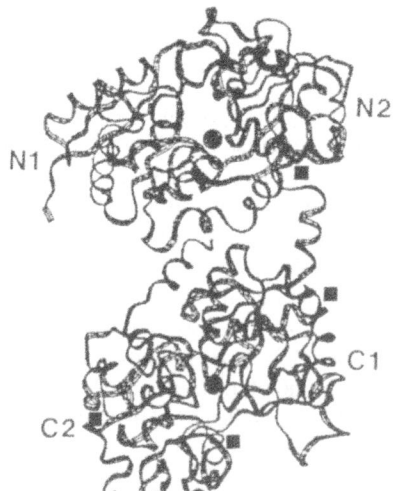

**Figure 1.** Ribbon diagram showing the polypeptide chain conformation of bovine lactoferrin. Iron atoms shown as filled circles, and glycosylation sites marked by squares. Note the somewhat more closed N-lobe (upper) compared with the human structure (Baker *et al*, this volume).

folding, nor are there any obvious differences at the glycosylation sites, even though three of these (at Asn 233, Asn 368, and Asn 545) are at positions not glycosylated in human lactoferrin.

The major difference in the structure of bovine lactoferrin is that the relative orientations of the lobes and domains are somewhat different. If the N-lobes of human and bovine lactoferrins are superimposed, the C-lobes are out of register and a 12° rotation of the bovine C-lobe is then required to bring it into correspondence with the human C-lobe. The significance of this difference in lobe orientations (if any) is not clear, but it could influence interactions with cells, cellular receptors and other molecules, if both lobes make contact. Likewise the domains of the N-lobe of bovine lactoferrin are slightly more closed over the iron site than in human lactoferrin.

Differences in metal and anion binding properties of bovine lactoferrin may be related to the following changes:

(i) Alterations in the interactions between the two domains of the N-lobe, which may change the balance between 'open' and 'closed' configurations. In bovine lactoferrin residue 217 is changed to Asn, resulting in the loss of a salt bridge Asp 217...Lys 295 which links the N1 and N2 domains in human lactoferrin. Further, residue 210, at the back of the iron site, is changed from Arg in human lactoferrin to Lys in bovine lactoferrin. Although the effect of this change in bovine lactoferrin is not yet clear, the same change in serum transferrin appears to weaken interdomain interactions (Baker and Lindley, 1992).

(ii) Alterations near the back of the C-lobe iron site (i.e., in the hinge region). In particular, Tyr 415 in human lactoferrin (an invariant residue in other transferrins) is changed to Arg in the bovine protein, and there is a second change nearby with the presence of a new glycosylation site at Asn 545. The Arg sidechain could provide a site for binding anions such as citrate, while the presence of a carbohydrate chain near the hinge region could again influence the 'open' to 'closed' structural transition.

The precise details of these changes, and their functional consequences, have still to be clarified by further crystallographic refinement and structural comparisons. It is clear at this stage, however, that the bovine lactoferrin structure analysis confirms and extends the functional insights derived from structural studies on the human protein.

## ACKNOWLEDGEMENTS

Support is gratefully acknowledged from the U.S. National Institute of Child Health and Human Development (Grant HD-20859), the Health Research Council of New Zealand, the Wellcome Trust, and the New Zealand Dairy Research Institute. We also thank Dr. N. Sakabe for help with data collection at the Photon Factory.

## REFERENCES

Aisen, P. & Leibman, A. (1972). *Biochim. Biophys. Acta.* 257, 314–323.

Baker, E.N. & Lindley, P.F. (1992). *J. Inorg. Biochem.* 47, 147–160.

Baker, E.N., Anderson, B.F., Baker, H.M., Haridas, M., Jameson, G.B., Norris, G.E., Rumball, S.V. & Smith C.A. (1991). *Int. J. Biol. Macromol.* 13, 122–129

Legrand, D., Mazurier, J., Colavizza, D., Montreuil, J. & Spik, G. (1990). *Biochem. J.* 266, 575–581.

Mead, P.E. & Tweedie, J.W. (1990). *Nucl. Acids. Res.* 18, 7167.

Norris, G.E., Anderson, B.F., Baker, E.N., Baker, H.M., Gartner, A.L., Ward, J. & Rumball, S.V. (1986). *J. Mol. Biol.* 191, 143–145.

Norris, G.E., Baker, H.M. & Baker, E.N. (1989). *J. Mol. Biol.* 209, 329–331.

Pierce, A., Colavizza, D., Benaissa, M., Maes, P., Tartar, A., Montreuil, J. & Spik, G. (1991). *Eur. J. Biochem.* 196, 177–184.

# BINDING OF PORCINE MILK LACTOFERRIN TO PIGLET INTESTINAL LACTOFERRIN RECEPTOR

Jóhannes Gíslason,[*] Suhasini Iyer,[†] Gordon C. Douglas,[‡]
T. William Hutchens,[◊] and Bo Lönnerdal[†]

[*]Department of Biochemistry
Faculty of Medicine
University of Iceland
Reykjavik, Iceland

[†]Department of Nutrition
University of California
Davis, California

[‡]Department of Human Anatomy and Cell Biology
University of California, Davis

[◊]Department of Pediatrics
Baylor College of Medicine
Children's Nutrition Research Center
Houston, Texas[*]

## INTRODUCTION

Lactoferrin (Lf)[**] is abundant in human milk. The physiological role of milk Lf is still debated but it is believed to exhibit a bacteriostatic effect in the small intestine. It has also been speculated whether Lf is actively taking part in iron absorption during infancy (1). Cox et al. (2) showed that human lactoferrin (HLf) has the ability to deliver iron to mucosal human intestinal biopsies. Recent evidence for the presence of specific Lf receptors in the small intestine has been provided in rhesus monkeys (3), mice (4) and humans (5) using brush border membranes. It should be noted that all these species have a relatively high Lf concentration in their milk. This may suggest that Lf plays a role in receptor mediated iron absorption in these species.

---

[*] Present address: Department of Food Science and Technology, University of California, Davis, California.

[**]Lf: Lactoferrin; PLf: porcine lactoferrin; PTf: porcine transferrin; BSA: bovine serum albumin; Tf: transferrin; BLf: bovine lactoferrin; BBMV: brush border membrane vesicles.

*Lactoferrin: Structure and Function*
Edited by T.W. Hutchens *et al.*, Plenum Press, New York, 1994

In order to answer questions regarding the role of interactions between milk Lf and its intestinal receptor, a reliable animal model has to be found where the interaction between viable enterocytes and homologous Lf can be studied.

Similar to human milk, Lf in sow's milk is a major iron binding protein. We studied the kinetics of Lf binding to brush border membranes (0–22 days of age) and viable enterocytes isolated from piglet intestine (7–20 days of age). We also studied the ontogeny, specificity and intestinal distribution of the piglet Lf receptor.

## METHODS

### Labelling of Proteins

$^{59}Fe$-citrate was added in an amount sufficient to saturate the protein and the solution was left at room temperature overnight to ensure maximum binding of iron. Unbound $^{59}Fe$ was removed by passing the solution through an Excellulose GF-5 desalting column (Pierce, Rockford, IL)

Porcine lactoferrin(PLf) was labelled with $^{125}I$ using Iodo-Gen from Pierce Chemical Co, Rockford, IL.

### Preparation of Brush Border Membrane Vesicles (BBMV)

Piglets were killed by intraperitoneal injection of pentobarbital. Small intestines were immediately dissected and flushed with ice cold saline and cut into three equal segments. Each part was slit open and the mucosa was scraped with a glass slide. BBMVs were prepared from 1.5–2.5 g of mucosa by differential centrifugation and magnesium precipitation techniques described by Muir et al. (6) and modified by Davidson and Lönnerdal (3).

### Cell Preparation

Pieces of duodenum (15 cm) were inverted upon a glass pipette which was hooked on Vibromixer Model E1 (Chemap AG, Zrich, Switzerland) (four at a time) in Dulbeccos PBS pH 7.4 (without $Ca^{2+}$ and $Mg^{2+}$) containing 3 mM EDTA, 10 mM glucose, 0.5 mM DTT and 0.1% bovine serum albumin (BSA). Dead cells and other particles were shaken off by gently vibrating for 2 min. After changing to fresh PBS, intestines were vibrated for 8 min. Viable epithelial cells were shaken off the tissue into the medium. The cell preparation was washed 3 times in Krebs-Henseleit buffer pH 7.4 containing 10 mM glucose and 1% BSA.

### Binding Assays

BBMVs were suspended at a concentration of 1 mg protein/ml. 20 μl of this suspension were incubated at 37°C with varying concentrations of labelled ligand, ranging from 0.1–1.25 μM and for various time points, ranging from 15 sec to 10 min. The reaction was stopped by addition of ice-cold saline. The mixture was immediately vacuum-filtered onto prewetted 0.22 μm Millipore filters, rinsed 3 times with 1 ml saline and counted in a gamma counter. Nonspecific binding of $^{59}Fe$-Lf to BBMVs was determined and corrected for in all experiments. Non-specific binding of $^{59}Fe$ to the filter membranes was monitored by incubation of $^{59}Fe$-labelled protein in absence of BBMVs. All incubations were done in triplicates. Incubations were carried out using $^{59}Fe$-labelled PLf and PTf, bovine lactoferrin (BLf) and HLf.

Ligand binding to suspended cells was measured by incubating approximately $2 \times 10^6$ freshly prepared cells with increasing concentrations of $^{125}I$-labelled PLf (0.1, 0.5, 1.0, 2.5, 5.0, and 10 μM). Cells were suspended at 4°C in 300 μl KrebsHenseleit buffer (pH 7.4),

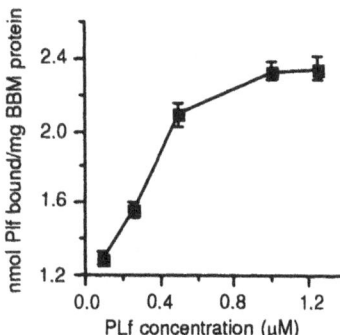

**Figure 1.** Saturation curve for PLf binding to BBMV. PLf (0.1–1.25 M) was incubated with 20 mg of membrane protein for 5 min at 37°C. Values are means of 3 experiments done in triplicates. Bars indicate SD.

containing 10 mM glucose and 1% BSA in Eppendorf tubes and vortexed at low speed every 15 min during incubation. Non-specific binding was assessed by including an excess of cold PLf (500 μM) in a parallel series of tubes. All series were done in duplicates. After 2.5 h of incubation, cells were spun down in a table top centrifuge and the cell pellets were washed three times with ice-cold Krebs–Henseleit buffer and two times with ice cold Dulbecco's PBS. During the last wash, cells were transferred to new tubes. Cells were solubilized in 1.5% sodium dodecyl sulfate (SDS) (0.5 ml) and counted in a Beckman gamma counter (Gamma 8500, Beckman, Fullerton, CA). After incubation at 15°C for 20 min, the solubilized cells were vortexed vigorously and an aliquot was taken for protein determination.

## RESULTS

Enrichment of sucrase activity in the brush-border membranes was usually 17–20 fold over that of the original homogenate. There was no detectable Na-K-ATPase activity in the BBM suspension. Recoveries were generally 20–30%. Viability of the isolated cells was typically higher than 85–90% as measured by Trypan blue exclusion but went down to 50–60% overnight at 4°C.

When using a Lf concentration of 0.1–1.25 μM and 0.1–10 mM, saturation of the receptor was observed at concentrations of 1 μM (BBMVs) and 5 mM (enterocytes) respectively (Figs 1 and 2).

Scatchard plot analysis was done for all ages and locations to determine receptor number and affinity constants for binding as calculated from the intercept and the slope of the plot, respectively. Figs 3 and 4 are representatives of the several Scatchard plots. There were no significant differences in binding affinity or receptor numbers between the three locations in

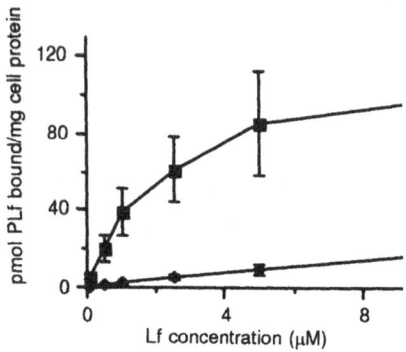

**Figure 2.** Saturation curve showing specific and unspecific binding of PLf to enterocytes. PLf (0.10–10 M) was incubated with approximately $2 \times 10^{-6}$ cells on ice for 2.5 h. Values are means of 3 experiments each of which was done in duplicates. Bars indicate SD. The upper line represents specific binding. The lower line represents nonspecific binding.

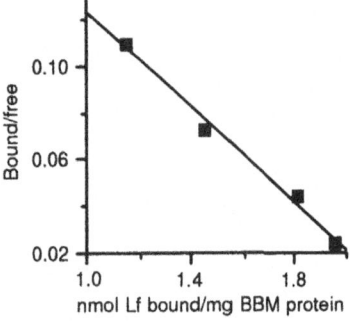

**Figure 3.** Scatchard plot analysis of PLf binding to BBMVs isolated from intestine of 10 days old piglet. Values are means of triplicates. Apparent Kd: $0.3 \times 10^{-6}$ M.

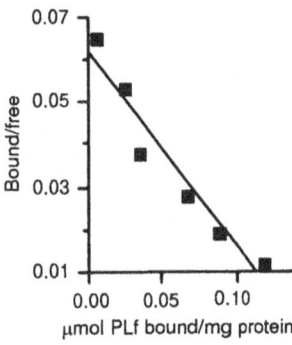

**Figure 4.** Scatchard plot analysis of PLf binding to enterocytes isolated from piglet intestine. Values are means of duplicates. Apparent Kd: $2 \times 10^{-6}$ M.

the intestine (Table 1) or between the different age groups (Table 2). The binding affinity expressed as $K_d$ was about 0.3 μM (BBMWs) and 2 μM (enterocytes).

When enterocytes were incubated with excess concentrations of PTf, PLf and BLf, only a moderate inhibition of $^{125}$I-PLf binding was observed for very high concentrations of PTf while a pronounced inhibition by PLf and BLf was observed (Fig 5).

## SUMMARY

The saturation kinetics, the competitive binding studies and the Scatchard plots clearly demonstrate the presence of a specific lactoferrin receptor in piglet's small intestine.

Like in mice, humans and rhesus monkeys, the intestinal porcine lactoferrin receptor is a low affinity receptor. Because of the high lactoferrin concentration in the milk of these species, a physiological role of the receptor can be assumed.

**Table 1.** Binding Kinetics of PLf Binding to BBMVs Prepared from Different Locations within the Piglet Small Intestine

| Location (cm from pylorus) | Dissociation constant (M) | Number of binding sites ($\times 10^{14}$/mg protein) |
|---|---|---|
| 0–120 | $0.31 \pm 0.8$ | $15 \pm 1.1$ |
| 240–360 | $0.36 \pm 0.6$ | $16 \pm 1.1$ |
| 480–600 | $0.32 \pm 0.4$ | $15 \pm 1.1$ |

**Table 2.** Binding Kinetics of PLf Binding to BBMVs Prepared from Piglets of Different Age Groups

| Age of piglet (days) | Dissociation constant (M) | Number of binding sites ($\times 10^{14}$/mg protein) |
| --- | --- | --- |
| 0 | $0.36 \pm 0.6$ | $15 \pm 1.1$ |
| 10 | $0.32 \pm 0.7$ | $12 \pm 1.0$ |
| 20 | $0.42 \pm 0.6$ | $15 \pm 1.2$ |

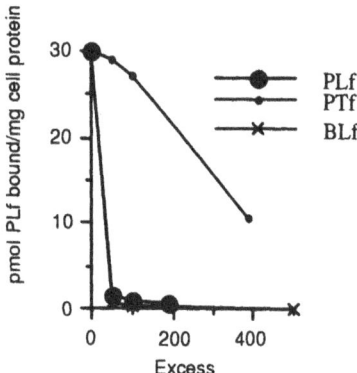

**Figure 5.** Competition curve showing inhibition of the binding of $^{125}$I-labelled porcine lactoferrin to enterocytes, by addition of increasing molar concentration of porcine lactoferrin, bovine lactoferrin and porcine transferrin.

There are reasons to suggest that only a small amount of milk casein is digested in the stomach of a suckling piglet. In species with milk high in casein (e.g., sow's milk), this will result in a relatively high concentration of incompletely digested casein causing immobilization and decreased availability of iron at the site where it is most efficiently absorbed (duodenum). The presence of lactoferrin receptors in the jejunum and ileum may facilitate absorption of iron, gradually being released from partly digested casein in the distal parts of the small intestine.

Our results indicate that the isolated piglet enterocyte might be a suitable model for further studies on the interaction between the enterocyte and species specific lactoferrin.

## ACKNOWLEDGMENTS

This work was supported by the Swedish Council for Forestry and Agricultural Research, The Swedish Nutritional Foundation, the Medical Faculty at the University of Uppsala and NIH grant DK-47850.

The authors thank Dr. Bernt Jones For his valuable guidance and practical help with housing of animals and tissue collection.

## REFERENCES

1. Lönnerdal, B. (1984) Iron in breast milk. In: Iron in Infancy and Childhood (Stekel, A., ed.), Nestle, Vevey/Raven Press, New York, Nestle Nutrition Workshop Series 4: 95–118.

2. Cox, T.M., Mazurier, J., Spik, G., Montreuil, J & Peters, T.J. (1979) Iron binding proteins and influx of iron across the duodenal brush border. Evidence for specific lactotransferrin receptors in the human intestine. Biochim. Biophys. Acta, 588: 120–128.

3. Davidson, L.A. & Lönnerdal, B. (1988) Specific binding of lactoferrin to brush border membrane: ontogeny and effect of glycan chain. Am. J. Physiol., 254: G580–585.

4. Hu, W.L., Mazurier, J., Sawatzki, G., Montreuil, J. & Spik, G. (1988) Lactoferrin receptor of mouse small-intestinal brush border. Biochem. J., 249: 435–441.

5. Kawakami, H. & Lönnerdal, B. (1991) Isolation and function of a receptor for human lactoferrin in human intestinal brushborder membranes. Am. J. Physiol., 261: G841–846.

6. Muir, W.A., Hopfer, U. & King, M. (1984) Iron transport across brush-border membranes from normal and iron-deficient mouse upper small intestine. J. Biol. Chem., 253: 4896–4903.

# LACTOFERRIN–RECEPTOR INTERACTION

## Effect of Surface Exposed Histidine Residues

Suhasini Iyer,[*] Tai-Tung Yip,[†] T. William Hutchens,[†] and Bo Lonnerdal[*]

[*]Department of Nutrition
University of California
Davis, California

[†]Department of Food Science and Technology
University of California
Davis, California

## ABSTRACT

Specific receptors for lactoferrin (Lf) have been identified in the small intestine brush-border membrane and on lymphocytes. Little is known about the mechanism of Lf interaction with its receptor protein. Because of the pH dependence of this interaction, the possible involvement of surface exposed histidine residues was explored. His residues on purified human milk Lf were progressively modified by carboxyethylation with diethyl pyrocarbonate; the reaction was monitored by UV absorbance and by affinity chromatography on immobilized Cu(II) ions. Human infant brush-border membrane vesicles (BBMVs) were prepared by differential centrifugation and precipitation methods. Human peripheral lymphocytes were prepared by Ficoll centrifugation and stimulated by mitogen (PHA) exposure in culture. A competitive receptor binding assay was performed with $^{59}$Fe- and/or $^{125}$I-labelled native Lf and His-modified Lf. Binding assays with labelled native Lf were conducted in the presence of a 10–200 fold molar excess of DEPC-Lf, the DEPC-Lf was not competitive. A 100-fold molar excess of native Lf completely blocked the interaction of labelled Lf with the receptor. Similar results were obtained for receptor proteins on BBMVs and lymphocytes. Thus, Lf receptors on these two different cell types appear to require His residues for Lf binding.

## INTRODUCTION

Lactoferrin is an iron-binding glycoprotein, closely related in structure to the transferrins. Unlike transferrin, however, only trace amounts of lactoferrin are present in circulation. Instead it is mainly found in milk and other mucous secretions and in the secondary granules of neutrophils (Masson et al 1966, Baggiolini et al 1970). The structure of lactoferrin has been examined in detail, using X-ray crystallography (Anderson et al 1987, 1989). The molecule is folded into two globular lobes, the N lobe comprising the N terminal half of the polypeptide chain (residues 1–333) and the C-lobe comprising the C terminal half of the polypeptide chain (residues 345–691 ). The two lobes are connected by a three-turn α helix (residues 334–344). Lactoferrin, like transferrin, reversibly binds two ferric ions, synergistically with an anion, usually bicarbonate or carbonate. The affinity of lactoferrin for iron is however, ~300 times stronger than that of transferrin for iron even at a low pH (pH 3). A conformational change apparently accompanies iron binding, with the structure becoming more compact (Aisen and Listowsky 1980; Grossman et al 1992). Spectroscopic analysis of the metal-ion complexes of human lactoferrin has revealed that iron binding to lactoferrin results in an increased conformational stability of Fe(III)-lactoferrin in urea when compared to apo-lactoferrin.

Lactoferrin is a basic protein with a large number and a variety of solvent-exposed electron donor groups on its surface which enable it to bind other metal ions in a nonspecific manner (Smith et al 1991). Iron binding to lactoferrin increases the affinity of lactoferrin to immobilized Cu(II) at sites that are distinct from the two known high-affinity metal-binding sites. Further, due to these solvent-exposed groups, lactoferrin is likely to bind other metal ions, like Mn (II) and Zn on the surface regions. This, however, has not been evaluated in terms of number of metal ions or their binding affinities.

Although the precise biological role of lactoferrin has not yet been established, lactoferrin has been reported to bind specifically to receptors in the small intestine of the mouse (Hu et al 1988), rabbit (Mazurier et al 1985), monkey (Davidson and Lonnerdal 1988), piglet (Gislason et al 1992) and the human fetus (Kawakami and Lonnerdal 1991 ) in a species-specific manner. Lactoferrin also binds specifically to receptors on phytohemagglutinin-stimulated lymphocytes (Mazurier et al 1989, Rochard et al 1989) and monocytes (Birgens and Kristensen 1990). Although the reported binding affinities of about $10^{-7}$ M are considered to be low, a high-affinity binding site has been recently described in lymphocytes (Birgens et al 1991). The binding characteristics and factors that contribute to this highly specific binding have not been evaluated so far. Lactoferrin binding to the receptor is shown to be temperature and pH dependent (Kawakami and Lonnerdal 1991), the optimum pH being 6.5–7.5 in the small intestinal brush border membrane and 6.0–7.2 in lymphocytes (Birgens et al 1983). Because of this pH dependence, we hypothesized that the surface-exposed histidine residues on the lactoferrin molecule, which are 7 in number, play a role in the interaction between lactoferrin and its receptor. In this study, we modified the surface-exposed histidine residues on lactoferrin by carboxyethylation using diethylpyrocarbonate (DEPC) and examined its effect on receptor binding in terms of binding kinetics and specificity.

## METHODS

### Modification of Lactoferrin with Diethylpyrocarbonate (DEPC)

Human Fe-saturated lactoferrin in sodium phosphate buffer (60 mM, pH 6) was mixed with DEPC at 10:1 and 50:1 molar ratios, DEPC:Lf, for 1 h at room temperature. The reaction was stopped by exchanging the buffer using a PD-10 column (Pharmacia, Piscataway, NJ). The extent of modification was monitored by observing the increase in absorbance at 240 nm relative to the unmodified protein. Histidine residue modification was verified by the inability of the DEPC modified protein to bind to an affinity column of immobilized copper.

### Preparation of Brush Border Membranes

Small intestine from human infant (6 m old), flash frozen and stored at −80°C was obtained from the International Institute for the Advancement of Medicine. The inside of the intestine was washed with ice-cold saline and the mucosa was scraped with a glass slide. Brush border membrane vesicles (BBMVs) were prepared by magnesium precipitation and differential centrifugation, as described previously (Davidson and Lonnerdal 1988). The final vesicle preparation was used immediately for binding assays.

The purity of the BBMV preparation was monitored by measuring the protein concentration and sucrase activity, in the initial homogenate as well as the final vesicle suspension. Protein concentration was measured by a modification of the Lowry assay (Peterson 1977) using bovine serum albumin (BSA) as standard. Sucrase activity was measured as the amount of glucose liberated on the addition of sucrose (Conklin et al 1975). Contamination of basolateral membranes in the BBMV preparations was assessed by measuring $Na^+$-$K^+$-ATPase activity (Murer et al 1976).

### Lymphocyte Cell Isolation and Stimulation

Blood from healthy adult volunteers was incubated at 37°C in the presence of dextran T-500. The leukocyte rich plasma was layered onto Ficoll–Paque density separation medium (Pharmacia, Piscataway, NJ) and centrifuged at 400 x g for 40 min. Lymphocytes were removed from the plasma/Ficoll–Paque interface and washed twice in Hank's balanced salt solution. The number of cells was determined using a hemocytometer and cell viability was typically greater than 95% as determined by Trypan blue exclusion.

Lymphocytes were cultured in 75 $cm^2$ flasks in RPMI Medium 1640 with 25 mM HEPES containing L-glutamine; 10% fetal calf serum; 100,000 U penicillin/L; 50 mg streptomycin/L and 20 mg phytohemagglutinin/L. On every second day, 1/3 of the nonadherent cells were removed and subcultured. The non-adherent cells used for the binding assays were 94–99% lymphocytes with no non-specific esterase staining cells detected; viability was 85–95%.

### Labelling of Lactoferrin

Radio-iodination of lactoferrin using Iodogen was carried out according to the instruction manual from Pierce (Rockford, IL). Two mg Iodogen dissolved in chloroform was precoated into 10 ml-scintillation vials and air-dried. One milligram of lactoferrin or DEPC-modified lactoferrin dissolved in 50 mM sodium phosphate buffer (pH 7.4) containing 0.15 M NaCl was added. The mixture was then spiked with 0.7 mCi $Na^{125}I$ and left at room temperature for 10 min. The reaction contents were then transferred directly onto an Excellulose GF-5 desalting column to remove unbound $^{125}I$. The specific activity routinely obtained was approximately 0.5 mCi/mg lactoferrin.

Lactoferrin was labelled with $^{59}Fe$ using $^{59}FeCl_3$. Lactoferrin (1 mg) was dissolved in Tris/HCl buffer containing 10 mmol/L $NaHCO_3$. 5 μCi $^{59}FeCl_3$ was added along with 9.8 μg ferrous ammonium sulfate dissolved in Tris/HCl buffer. The mixture was incubated at 37°C for 30 min. Labelled lactoferrin was routinely stored at 4°C for up to one week.

### Binding Assays (BBMVs)

Purified BBMVs were suspended at a protein concentration of 1 mg/ml. Assays were performed in triplicate by incubating labelled DEPC-lactoferrin with 20 μg BBMV protein. The assay medium consisted of 40 mM tris (hydroxymethyl) aminomethane-N-2hydroxyethylpiperazine-N'-2-ethanesulfonic acid (Tris-HEPES) buffer (pH 7.4) containing 0.1 M D-mannitol, 0.1 M NaCl and 2 mM D-glucose to a final volume of 100 μl as described previously (Davidson and Lonnerdal 1988). The concentration of $^{125}I$-DEPC lactoferrin was varied from 100 nM to 5′ μM. The incubation was carried out in a water bath at 37°C for various time points that ranged from 30 sec. to 30 min. and the reaction was terminated by the addition of 1 ml of ice-cold saline. The mixture was immediately vacuum filtered onto prewetted 0.22 μM hydrophilic membrane filters (Millipore, Bedford, MA) and rinsed 3 times with 1 ml ice-cold saline. The filters were counted in a gamma-well scintillation counter (Gamma 8500, Beckman, Fullerton, CA) to determine the amount of $^{125}$I associated with the membranes. Non-specific binding was determined by the addition of a 100-fold molar excess of unlabelled lactoferrin to the incubation mixture.

### Binding Studies (Lymphocytes)

Mitogen-stimulated lymphocytes cultured for 6 days were washed twice in Hank's balanced salt solution and resuspended in RPMI 1640 without fetal calf serum and phytohemagglutinin. $^{59}$Fe-labelled lactoferrin, over a concentration range of 20 nM to 5 μM was incubated with the cells at 24°C for 1.5 h. The lymphocytes were transferred to a 96-well microfiltration plate with a 0.65 μM PVDF membrane (Millipore, Bedford, MA). Vacuum was applied to remove unbound lactoferrin; the cells/filters were washed 5 times with ice-cold Hank's balanced salt solution and counted in a gamma scintillation counter to determine the amount of lactoferrin associated with the cells. Non-specific binding was determined in the presence of a 100-fold molar excess of unlabelled lactoferrin.

### Competitive Inhibition Assays

The binding of $^{125}$I-lactoferrin (unmodified) to brush border membranes at 1.25 μM concentration was evaluated in the presence of (a) no competitor; (b) a 10–200 fold molar excess of unlabelled DEPC-Lf; (c) a 100 fold molar excess of unlabelled native human lactoferrin (unmodified); (d) a 50–100 fold molar excess of unlabelled human transferrin and (e) a 50–100 fold molar excess of unlabelled bovine serum albumin (BSA). The binding of $^{59}$Fe to phytohemagglutinin-stimulated lymphocytes at 1 μM concentration was evaluated in the presence of (a) a 2–100 fold molar excess of unlabelled human lactoferrin and (b) a 2–100 fold molar excess of DEPC-modified lactoferrin at 10:1 and 50:1 molar ratios of DEPC: human lactoferrin.

## RESULTS

### Binding of DEPC-lactoferrin to BBMVs

The saturation curves and non-specific binding of $^{125}$I-DEPC-lactoferrin to the BBMVs are shown in Figure 1. The binding of $^{125}$I-DEPC-lactoferrin increased with increasing concentration. However, it was not saturable at concentrations of 0.1–10 μM. Approximately 50% of the bound protein was non-specifically bound or not displacable with excess unlabelled DEPC-lactoferrin.

### Competitive Inhibition Assay (BBMVs)

Figure 2 shows the results of the competitive binding assay for BBMVs. DEPC-modified lactoferrin (50:1; DEPC:lactoferrin) did not compete with native human lactoferrin for receptor binding. A 200-fold molar excess exhibited only an approximately 50% inhibition of native Lf binding. Excess unlabelled human lactoferrin could displace bound labelled human lactoferrin. Human transferrin and bovine serum albumin did not compete.

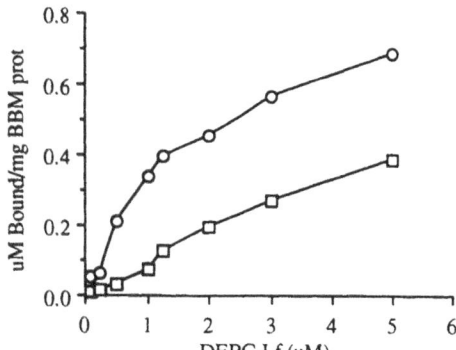

**Figure 1.** Binding of $^{125}$I-DEPC-lactoferrin to human brush-border membrane vesicles (37°C, pH 7). Total binding is shown in the upper curve (open circles). Non-specific binding (open squares) was measured in the presence of 100-fold molar excess of unlabelled DEPC-lactoferrin. Each point is the mean of 3 experiments doen in triplicate.

## Competitive Inhibition Assays (Lymphocytes)

Figure 3 shows the results of the competitive binding assays for lymphocytes. An inhibition pattern similar to that of BBMVs was observed, where a 100-fold molar excess of DEPC-lactoferrin (50:1) resulted in approximately a 50% inhibition of human lactoferrin binding to the cells. DEPC-lactoferrin (10:1), however, did compete more effectively for binding. Excess unlabelled human lactoferrin was found to displace bound human lactoferrin.

## DISCUSSION

Lactoferrin receptors were first described in rabbits by Mazurier et al. (1985) who demonstrated a specific, saturable binding of human lactoferrin to brush-border membrane vesicles. Subsequent studies by Davidson and Lonnerdal (1988)in infant rhesus monkeys reported that rhesus lactoferrin and human lactoferrin bound to a lactoferrin receptor in a saturable manner. Lactoferrin receptors have also been documented in human monocytes/macrophages (Bennett and Davis 1981; Birgens et al 1983; Moguilevsky et al 1987) and in mitogen stimulated peripheral lymphocytes (Mazurier et al 1989). However, the role of lactoferrin binding to the receptor is yet unclear. A number of hypotheses have been put forward. Since lactoferrin binds iron with high affinity, the lactoferrin–lactoferrin receptor system is suggested to play a role in iron absorption during infancy (Davidson and Lonnerdal 1988); as an iron-scavenger in circulation and therefore in protection against free-radical damage (Rochard et al 1989); as a growth factor resulting in cell differentiation (Nichols et al 1990) and as a modulator in immune response in case of infection (Crouch et al 1992). More

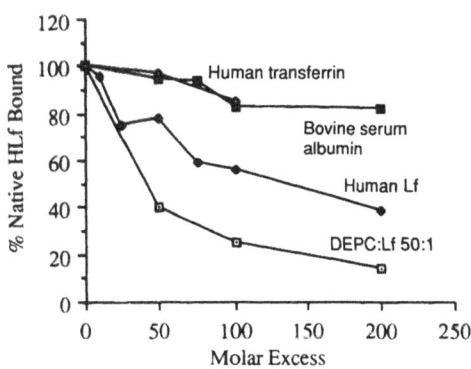

**Figure 2.** Competitive inhibition of $^{125}$I-lactoferrin binding to brush-border membrane vesicles at 1 µM by DEPC:Lf 50:1 (open squares), human Lf (open circles), human transferrin (open triangles) and bovine serum albumin (closed triangles). Concentration of competitor ranged from a 10 to 200-fold molar excess. Each point is the mean of triplicate values.

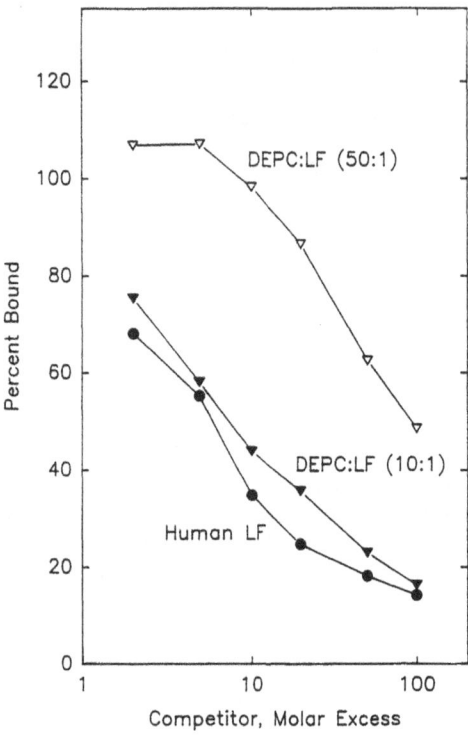

**Figure 3.** Competitve inhibition of $^{59}$Fe-lactoferrin binding to cultured (6 day) phytohemagglutinin stimultated lymphocytes at 1 μM by DEPC-Lf 50:1 (open triangles), DEPC:Lf 10:1 (filled triangles) and human lactoferrin (filled circles). Each point is the mean of triplicate values.

recently, lactoferrin has been shown to cause strand breakage in supercoiled DNA due to possible free-radical damage caused by surface bound metal ions (unpublished).

Based on current information, however, it is evident that lactoferrin binds to a receptor in a species-specific manner. Davidson and Lonnerdal (1988) reported that bovine lactoferrin and human serum transferrin showed virtually no specific binding to the rhesus monkey brush border membrane. This is consistent with our finding (Gislason et al 1992) that sow milk lactoferrin binds to the piglet small intestinal BBM receptor in a species-specific manner and that bovine lactoferrin, human lactoferrin and human transferrin showed no binding.

The specificity of the binding of lactoferrin to cells from the monocyte/macrophage line has been studied in competitive binding studies with various proteins that include human transferrin, monomeric and aggregated IgG (Van Snick and Masson 1976, Bennett and Davis 1981, Birgens et al 1983), bovine albumin (Campbell 1982) and cytochrome c (Van Snick and Masson 1976, Birgens et al 1983). Human transferrin showed no inhibitory effect on binding of $^{125}$I-lactoferrin binding and neither did bovine albumin or cytochrome c. This is interesting in the sense that lactoferrin and transferrin exhibit 60% homology in structure and sequence and both proteins are present in the circulation but show no competition in binding to the monocyte receptor.

The physical conditions as well as the structural features of lactoferrin responsible for the above specificity have been under investigation. Apo-lactoferrin has been reported to bind the receptor, albeit with a somewhat lower affinity than Fe-saturated lactoferrin (McAbee and Esbensen 1991). Lactoferrin is a glycoprotein with 2 oligosaccharide chains, which contain terminal sialic acid, fucose and galactose residues. Deglycosylation of lactoferrin using Peptide-N Glycosidase F seemed to have no effect on binding (Kawakami and Lonnerdal 1991). However, complete enzymatic deglycosylation and sensitive techniques to determine the degree of deglycosylation are difficult to achieve. Recent studies using recombinant human lactoferrin which contained mutated glycosylation sites indicate that the glycan chains do not appear to play a role in receptor binding (unpublished data).

We have demonstrated in this study that modification of surface-exposed His residues on the lactoferrin molecule results in a loss of specificity of binding. DEPC-lactoferrin did not completely inhibit the binding of native human lactoferrin either to the brush-border membrane or to PHA-stimulated lymphocytes, both of which are known to have specific receptors for lactoferrin. It is interesting that DEPC-lactoferrin alone, incubated with the membranes or lymphocytes exhibited non-saturable binding rather than a total loss of binding. A possible reason could be that Lf is a "sticky" protein and is known to bind many proteins and materials like glass and plastic (Birgens 1991, Hekman 1971). Almost 50% of the bound DEPC-Lf was non-specific in nature and could not be displaced by native human lactoferrin.

DEPC is a compound that neutralizes the positive charge on His residues by complexing the charged amino groups but leaves other amino acids unchanged, as long as the incubation time and molar ratios are kept optimum. We did not determine the exact number of surface-exposed His residues modified by our method. However, a 10:1 molar ratio of DEPC:lactoferrin was more competitive than a 50:1 molar ratio in the competitive binding assays with native lactoferrin. Hence, the observed loss in specificity could be attributed to a loss in positive charge on the surface of the molecule. This observation is in agreement with that of Ziere et al (1992) who reported that modifying the 4-clustered arginine residues at the N-terminus of lactoferrin with 1,2-cyclohexanedione by neutralizing the positive charge, resulted in a loss of uptake of bovine lactoferrin by rat liver parenchymal cells. Further, modification of Arg with 1,2-cyclohexanedione abolished its capacity to compete with $^{125}$I-lactoferrin binding to isolated parenchymal ceils.

We have observed similar results in two different cell types, brush-border membrane from infant intestinal cells and PHA-stimulated lymphocytes. This indicates that receptor binding in both these cell types shares a common mechanism. In conclusion, the positive charge on the lactoferrin molecule may be one of the factors that regulate specific binding to the receptor.

## REFERENCES

Aisen P, Listowsky I (1980). Iron transport and storage proteins. Ann. Rev. Biochem. 49, 357–393.

Anderson B F, Baker H M, Dodson E J, Norris G E, Rumball S V, Waters J M, Baker N (1987). Structure of human lactoferrin at 3.2 Å resolution. Proc. Nat. Acad. Sci. USA 84, 1769–1773.

Anderson B F, Baker H M, Norris G E, Rice D W, Baker E N (1989). Structure of human lactoferrin: crystallographic structure analysis and refinement at 2.8 Å resolution. J. Mol. Biol. 209, 711–734.

Baggiolini M, de Duve C, Masson P L, Heremans J F (1970). Association of lactoferrin with specific granules in rabbit heterophil leukocytes. J. Exp. Med. 131, 559–570.

Bennett R M, Duve C, Masson P L., Heremans J F (1970). Association of lactoferrin with specific granules in rabbit heterophil leukocytes. J. Exp. Med. 131, 559–570.

Bennett R M, Davis J (1981). Lactoferrin binding to human peripheral blood cells: An interaction with B-enriched population of lymphocytes and a subpopulation of adherent mononuclear cells. J. Immunol. 127, 1211–1216.

Bennett R M, Davis J D (1981). Lactoferrin binding to human peripheral blood cells: an interaction with B-enriched population of lymphocytes and a subpopulation of adherent mononuclear cells. J. Immunol. 127, 126–127.

Birgens H S (1991). The interaction of lactoferrin with human monocytes. Dan. Med. Bull. 38, 244–252.

Birgens H S, Hansen N E, Karle H, Kristensen L. O. (1983). Receptor binding of lactoferrin by human monocytes. Br. J. Haematol. 54, 383–391.

Birgens H S, Kristensen L O (1990). Imparied receptor binding and decrease in iseolectric point of lactoferrin after interaction with human monocytes. Eur. J. Haematol. 45, 31–35.

Campbell E J (1982). Human leukocyte elastase, cathepsin G and lactoferrin: Family of neutrophil granule glycoproteins that bind to an alveolar macrophage receptor. Proc. Natl. Acad. Sci. 79, 6941–6945.

Conklin K A, Yamashiro K M, Gray G M (1975). Human intestinal sucrase-isomaltase. J. Biol. Chem. 250, 5735–5741.

Crouch S P M, Slater K J, Fletcher J (1992). Regulation of cytokine release from mononuclear cells by the iron- binding protein lactoferrin. Blood 80, 235–240.

Davidson L A, Lonnerdal B (1988). Specific binding of lactoferrin to brush-border membranes: ontogeny and effect of glycan chain. Am. J. Physiol. 254 (Gastrointest. Liver Physiol. 17), G580–G585.

Gislason J, Iyer S, Hutchens T W, Lonnerdal B (1992). Lactoferrin receptors in piglet small intestine: binding kinetics, specificity, ontogeny and regional distribution. J. Nutr. Biochem. 4, 528–533.

Grossman J G, Neu M, Pantos E, Schwab F J, Evans R W, Townes-Andrews E, Lindley P F, Appel H, Theis W G, Hasnain S S (1992). X-ray solution scattering reveals conformational changes upon iron uptake in lactoferrin, serum and ovotransferrins. J. Mol. Biol. 225, 811–819.

Hekman A (1971). Association of lactoferrin with other proteins. as demonstrated by changes in electrophoretic mobility. Biochim. Biophys. Acta 251, 380–387.

Hu W L, Mazurier J, Sawatzki G, Montreuil J, Spik G (1988). Lactotransferrin receptor of mouse small intestinal brush border. Biochem J. 248, 435–441.

Kawakami H, Dosako S, Lonnerdal B (1990). Iron uptake from transferrin and lactoferrin by rat intestinal brush border membrane vesicles. Am. J. Physiol. 258 (Gastrointest. Liver Physiol. 21), G535–G541.

Kawakami H, Lonnerdal B (1991). Isolation and function of a receptor for human lactoferrin in human fetal intestinal brush border membranes. Am. J. Physiol. 261 (Gastrointest Liver Physiol 24), G841–G846.

Masson P L, Heremans J F, Dive C (1966). An iron binding protein common to many external secretions. Clin. Chim. Acta 14, 735–739.

Mazurier J, Legrand D, Hu W L, Montreuil J, Spik G (1989). Expression of human lactotransferrin receptors in phytohemagglutinin stimulated human peripheral blood lymphocytes. Isolation of the receptors by antiligand affinity chromatography Eur. J. Biochem. 179, 481–487.

Mazurier J, Montreuil J, Spik G (1985). Visualization of lacto transferrin brush-border receptors by ligand blotting. Biochim. Biophys. Acta 821, 453–460.

McAbee D D, Esbensen K (1991 ). Binding and endocytosis of Apo- and Holo-lactoferrin by isolated rat hepatocytes. J. Biol. Chem. 266. 23624–23631.

Moguilevsky N. Masson P L, Courtoy P J (1987). Br. J. Haematol. 66, 129–136.

Murer H, Amman E, Biber J, Hopfer U (1976). The surface membrane of the small intestinal epithelial cell. Biochim. Biophys. Acta 433, 409–519.

Nicols B L, McKee K S, Heubers H A (1990). Iron is not required in the lactoferrin stimulation of thymidine incorporation into the DNA of rat crypt enterocytes. Pediatr. Res. 27, 525–528.

Peterson G L (1977). A simplification of the protein assay method of Lowry et al. which is more generally applicable. Anal. Biochem. 83, 346–356.

Rochard E, Legrand D, Mazurier J, Montreuil J, Spik G (1989). The N-terminal domain of human lactotransferrin binds specifically to phytohemagglutinin stimulated peripheral blood human lymphocyte receptors. FEBS Lett. 255, 201–204.

Smith C A, Baker H M, Baker E N (1991). Preliminary crystallographic studies of copper(II)- and oxalate-substituted human lactoferrin. J. Mol. Biol. 219, 155–159.

Van Snick J L, Masson P L (1976). The binding of lactoferrin to mouse peritoneal cells. J. Exp. Med. 144, 1568–1580.

Ziere G J, Van Dijk M C M, Bijsterbosch M K, Van Berke T J C (1992). Lactoferrin by the rat liver. Characterization of the recognition site and effect of selective modification of the arginine residues. J. Biol. Chem. 267, 11229–11235.

# KINETIC PARAMETERS FOR THE HEAT DENATURATION OF BOVINE LACTOFERRIN IN MILK, AND ITS EFFECT ON INTERACTION WITH MONOCYTES

Lourdes Sánchez,[*] José María Peiró,[*] Rosa Oria,[*] Helena Castillo,[*]
Jeremy H. Brock,[†] and Miguel Calvo[*]

[*]Tecnología de los Alimentos
Facultad de Veterinaria
Universidad de Zaragoza
Miguel Servet 177
50013 Zaragoza, Spain

[†]Department of Immunology
Western Infirmary,
Glasgow G11 6NT
Scotland, United Kingdom

## 1. INTRODUCTION

The high concentration of lactoferrin in human milk throughout lactation suggests that this protein fulfils some biological role in the development of the newborn infant. Various functions have been proposed, including regulation of iron absorption and development of the gut flora, stimulation of the immune system, and maturation of the gastrointestinal tract (reviewed by Sánchez et al, 1992a).

Human breast-feeding is commonly substituted by formula-milks which are manufactured from cow's milk. Due to the low concentration of lactoferrin in bovine milk, there is a considerable interest in supplementing artificial milks with this protein. Since these milks are subjected to heat-treatment for preservation it is necessary to know what effect such treatments would have on lactoferrin structure and activity. Furthermore, it should be taken into account that lactoferrin, due to its basic character, interacts with other milk proteins, such as albumin, casein and β-lactoglobulin (Lampreave et al 1990). These interactions might affect the susceptibility of lactoferrin to heat-denaturation and/or interfere with its biological activity.

The aim of this work has been to study the effect of heat treatment on lactoferrin denaturation and also on its interaction with a monocytic cell line.

*Lactoferrin: Structure and Function*
Edited by T.W. Hutchens *et al.*, Plenum Press, New York, 1994

## MATERIALS AND METHODS

### 1. Materials

Bovine lactoferrin was provided by Fina Research (Seneffe, Belgium) and bovine β-lactoglobulin by Sigma (Poole, Dorset, England). Fresh cow's milk (Tauste Ganadera, Zaragoza, Spain) was skimmed by centrifugation at 2000 x g for 15 min at 4°C.

### 2. Methods

**2.1. Heat Treatment.**   Full details are given elsewhere (Sanchez *et al*, 1992b). Briefly, solutions of lactoferrin (apo or iron-saturated) in either Phosphate-buffered saline, pH 7.4 or in milk were subjected to heat treatment in capillary tubes at various temperatures, and denaturation assessed by radial immunodiffusion. Lactoferrin, β-lactoglobulin or a mixture of both proteins (1:1 by weight) were subjected to differential scanning calorimetry from 35 to 110°C and the results analysed on a DuPont model DSC 10 thermal analyser.

**2.2. Cellular Studies.** ⁺ These studies were carried out using the human promonocytic cell line U937. Full details are given elsewhere (Iturralde et al 1992; Brock et al, this volume). Briefly, ligand-binding studies were carried out by incubating $10^6$ log-phase cells with [125]I-labelled bovine Fe-lactoferrin (10 μg/ml) and unlabelled competing protein (50 μg/ml) for 30 minutes at 4°C. Proliferation was determined by uptake of [3]H-thymidine. Cultures contained 50 μg/ml native or heated apo- or Fe-lactoferrin.

## RESULTS AND DISCUSSION

The concentration of undenatured lactoferrin remaining after heat treatment at various temperatures and different times was subjected to kinetic analysis, which showed that the denaturation of lactoferrin followed first-order kinetics. The temperature of treatment (T) and the rate constant (k) in a denaturation process are related according to the Arrhenius equation ($k = K \cdot e^{-E_A/RT}$). The values for k at different temperatures and conditions (Table 1) showed that the rate constant for lactoferrin denaturation increased with temperature and was greater for apolactoferrin than for iron-saturated lactoferrin. This probably reflects the more stable conformation of lactoferrin with bound iron (Anderson *et al*, 1990). The rate constants of both forms of lactoferrin were lower when treated in milk than in phosphate buffer. This might be due to an increase with temperature in the interactions of lactoferrin with caseins and whey proteins, which may increase its heat sensitivity. In addition, the pH of milk decreased more rapidly with increasing temperature, compared with phosphate buffer, and this might also contribute to the low thermoresistence of lactoferrin when treated in milk.

The values of enthalpy of activation for the denaturation of lactoferrin were high (Table 1) indicating that during denaturation lactoferrin experiences a large conformational change with breakage of a large number of bonds. The change in the enthalpy of activation was constant at the temperatures studied. However, the values were still lower for apolactoferrin than for the iron-saturated form and lower when treated in milk than in phosphate buffer (Table 1).

Results of differential scanning calorimetry are shown in Figure 1. The endotherm of lactoferrin denaturation gave two peaks with temperatures of maximum heat absorption at 74°C and 86.5°C. These two peaks suggest that the two lobes of lactoferrin denature independently and have different heat sensitivity. This differs from the situation in transferrin, in which independent denaturation of the two lobes only occurs in the presence of

**Table 1.** Kinetic Parameters for the Denaturation of Iron-Saturated Lactoferrin (Sat) and Apolactoferrin (Apo) in Phosphate Buffer and in Milk.

| | Lactoferrin dissolved in phosphate buffer | | | | Lactoferrin dissolved in milk | | | |
| | k | | $\Delta H^{\#}$ (kJ/mol) | | k | | $\Delta H^{\#}$ (kJ/mol) | |
| Temperature | Sat | Apo | Sat | Apo | Sat | Apo | Sat | Apo |
| --- | --- | --- | --- | --- | --- | --- | --- | --- |
| 72°C | 1.2 | 4.6 | 199.92 | 134.05 | 4.7 | 12.2 | 181.10 | 154.64 |
| 77°C | 2.5 | 5.2 | 199.88 | 134.01 | 8.1 | 20.4 | 181.06 | 154.60 |
| 81°C | 6.6 | 19.1 | 199.85 | 133.98 | 18.4 | 54.4 | 181.03 | 154.57 |
| 85°C | 18.0 | 22.5 | 199.81 | 133.94 | 56.9 | 91.2 | 180.99 | 154.53 |

k = rate constant.
$\Delta H^{\#}$ = enthalpy of activation.

the chaotropic agent sodium perchlorate (Donovan 1977). The β-lactoglobulin endotherm showed only one peak with a denaturation temperature of 73.8°C, while a mixture of β-lactoglobulin and iron-saturated lactoferrin gave two peaks, the first one corresponding to the first of the lactoferrin endotherm and that of β-lactoglobulin, and the second one to the second peak of lactoferrin. In addition, estimation of the enthalpy of denaturation of a mixture of β-lactoglobulin and iron-saturated lactoferrin by integration of the endotherms gave a value of 26.8 J/g, which was almost identical to the sum of the enthalpies of denaturation for each individual protein (21 J/g for lactoferrin and 6.4 J/g for β-lactoglobulin. Thus although lactoferrin can form a complex with β-lactoglobulin, this does not perceptibly alter its heat sensitivity, suggesting that complex formation does not result in any major change in the conformation of lactoferrin.

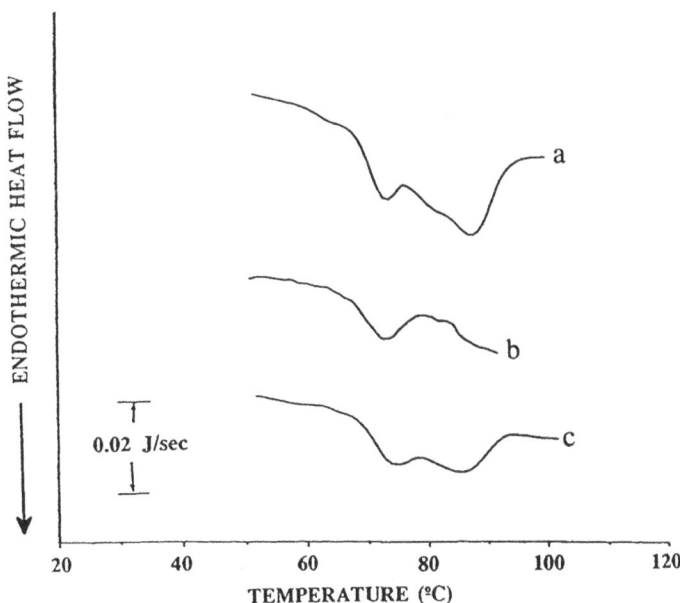

**Figure 1.** Differential scanning calorimetry thermograms of iron-saturated lactoferrin (a), β-lactoglobulin (b) and a mixture of both (c).

**Table 2.** Effect of Heat-Treated Bovine Lactoferrin on the Binding of
Bovine Fe-lactoferrin to U937 Promonocytic Cells. Relative Binding
in the Presence of Fe-lactoferrin or Apolactoferrin.

| Treatment | Fe-lactoferrin | Apolactoferrin |
|---|---|---|
| None (native) | 50 ± 9 | 64 ± 14 |
| 72°C for 20 sec | 82 ± 11 | 80 ± 10 |
| 85°C for 20 min | 84 ± 9 | 79 ± 9 |
| 135°C for 8 sec | 83 ± 6 | 80 ± 6 |

Cells were incubated with 10 µg/µl $^{125}$I-labelled bovine Fe-lactoferrin. Competing
proteins were added at 50 µg/µl.
Results are mean ± s.d., n=4, and are expressed as percentage of binding occurring
*in* the absence of any competing protein.

**Table 3.** Effect of Native and Heat-Treated Bovine Lactoferrin on the
Proliferation of U937 Promonocytic Cells. Relative Proliferation in the
Presence of Fe-lactoferrin and Apolactoferrin.

| Treatment | Fe-lactoferrin | apolactoferrin |
|---|---|---|
| None (native) | 204 ± 91 | 186 ± 67 |
| 72°C for 20 sec | 150 ± 39 | 169 ± 42 |
| 85°C for 20 min | 76 ± 4 | 175 ±3 7 |
| 135°C for 8 sec | 151 ± 36 | 130 ± 26 |

Cells were incubated with 50 µg/µl of each lactoferrin, and proliferation assessed
by $^3$H-thymidine incorporation.
Results are mean ± s.d., n=4, and are expressed as percentage of proliferation
occurring in the absence of lactoferrin.

Binding of labelled bovine Fe-lactoferrin to U937 cells was inhibited by an excess of both
native apo- and Fe-lactoferrin (Table 2). Lactoferrin heat-treated at 72°C for 20 sec, 85°C for
20 min, or at 135°C for 8 sec was somewhat less effective at inhibiting binding of labelled
native lactoferrin, though there was no noticeable difference between any of the three heat
treatments.

Bovine lactoferrin caused a marked increase in $^3$H-thymidine incorporation by U937
cells compared with cells incubated with medium alone (Table 3). This agrees with our
earlier findings using the murine macrophage cell line P388D1 (Oria *et a*l, 1988). Apo-
lactoferrin and Fe-lactoferrin were equally effective. In most cases heat treatment of
Fe-lactoferrin did not abolish the stimulatory effect, although Fe-lactoferrin heated at 85°C
for 20 min lost its stimulatory properties and became somewhat inhibitory. The reasons
for this are not clear, but might be due to destabilisation of the metal-binding site, which
could allow incompletely liganded iron to catalyse potentially harmful free radical reactions
at the cell surface.

Overall, these results show that apolactoferrin in particular is sensitive to heat denaturation, and that sensitivity is increased by other milk components. However, such treatment does
not greatly affect the ability of lactoferrin to interact with monocytic cells, or impair its ability
to enhance cell proliferation.

# REFERENCES

Anderson BF, Baker HM, Norris GE, Rumball SV, Baker EN (1990) Apolactoferrin structure demonstrates ligand-induced conformational change in transferrins. *Nature (London)* 344: 784–787.

Donovan JW (1977) Differences between ovotransferrin and human serum transferrin in structural and metal-binding cooperativity. In *Proteins of iron metabolism* (Ed Brown EB, Aisen P, Fielding J, Crichton RR), Grune and Stratton, New York, pp 179–186.

Iturralde M, Vass JK, Oria R, Brock JH (1992) Effect of iron and retinoic acid on the control of transferrin receptor and ferritin in the human promonocytic cell line U937. *Biochim Biophys Acta* 1133: 241–246.

Lampreave F, Piñeiro A, Brock JH, Castillo H, Sánchez L, Calvo M (1990) Interaction of bovine lactoferrin with other proteins of milk whey. *Int J Biol Macromol* 12: 2–5.

Oria R, Alvarez-Hernández X, Licéaga J, Brock JH (1988) Uptake and handling of iron from transferrin, lactoferrin and immune complexes by a macrophage cell line. *Biochem J* 252: 221–225

Sánchez L, Calvo M, Brock JH (1992a) Biological role of lactoferrin. *Arch Dis Child* 67: 657–661.

Sánchez L, Peiró JM, Castillo H, Pérez MD, Ena JM, Calvo M (1992b) Kinetic parameters for denaturation of bovine milk lactoferrin. *J Food Sci* 57: 873–879.

# CLONING AND EXPRESSION OF THE C-TERMINAL LOBE OF HUMAN LACTOFERRIN

Bhavwanti Sheth, Kathryn M. Stowell, Catherine L. Day, Edward N. Baker, and John W. Tweedie

Department of Chemistry and Biochemistry
Massey University
Palmerston North
New Zealand

## INTRODUCTION

Three dimensional studies of human lactoferrin (Anderson et al, 1989) have shown that like all other members of the transferrin family, lactoferrin is divided into two lobes; the N-terminal and the C-terminal lobes. Each lobe is capable of synergistically binding one $Fe^{3+}$ ion and one $CO_3^{2-}$ anion. The cloning of the cDNA for human lactoferrin (hLf) and its subsequent expression in mammalian cells (Stowell et al, 1991) has provided an excellent system to probe the structure and function of hLf by site-directed mutagenesis. The first mutant to be cloned and expressed using this system was the N-terminal lobe ($Lf_N$) of hLf (Day et al, 1992). Recombinant protein concentrations of up to 30 mg/l in the tissue culture medium have been obtained.

Iron-binding studies of the N-lobe protein have shown no major differences between the half molecule and the full length lactoferrin. However, iron-release experiments have shown that 50% of the iron is lost from the recombinant N-lobe at a pH of 4.8 whereas full length human lactoferrin does not release its iron until much lower pH values ($pH_{0.5} \sim 3$); suggesting that in the whole molecule interaction between the two lobes may contribute to the affinity for iron of Lf at low pH values.

The studies discussed below have addressed the cloning and expression of the C-terminal lobe of human lactoferrin. Until the N-lobe was cloned, the only means available to study the two individual lobes was by proteolysis. This results in an asymmetrical cleavage of the protein and consequently the two lobes have therefore never been available as totally independent moieties. The availability of recombinant C-terminal protein will allow us and other groups to examine the physical and functional differences between the two lobes. With pure recombinant protein for the N- and C-lobes we can also investigate the significance of the interactions between these two lobes of human lactoferrin.

*Lactoferrin: Structure and Function*
Edited by T.W. Hutchens *et al.*, Plenum Press, New York, 1994

## EXPERIMENTAL

### Synthesis of a cDNA Construct for Expression of the C-terminal lobe

The initial hurdle in making a construct for the C-lobe was to join the signal peptide sequence for hLf to the cDNA sequence of the C-terminal half of lactoferrin. This was achieved by PCR of a 1.5 kb truncated fragment of hLf cDNA. The 5'-primer for this reaction was designed to have most of the coding sequences for the signal peptide, including a unique ApaI restriction site and the first 24 bases of the C-terminal lobe, starting at position 341 in the amino acid sequence. The 3'-primer was specific for the 3' untranslated region of hLf cDNA.

The PCR product (~1.2 kb) was then digested with ApaI and BgIII, a unique restriction site within the C-terminal cDNA sequence. This gave a 0.5 kb fragment which was cloned into the plasmid pGEM:hLf, which had also been digested with ApaI and BgIII. Digestion of pGEM:hLf with these restriction enzymes resulted in the complete removal of the N-terminal sequences. After verification by DNA sequencing, the hLf$_C$ cDNA was cleaved from pGEM by restriction enzymes BamHI and HindIIi. This 1.2 kb fragment was then cloned into the SmaI sites of pNUT, an expression vector. Since this was a blunt-ended ligation the fragment was cloned in both orientations; pNUT:hLf$_C$-1 was cloned in the correct orientation for transcription and translation, whilst pNUT:hLf$_C$-2 was cloned in the opposite orientation. pNUT:hLf$_C$-2 was used as a control that would not express the C-terminal protein.

Both pNUT:hLf$_C$ constructs were then transfected into baby hamster kidney (BHK) cells by the calcium phosphate method of DNA co-precipitation. The transformants were selected in medium containing 0.5 mM methotrexate and the presence of the C-terminal protein was tested for by immunoprecipitation and/or western blotting using affinity purified antibodies to human lactoferrin.

A second hLf$_C$ construct was made using a different 5'-primer at the PCR stage. As before this primer contained the sequences corresponding to the signal peptide, including the ApaI restriction site. The first 15 nucleotides of the C-lobe however were altered such that the sequence would code for the first 5 amino acids at the N-terminal of hLf. This strategy was designed to produce a recombinant C-terminal pre-protein which would have the same signal peptidase cleavage site as in the whole molecule and the N-lobe. The cloning and transfection into BHK cells was as described for the first construct.

## RESULTS

DNA sequencing of the C-lobe constructs showed that in both cases the signal peptide sequences had been successfully attached to the cDNA for the C-lobe of hLf. However, when the medium of the transfected cells was checked for secretion of the C-lobe protein, no protein was detected by immunoprecipitation. The cells were therefore checked for the presence of cDNA for the C-lobe. This was demonstrated by PCR of the DNA isolated from the cells using specific primers. BHK cells transfected with the C-lobe construct in either orientation, produced a PCR product of ~1.2 kb which was not evident in the control reactions (Fig 1). RNA was then isolated from these cells in order to determine whether the DNA was being transcribed by the BHK cells. The RNA was separated by electrophoresis and blotted onto nitrocellulose membrane. This blot was probed with cDNA from hLf and the RNA from the transfected cells indicated that a mRNA transcript of ~1.2 kb was hybridising specifically to the cDNA (Fig 2). The transfected cells were therefore successfully transcribing the cDNA into mRNA for the C-lobe. Finally, we checked the BHK cell lysate for protein that may be

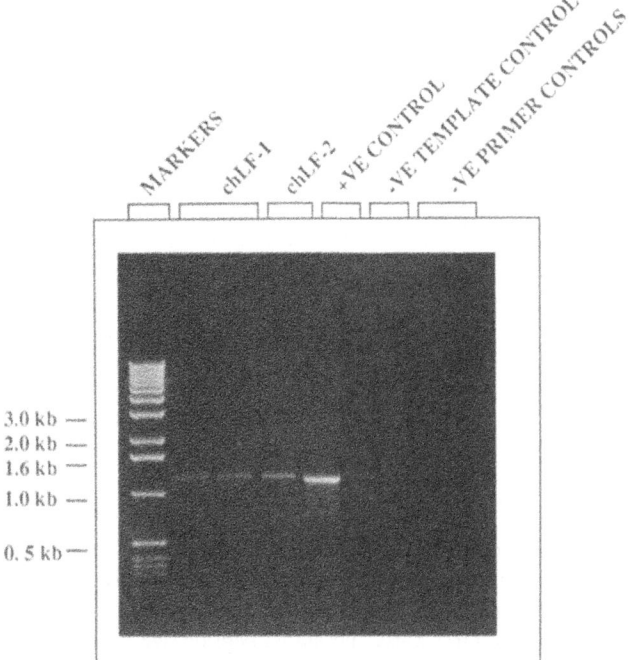

**Figure 1.** PCR analysis of DNA extracted from transfected BHK cells.

trapped inside the cells. The cells were lysed in an NP-40 buffer and the nuclear debris centrifuged. The supernatant was then immunoprecipitated using the polyclonal antibody to hLf. The immunoprecipitate produced was washed and loaded onto an SDS-polyacrylamide gel. Figure 3, a Western blot of this gel indicates a weak doublet at a molecular weight of ~40 kD in the lysate of cells transfected with cDNA for the C-lobe in the correct orientation (hLf$_C$-1) but not in the lysate of cells transfected with pNUT:hLf$_C$-2.

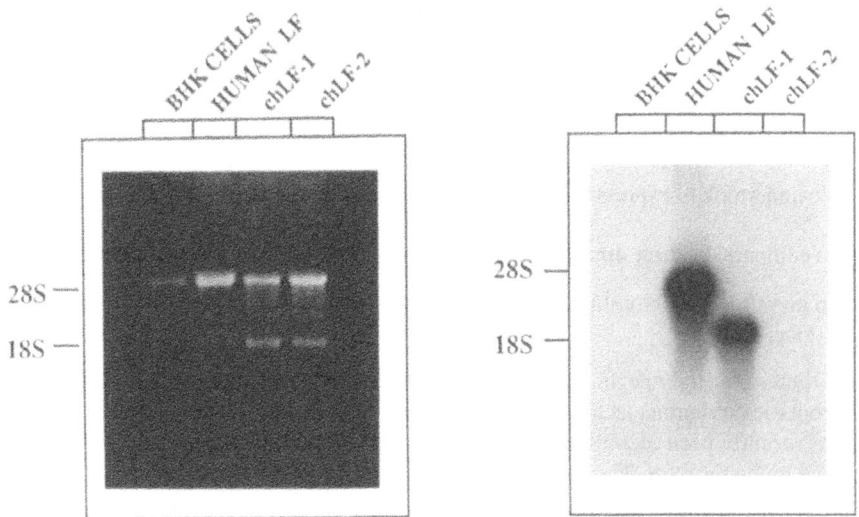

**Figure 2.** Northern blot analysis of RNA isolated from transfected cells.

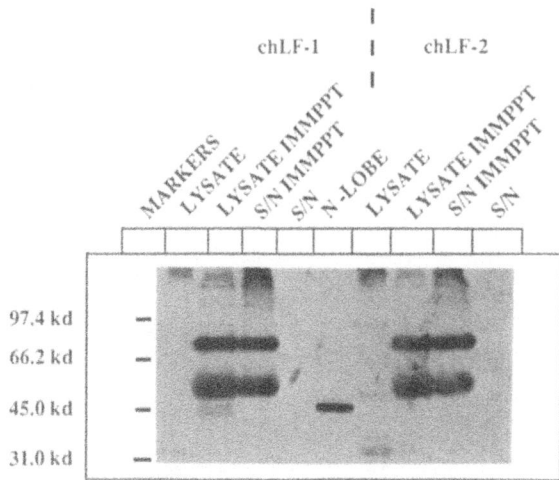

**Figure 3.** Western blot probed with rabbit anti-human Lf antibody.

We hypothesised at this stage that the C-lobe protein may be found in small quantities inside the cells because the signal peptidase of the BHK cells was unable to cleave the signal peptide. Since the full length hLf and the N-lobe protein were both successfully expressed in BHK cells with this particular signal peptide, we designed the second construct such that the first 5 amino acids of this C-lobe would duplicate those of the whole molecule. The first alanine at position 341 was changed to a glycine whilst the second alanine at position 344 was changed to an arginine. A comparison of the first 7 amino acids at the N-terminal and the C-terminal of hLf is shown below:

N-terminal:    $^{1}$Gly Arg Arg Arg Arg Ser Val

C-terminal:    $^{341}$**Ala** Arg Arg **Ala** Arg Val Val

When the second construct was transfected into BHK cells, the C-lobe protein was still not secreted into the medium. FITC-labelled antibodies to hLf were used to check if the protein was inside the cells and these results were also negative.

## CONCLUSIONS

We undertook to express the C-terminal protein of human lactoferrin for several reasons:

i)  To compare the three dimensional structure of the two halves when expressed separately.

ii)  To investigate the iron-binding and release properties of the C-lobe in the absence of the N-lobe.

iii)  To determine the role, if any, of the C-terminal in the microbiocidal and receptor binding properties of human lactoferrin. The presence of an anti-bacterial domain on the N-lobe has recently been identified (Bellamy et al, 1992) and the N-terminal is also thought to bind to the receptor. The role of the C-lobe however is not known.

iv)  To make chimeric molecules of lactoferrin and other transferrins as probes to further investigate the functions of these proteins.

The experiments described above show that we have successfully cloned the cDNA for the C-lobe. However, from the two cDNA constructs made here, neither have been successful in the production of secreted hLf$_C$ protein. The presence of the cDNA in the BHK cells has been demonstrated by PCR (Fig 1) and on induction of the gene, the cDNA is transcribed to produce mRNA of the expected size (Fig 2). Immunoprecipitates prepared from the transfected cells indicate that in the first construct very small amounts of the protein were present inside the cells (Fig 3). These results suggest that the mRNA produced by the BHK cells is not being translated by the ribosomes or that if translation is occurring the resulting protein is not processed correctly and is rapidly being degraded within the cells.

## ACKNOWLEDGEMENTS

This work was supported by the Health Research Council of New Zealand.

## REFERENCES

Anderson, B.F., Baker, H.M., Norris, G.E., Rice, D.W. and Baker, E.N. (1989) Structure of human lactoferrin: Structure analysis and refinement at 2.8 Å resolution. *J. Mol. Biol.* 209 711–734

Bellamy, W., Takase, M., Yamauchi, K., Wakabayashi, H., Kawase, K. and Tomita, M. (1992). Identification of the bactericidal domain of lactoferrin. *Biochimica et Biophysica Acta* 1121 130–136

Day, C.L., Stowell, K.M., Baker, E.N. and Tweedie, J.W. (1992) Studies of the N terminal half of human lactoferrin produced from the cloned cDNA demonstrate that interlobe interactions modulate iron release. *J. Biol. Chem.* 267 13857–13862

Stowell, K.M., Rado, T.A. Funk, W.D. and Tweedie, J.W. (1991) Expression of cloned human lactoferrin in baby-hamster kidney cells. *Biochem. J.* 276 349–355

# CRYSTALLOGRAPHIC STUDIES ON METAL AND ANION SUBSTITUTED HUMAN LACTOFERRIN

Clyde A. Smith, Heather M. Baker, Musa S. Shongwe, Bryan F. Anderson, and Edward N. Baker

Department of Chemistry and Biochemistry
Massey University
Palmerston North, New Zealand

The metal and anion binding function of the transferrins has been well characterised by a host of spectroscopic methods, and the members of this family of proteins are known to bind a wide variety of metal ions including the first, second and third row transition metals, the group 13 metals, the lanthanides and some of the actinides (Aisen and Harris, 1989). Crystallographic studies, on the other hand, have concentrated primarily on diferric and apolactoferrin (Anderson et al., 1989; 1990) and diferric rabbit transferrin (Bailey et al., 1988; Sarra et al., 1990). These crystallographic studies have defined the polypeptide chain folding and the metal (iron) and anion (carbonate) sites (Baker et al., this volume).

Schlabach and Bates (1975) showed that the synergistic anion requirement of the transferrins can be met with anions other than carbonate, including oxalate, malonate, thioglycolate and nitrilotriacetate. This study led to the so-called "interlocking sites" model for the interaction of the anion with the metal and the protein (Figure 1a). In order to display the properties of a synergistic anion, the anion must possess a carboxylate group and a second electron donor group no more than 7 Å away (Figure 1b).Recently, electron spin echo envelope modulation (ESEEM) experiments on a variety of copper- and vanadyl-anion-transferrin complexes have indicated that the "interlocking sites" model may be an oversimplification and that the mode of anion coordination is dependent on the electron donating groups on the anion (Dubach et al., 1991).

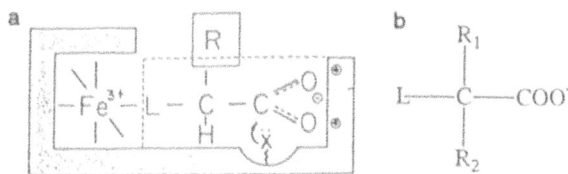

**Figure 1.** (a) The "interlocking sites" model of synergistic anion binding to the transferrins, as proposed by Schlabach and Bates (1975); (b) The general structure of a synergistic anion, where L is a second donor atom which can interact with the metal ion.

*Lactoferrin: Structure and Function*, Edited by T.W. Hutchens, B. Lönnerdal, and
S. Rumball, Plenum Press, New York, 1994

X-ray structural studies demonstrate the existence of conformational flexibility in the protein. It can be inferred from the diferric and metal-free human lactoferrin structures that iron binding is associated with a large scale conformational change, as illustrated by the 54° rotation seen for the N2 domain. Small angle X-ray scattering studies on apoLf and various metallo-lactoferrin complexes indicate that a similar conformational change occurs upon copper binding as seen with iron (Grossman et al., 1992). In order to understand the structure–function relationships, crystallographic studies involving other metals and anions were conducted.

## METHODS, RESULTS AND DISCUSSION

ApoLf was prepared using the method described by Norris et al. (1989). The dicupric, $Cu_2(CO_3)_2Lf$, dicupric monooxalate, $Cu_2(CO_3)(C_2O_4)Lf$ and diferric dioxalate, $Fe_2(C_2O_4)_2$, complexes were prepared as shown in Figure 2 and characterised by UV-visible and ESR spectroscopy. It is interesting to note that while the dicupric-oxalate complexes can be prepared by reacting $Cu_2(CO_3)_2Lf$ with excess sodium oxalate, the corresponding reaction with $Fe_2(CO_3)_2Lf$ is not possible. It appears that the carbonate ion is bound in the iron ternary complex much more tightly than in the copper complex and that oxalate is unable to remove it. These anion competition reactions clearly show the synergistic relationship between the metal and anion and the inequivalence of the two sites.

Crystals of $Cu_2(CO_3)_2Lf$, $Cu_2(CO_3)(C_2O_4)Lf$ and $Fe_2(C_2O_4)_2Lf$ were grown by dialysis from phosphate buffer using methanol as the precipitant. Data sets for the first two complexes

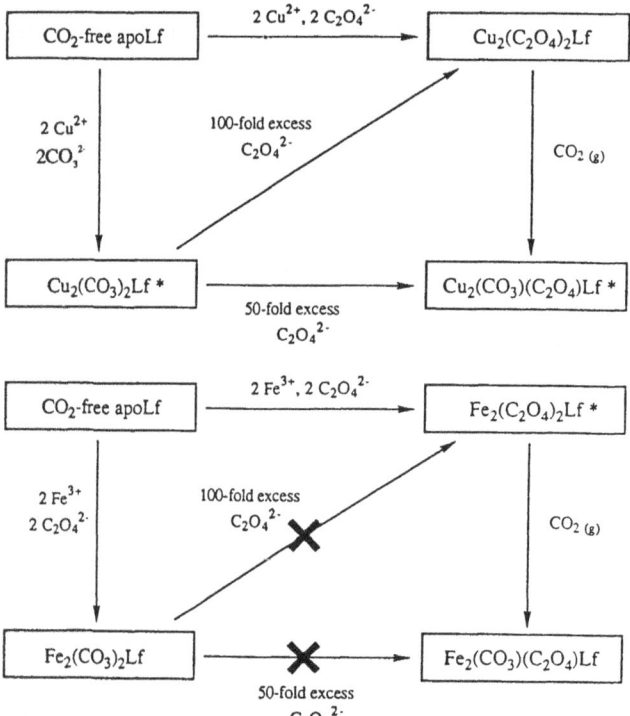

**Figure 2.** Preparative schemes for the iron and copper complexes of lactoferrin with carbonate and oxalate as the synergistic anions. * identifies the complexes studied by crystallography.

were collected using an Enraf-Nonius CAD4 diffractometer and the Synchrotron Radiation Source at Daresbury, UK, while data from the latter crystals were collected at the Photon Factory at Tsukuba, Japan. The space group is $P2_12_12_1$ for all three complexes, and they are isomorphous with $Fe_2(CO_3)_2Lf$.

The structures were solved using the $Fe_2(CO_3)_2Lf$ atomic coordinates as the starting model, with the iron atoms, carbonate ions, protein ligands, solvent molecules and carbohydrate chains removed. Initial R-factors were 0.34, 0.29 and 0.28 for the three complexes respectively. Table 1 gives some of the refinement statistics for the complexes. It should be noted that the refinement of the $Cu_2(CO_3)(C_2O_4)Lf$ and $Fe_2(C_2O_4)_2Lf$ structures are not yet complete, while the final structural details for $Cu_2(CO_3)_2Lf$ have been previously reported (Smith et al., 1992).

These structural studies show that $Cu^{2+}$ and $C_2O_4^{2-}$ bind in the specific binding sites of human lactoferrin and do not lead to any major structural changes except in the immediate vicinity of the metal. In the first complex, with the two iron atoms replaced by copper, the only apparent changes are in the position and the coordination geometry of the N-lobe metal ion. The copper atom has moved about 1 Å from the position occupied by iron and this movement, along with a small rotation of the anion, results in a decrease in the copper coordination number from 6 to 5. The anion is monodentate and one of the Cu–tyrosyl oxygen bond distances has lengthened to 2.7 Å.

When compared to small molecule copper(II) complexes, the geometry observed in the N-lobe is favoured over the 6-coordinate geometry in the C-lobe, as copper (II) complexes are subject to Jahn–Teller distortion and are rarely regular. The rearrangement of the copper geometry is readily accommodated in the N-lobe due to the more flexible nature of this lobe relative to the C-lobe. This difference in flexibility of the two lobes has already been observed in the apoLf structure, where the N-lobe was open while the C-lobe remained closed (Anderson et al, 1990).

The first difference electron density maps calculated between the $Cu_2(CO_3)(C_2O_4)Lf$ and $Fe_2(CO_3)_2Lf$ data sets indicated that the oxalate had replaced the carbonate in the C-lobe site (Figure 3) and that this in turn resulted in a change in the positions of the sidechains of Arg465 and Tyr398. Refinement of the structure showed that the oxalate was coordinated in a symmetric 1,2-bidentate fashion to the copper atom and that the geometry in the N-lobe site was essentially identical to that observed in $Cu_2(CO_3)_2Lf$. Further analysis of the changes in the C-lobe shows that the new position of Arg465 is stabilised by a hydrogen bonding interaction with Tyr526, which may help to explain why substitution is favoured in this lobe over the N-lobe, where such an interaction is not possible.

**Table 1.** Selected Model and Refinement Statistics for $Cu_2(CO_3)_2Lf$, $Cu_2(CO_3)(C_2O_4)Lf$ and $Fe_2(C_2O_4)_2Lf$

| Complex | $Cu_2(CO_3)_2Lf$ | $Cu_2(CO_3)(C_2O_4)Lf$ | $Fe_2(C_2O_4)_2Lf$ |
| --- | --- | --- | --- |
| Unit cell edges [Å] | 155.9 97.0 56.0 | 155.8 96.7 55.9 | 155.5 96.9 55.9 |
| Protein atoms | 5321 | 5328 | 5322 |
| Water molecules | 308 | 307 | 122 |
| Resolution limit [Å] | 2.1 | 2.1 | 2.3 |
| R-factor | 0.196 | 0.215 | 0.22 |
| rms deviation (bonds) [Å] | 0.018 | 0.017 | 0.022 |

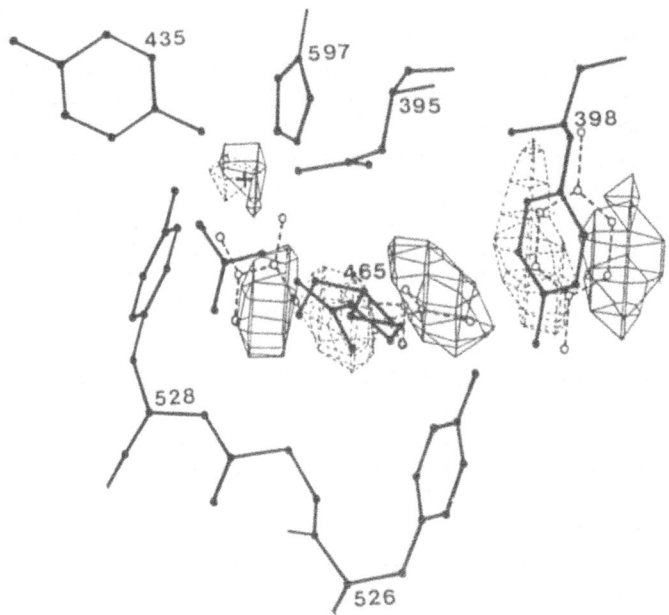

**Figure 3.** Difference electron density map in the vicinity of the C-lobe metal and anion binding site of $Cu_2(CO_3)(C_2O_4)Lf$, calculated with coeffcients, $\Delta F = |F_{Cu_2(CO_3)(C_2O_4)Lf}| - |F_{Cu_2(CO_3)_2Lf}|$. The $Cu_2(CO_3)(C_2O_4)Lf$ model is shown with bold lines, while the electron density is shown as thin lines (positive density) and dashed lines (negative density).

In $Fe_2(C_2O_4)_2Lf$, substitution takes place in both the N- and C-lobe sites, with the C-site virtually identical to the corresponding site in $Cu_2(CO_3)(C_2O_4)Lf$. In the N-lobe, as in the C-lobe, the arginine adjacent to the anion is pushed aside by the larger oxalate ion. There is a difference between the two sites, however, in that the anion in the N-lobe site is coordinated in an asymmetric fashion, with the two Fe–O (oxalate) bond distances being 2.0 and 2.6 Å. This contrasts with the C-lobe where oxalate coordination is symmetric, and with the carbonate structure where the anion is coordinated symmetrically in both lobes. The reason for the asymmetric oxalate coordination in the N-lobe is not clear, but it may be for this reason that oxalate binding in the N-lobe is less facile than in the C-lobe.

The coordination observed for oxalate contradicts the "interlocking sites" model of Schlabach & Bates (1975) in which the anion is bound to iron via the proximal electron donor group, while its carboxylate group interacts with the protein. In fact, in the case of oxalate, both the proximal ligand *and* the carboxylate bind to the metal ion. This can very likely be extended to other anions such as thioglycolate, for which EXAFS studies indicate that the sulphur ligand at least is directly bound to the iron (Schneider et al., 1984). These arguments lead to a revised general model for anion binding in the transferrins, based on a synergistic anion with the structure described by Schlabach and Bates (1975). This is shown in Figure 4.

Finally, the monodentate anion in the N-sites of both copper complexes, together with the hydrogen bonding networks, suggests that the form of the anion may be bicarbonate rather than carbonate. The observed monodentate coordination could also model the changes which might occur upon metal release, assuming protonation of the anion is the first step (Aisen et al., 1974; Sarra et al., 1990). Protonation would disrupt the anion hydrogen bonding and the subsequent movement of the anion and a change to monodentate coordination could lead to the destabilization of the metal-anion-lactoferrin ternary complex.

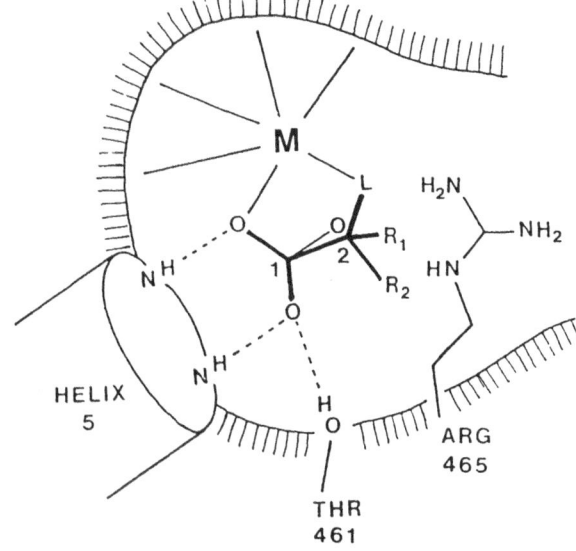

**Figure 4.** A revised model of the binding of a carboxylate synergistic anion to the transferrins (shown in the C-lobe binding site of lactoferrin). The carboxylate group interacts with the metal, the N-terminus of helix 5 and Thr461. The rest of the anion (heavy lines) extends past the third carbonate position (carbonate shown with thin lines) and the electron donor group, L, can then bind to the metal. The sidechain of Arg465 is necessarily displaced away from the anion.

## ACKNOWLEDGEMENTS

We gratefully acknowledge support from the U.S. National Institute of Child Health and Human Development (Grant HD-20859), The Health Research Council of New Zealand and the New Zealand Dairy Research Institute.

## REFERENCES

Aisen, P. & Harris, D.C. (1989) in Iron Carriers and Iron Proteins (T. M. Loehr, Ed.) pp 241-351, VCH Publishers, New York.

Aisen, P., Pinkowitz, R.A. & Leibman, A. (1974) *Ann. N.Y. Acad. Sci. 222,* 337.

Anderson, B.F., Baker, H.M., Norris, G.E., Rice, D.W. & Baker, E.N. (1989) J. Mol. Biol. 209, 711.

Anderson, B.F., Baker, H.M., Norris, G.E., Rumball, S.V., & Baker, E.N. (1990) *Nature (London) 344,* 784.

Bailey, S., Evans, R.W., Garratt, R.C., Gorinsky, B., Hasnain, S.S., Horsburgh, C., Jhoti, H., Lindley, P.F., Mydin, A., Sarra, R. & Watson, J.L. (1988) *Biochemistry 27,* 5804.

Dubach, J., Gaffney, B.J., More, K., Eaton, G.R. & Eaton, S.S. (1991) *Biophys. J. 59,* 1091.

Grossman, G., Appel, H., Hasnain, S.S., Neu, M., Schwab, F. & Thies, W.-G. (1991) Abstract from the *10th International Conference on Iron and Iron Proteins,* Oxford, U.K.

Norris, G.E., Baker, H.M. & Baker, E.N. (1989) *J. Mol. Biol. 209,* 329.

Sarra, R., Garratt, R., Gorinsky, B, Jhoti, H. & Lindley, P. (1990) *Acta Cryst. B46,* 763.

Schlabach, M.R. & Bates, G.W. (1975) *J. Biol. Chem. 250,* 2182.

Schneider, D.J., Roe, A.L., Mayer, R.J. & Que, L., Jr. (1984) *J. Biol. Chem. 259,* 9699.

Smith, C.A., Anderson, B.F., Baker, H.M. and Baker, E.N. (1992) *Biochemistry 31,* 4527.

# PROPOSED MECHANISMS FOR THE INVOLVEMENT OF LACTOFERRIN IN THE HYDROLYSIS OF NUCLEIC ACIDS

Xiao-Yan Zhao and T. William Hutchens

Protein Structure Laboratory
USDA/ARS Children's Nutrition Research Center,
Department of Pediatrics
Baylor College of Medicine
1100 Bates
Houston, Texas 77030

**Abstract**   Lactoferrin has recently been proposed to have ribonuclease activity in the absence of bound iron. We and others have demonstrated previously that lactoferrin interacts with DNA and will bind a number of transition metal ions via surface-exposed histidyl residues. In the present study, we investigated the possibility that surface-bound copper ions on lactoferrin may catalyze the production of active oxygen species responsible for the hydrolysis of nucleic acids. Purified lactoferrin (apo- and holo-forms) was incubated with $CuCl_2$ in solution to obtain lactoferrin with surface binding sites saturated by Cu(II) ions. The lactoferrin-Cu(II) complex was purified by Bio-Gel P-6 chromatography columns and tested for hydrolytic activity against DNA and RNA as analyzed by agarose gel electrophoresis. Incubation of lactoferrin-Cu(II) complexes with supercoiled plasmid Bluescript II SK DNA led to the rapid formation of relaxed open circular or linear forms of DNA characterized by changed electrophoretic mobility. Lactoferrin with bound Cu(II) also caused extensive degradation of yeast tRNA molecules in the presence of hydrogen peroxide. Covalent modification of surface-exposed histidyl residues by carboxyethylation with diethylpyrocarbonate abolished the lactoferrin-associated hydrolytic activity. These results indicate that lactoferrin-bound Cu(II) can indeed facilitate the hydrolysis of DNA and RNA molecules. Copper-binding sites on lactoferrin appear to serve as centers for repeated production of hydroxyl radicals via a Fenton-type Haber–Weiss reaction. Enhanced nuclease activity associated with elevated local concentrations of lactoferrin would promote microbial degradation.

## INTRODUCTION

Lactoferrin is a 78-kD glycoprotein found in human milk and other external secretions (1). It is a major constituent of the neutrophil specific granules (2) and has been associated with the immune cell modulation (3). Other functions have also been proposed for this molecule such as iron absorption and microbial degradation (4). Lactoferrin binds two Fe(III) with high affinity in the two specific metal-binding sites of the N and C lobes. Ligands for the specific metal binding in the case of ferric-lactoferrin are provided by an aspartic acid, a histidine, and two tyrosine residues, as well as by an anion (5). The iron binding protein, lactoferrin, was shown to mediate the formation of oxygen radicals (6–16) which are generally believed to be active species in the hydrolysis of nucleic acids (17). Furmanski et al. have recently indicated that lactoferrin possesses ribonuclease activity in the absence of bound iron (18). The mechanism for the observed lactoferrin-associated RNase activity is unknown.

Our earlier study demonstrated that lactoferrin binds to a number of transition metal ions (e.g., Cu) via surface-exposed histidyl residues (19). Copper (II) ion is an efficient catalyst of redox reactions that produce free radicals (20). Since lactoferrin interacts with DNA under physiologic conditions (21–24), we examined the possibility that surface-bound copper (II) ions on lactoferrin may catalyze the production of active oxygen species responsible for the hydrolysis of nucleic acids. A sensitive assay was utilized for detection of the free radical-induced strand scission in DNA and RNA molecules. We discovered that lactoferrin with bound copper ions has evident hydrolytic activities against nucleic acids in the presence of hydrogen peroxide. In this report, we present a novel approach to investigate the biological function of lactoferrin inferred from its structural characteristics. The mechanisms for the role of lactoferrin in the hydrolysis of nucleic acids are proposed.


## METHODS

Purification of human lactoferrin by DNA affinity chromatography. The purification procedure was described previously (24). Briefly, colostrum was first adjusted to pH 4 and subjected to ultracentrifugation for collecting the whey proteins. For hololactoferrin preparations, the whey proteins were readjusted to pH 7 or 8 followed by the addition of ferrous sulfate in the presence of bicarbonate. The whey samples were loaded into single-stranded DNA affinity column in the presence of 6 M urea. Lactoferrin was eluted with a linear gradient of NaCl (0–1.0 M) in 20 mM HEPES, pH 8.0, after the urea was removed. The high purity of the preparations was confirmed by immobilized metal-ion affinity HPLC, reverse phase HPLC, and SDS-PAGE followed by silver staining.

Chemical modification of lactoferrin by carboxyethylation with diethylpyrocarbonate (DEPC). As described previously (19), DEPC (Sigma) stock solution was prepared in anhydrous ethanol, and the concentration was determined before use by adding an aliquot of the stock to a known concentration of imidazole solution and measuring the absorbance at 240 nm. An aliquot of this stock was incubated for 30–40 min with a known concentration of hololactoferrin at room temperature. The final molar ratio of DEPC to lactoferrin in the reaction was 50:1. The reactions were monitored by UV absorbance. DEPC-lactoferrin complexes were isolated by affinity chromatographies on immobilized Cu(II) ions.

Isolation of supercoiled plasmid DNA by $CsCl_2$ gradient ultra-centrifugation. A single colony of bacteria DH-5 alpha carried the plasmid Bluescript II SK was inoculated and the overnight culture was subjected to alkali lysis followed by potassium acetate neutralization. The lysate was applied to two cycles of $CsCl_2$-ethidium bromide gradient ultra-centrifugation after the bacterial debris was removed. Supercoiled plasmid DNA was isolated and checked for its purity on agarose gel electrophoresis as previously published (25).

DNA and RNA strand scission assay. Hololactoferrin was passed through a Sephadex G-25M column (Pharmacia) for removing surface bound iron on lactoferrin. Lactoferrin (apo- and holo- forms) or DEPC-lactoferrin (1 eq) was incubated with 10 eq of $CuCl_2$ in 10 mM sodium phosphate buffer (pH 7.2) plus 145 mM NaCl at 37°C for 2 h to obtain lactoferrin-Cu(II) complex. The complexes were purified by chromatography over Bio-Gel P-6 columns (Bio-Rad). Strand scission reaction was carried out at 37°C for 2 h. The reaction mixture had a final volume of 24 μl, containing 54 μM base-pair DNA or 270 μM tRNA, 62.5 μM lactoferrin–copper complex, and various concentrations of hydrogen peroxide, 145 mM NaCl, 10 mM sodium phosphate buffer, pH 7.2. Reactions were terminated by adding 1 mM EDTA. The reaction mixtures were treated with Proteinase K at a concentration of 200 µg/ml for 15 min at 37°C followed by phenol/chloroform extraction. DNA or RNA samples were analyzed by 0.8% agarose gel electrophoresis.

## RESULTS

Lactoferrin-metal complexes. Lactoferrin was initially recognized as an iron-binding protein (1). Later reports from Ainscough et al. (26–29) demonstrated that the human lactoferrin can also form complexes with other metals in place of iron, such as chromium, manganese, cobalt, and copper. The occupancy of the specific metal-binding sites on lactoferrin results in dramatic changes regarding to the absorption spectrum. As shown in Figure 1, absorbance spectra of various lactoferrin-metal complexes were recorded from 300 nm to 700 nm. All the complexes show an increased absorbance at 300 nm (near 295 nm) and exhibit additional adsorption maxima between 400 nm and 500 nm, when compared with native lactoferrin (94% apo-form). The colorful lactoferrin-metal complexes are easily distinguished from apolactoferrin, which is colorless. The pink ferric-lactoferrin (Fe) and yellow cupric-lactoferrin (Cu) complexes display absorption peaks at 465 nm and 438 nm, respectively, which may result from phenolate-metal ion charge transfer transition (26). We have presented

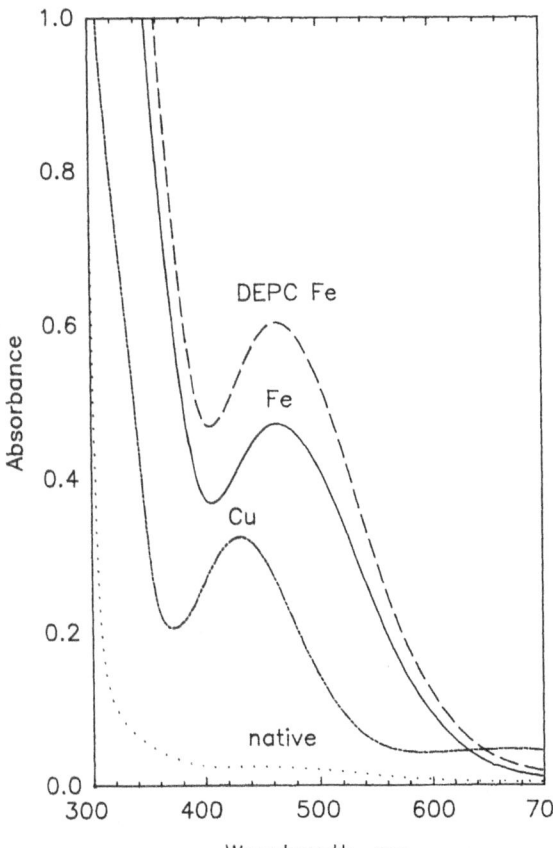

**Figure 1.** Optical absorption spectra of apolactoferrin (·····), cupric-lactoferrin (—·—··), ferric-lactoferrin (—), and DEPC-ferriclactoferrin (- - - - -). 1% protein solutions were prepared in 10 mM sodium phosphate buffer (pH 7.2) plus 145 mM NaCl.

previously that lactoferrin can also bind to a number of transition metal ions via surface-exposed histidyl residues (19). These surface-exposed histidyl residues of lactoferrin are accessible to modifications by chemicals such as carboxyethylation with diethylpyrocarbonate (DEPC). Covalent modification of surface-exposed histidyl residues by DEPC, however, does not change the spectrum characteristic of ferric-lactoferrin significantly. Thus, the absorption maxima of pink DEPC-lactoferrin (DEPC Fe) resembles that of ferric-lactoferrin (Fe), indicating that the specific metal binding sites of lactoferrin have not been altered by DEPC modification.

Degradation of RNA by lactoferrin-Cu(II). Copper ions are efficient catalysts of redox reactions that produce free radicals (20), which are known to be able to degrade nucleic acids (17). The question was raised whether lactoferrin could be involved in the production of free radicals in the presence of surface bound copper(II) ions. As shown in Figure 2 (lanes 1–6), lactoferrin-Cu(II) complex indeed had the hydrolytic activity towards yeast tRNA. Furthermore, the hydrolytic activity is dependent upon addition of hydrogen peroxide. Similar results were obtained with bacterial RNA (data not shown). Significant hydrolysis occurred in the presence of more than 250 µM of hydrogen peroxide (lanes 4–6). DEPC was used to modify the seven surface-exposed histidyl residues to abolish its copper-binding capacity. The modified lactoferrin (DEPC-lactoferrin) completely lost the hydrolytic activity (lane 7), suggesting that copper-binding ability is essential for lactoferrin to hydrolyze RNA. EDTA, the chelating agent for Cu(II) ions, totally inhibited tRNA hydrolysis (lane 8). As shown in Table 1, incubation of free copper with hydrogen peroxide hydrolyzed RNA, but incubation of copper plus 2-mercaptoethanol did not. The hydrolytic effect of free copper in the presence of hydrogen peroxide was contributed by copper-induced formation of hydrogen radical

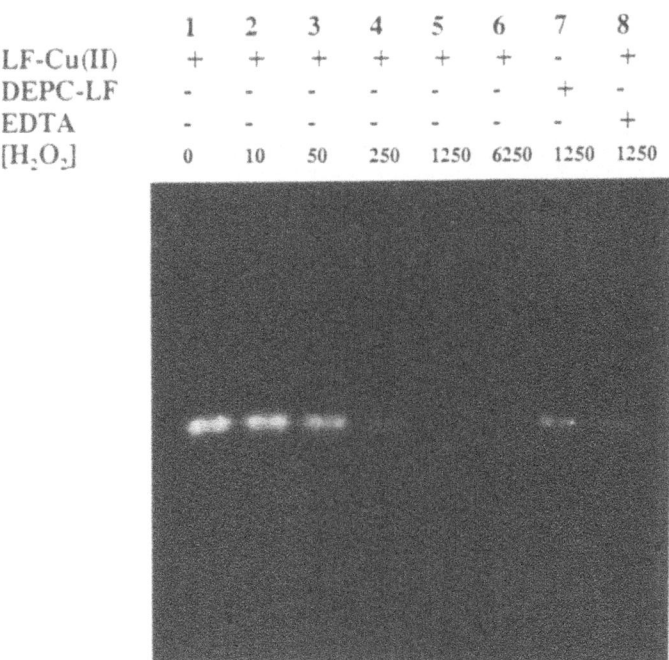

|           | 1 | 2 | 3 | 4 | 5 | 6 | 7 | 8 |
|-----------|---|---|---|---|---|---|---|---|
| LF-Cu(II) | + | + | + | + | + | + | - | + |
| DEPC-LF   | - | - | - | - | - | - | + | - |
| EDTA      | - | - | - | - | - | - | - | + |
| [$H_2O_2$] | 0 | 10 | 50 | 250 | 1250 | 6250 | 1250 | 1250 |

**Figure 2.** Degradation of yeast tRNA by lactoferrin-Cu(II). Samples containing yeast tRNA(270 µM), and lactoferrin-Cu(II) (62.5 µM) or DEPC-lactoferrin (62.5 µM) plus various concentrations of $H_2O_2$ were incubated for 2 h at 37°C.

**Table 1.** Hydrolytic Activity of Lactoferrin on RNA Molecules

| SAMPLE | HYDROLYTIC ACTIVITY |
|---|---|
| apo Lf | − |
| apo Lf-Zn(II) | − |
| apo Lf-Cu(II) | + |
| apo Lf-Cu(II) + EDTA | − |
| holo Lf before Sephadex G-25M chromatography | + |
| ferric Lf | − |
| ferric Lf-Cu(II) | + |
| DEPC-ferric Lf | − |
| Cu(II) + 2-mercaptoethanol | − |
| Cu(II) + $H_2O_2$ | + |
| RNase A | + |
| RNase A + EDTA | + |

(30–32) via Fenton-type Haber–Weiss reactions. Reduction of copper has been shown to be critical in the repeated production of free radical in Fenton reaction (20, 30–32). Reduction of copper in the presence of 2-mercaptoethanol, however, is apparently not sufficient to cleave RNA. It appears that lactoferrin may serve as a vehicle in carrying out the Cu(II)-catalyzed nuclease activity in the biological system.

To exclude the possibility of ribonuclease (RNase) contamination, we carried out two control experiments. First, we replace copper in the system with a redox inactive metal, zinc. Lactoferrin-Zn(II) can not hydrolyze tRNA molecules even in the presence of high concentration (6 mM) of hydrogen peroxide (Table 1). Second, we found that lactoferrin-Cu(II) associated RNase activity was very sensitive to EDTA (Table 1), which is known to have no effect on RNase (33). These results strongly suggest that lactoferrin-Cu(II) is indeed responsible for RNA hydrolysis in the present system.

The lactoferrin-associated RNase activity was not affected by the occupancy of the specific metal binding sites of lactoferrin. As shown in Table 1, either apolactoferrin-Cu(II) or ferric-lactoferrin-Cu(II) has hydrolytic activity on RNA molecules. In the hydrolysis of tRNA, hololactoferrin is as effective as lactoferrin-Cu(II) before the surface bound iron were removed by Sephadex G-25M chromatography (Table 1), indicating that the surface metal binding is important for the lactoferrin-associated nuclease activity.

Cu(II)-dependent hydrolytic effect of lactoferrin on DNA. To investigate the hydrolytic activity of lactoferrin-Cu(II) on DNA molecules, supercoiled plasmid Bluescript II SK DNA was isolated. As shown in Figure 3, supercoiled DNAs were nicked and converted into relaxed open circular DNAs after incubation with lactoferrin-Cu(II) in the presence of hydrogen peroxide (lanes 3–6). Open circular DNA molecules were detected on the agarose gel as a retarded band. The conversion of supercoiled DNA to open circular DNA was complete in the presence of 1250 μM of $H_2O_2$ (lane 5). Higher concentration of $H_2O_2$ (lane 6) resulted in appearance of linear DNA due to extensive DNA cleavage. DEPC-lactoferrin lost DNase activity even in the presence of high concentration of $H_2O_2$ (lane 7), suggesting that copper binding are essential for lactoferrin to cleave DNA. EDTA prevented DNA from cleavage by lactoferrin-Cu(II) in the presence of $H_2O_2$ (lane 8). Similarly to the hydrolysis of RNA, these results demonstrated that the lactoferrin-associated hydrolysis of DNA was dependent on the

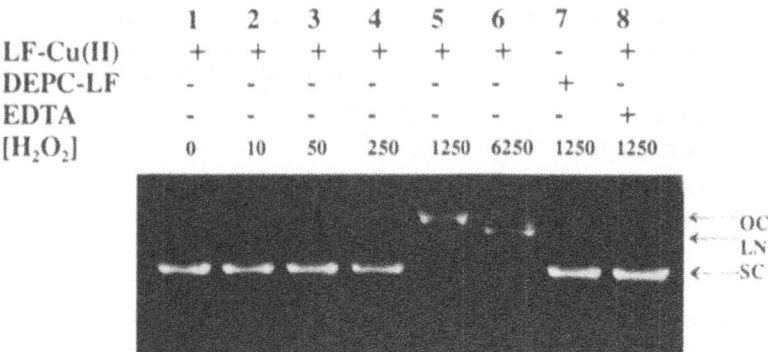

**Figure 3.** Cleavage of double-stranded DNA by lactoferrin-Cu(II). Samples containing plasmid Bluescript II SK DNA (54 μM base pair), and lactoferrin-Cu(II) (62.5 μM) or DEPC-lactoferrin (62.5 μM) plus various concentrations of $H_2O_2$ were incubated for 2 h at 370C. Open circular (oc), supercoiled (sc), and linear (ln) DNA band positions are indicated.

bound copper on the protein surface. Furthermore, the lactoferrin-associated DNA strand scission required an optimal concentration of hydrogen peroxide.

We noticed that the method for sample treatment after DNA strand scission assay has profound effects on the electrophoretic patterns. As summarized in Table 2, the DNA sample without treatment after scission assay was physically trapped in the gel wells due to DNA-lactoferrin aggregates. Thus, DNA bands on the gel wells were detected. Addition of 0.1% SDS can not dissociate the DNA-lactoferrin aggregates. Although DNAs can migrate into the agarose gel if the sample was boiled for 3 min in 0.1% SDS, multiple DNA bands were observed due to denaturation of DNA molecules by heat. The mentioned problems can be solved by phenol/chloroform extraction to remove bound lactoferrin on DNA molecules. But the extraction resulted in a low recovery of DNAs. We found that incubation of Proteinase K at 200 μg/ml for 15 min at 37°C prior to phenol/chloroform extraction gave satisfying results as shown in Figure 3.

## CONCLUSIONS

In summary, the human lactoferrin in the presence of bound copper ions has hydrolytic activity against RNA and DNA molecules. The hydrolytic effect is dependent upon the

**Table 2.** Effects of Post-reaction Sample Treatment on DNA Electrophoretic Patterns

| TREATMENT | OBSERVATION |
|---|---|
| none | DNA bands on the gel wells |
| addition of 0.1% SDS | DNA bands on the gel wells |
| boiling in 0.1% SDS | multiple DNA bands |
| phenol/chloroform extraction | very faint DNA bands |
| incubation with Protease K prior to extraction | patterns as shown in Fig. 3. |

concentration of hydrogen peroxide. The surface copper binding sites on lactoferrin are essential for the hydrolysis of nucleic acids.

## REFERENCES

1. Aisen P, and Listowsky I. 1980. Iron transport and storage protein. Ann Rev Biochem 49: 357–393.
2. Baggliolini MC, Masson DPL, and Heremans JF. 1970. Association of lactoferrin with specific granules in rabbit heterophil leukocytes. J Exp Med 131: 559.
3. Bagby Jr GC, Rigas VD, Bennet RM, Vandenbark AA. 1981. Interaction of lactoferrin, monocytes and T lymphocyte subsets in the regulation of steady-state granulopoiesis in vitro. J Clin Invest 68: 56–63.
4. Sanchez L, Calvo M, and Brock JH. 1992. Biological role of lactoferrin. Arch Dis Child 67: 657–661.
5. Shongwe MS, Smith CA, Ainscough EC, Baker HA, Brodie AM, and Baker EN. 1992. Anion binding by human lactoferrin: results from crystallographic and physicochemical studies. Biochem. 31: 4451–4458.
6. Ambruso DR and Johnston RB. 1981. Lactoferrin enhances hydroxyl radical production by human neutrophils, neutrophil particulate fractions, and an enzymatic generating sysstem. J. Clin. Invest. 67: 352–360.
7. Winterbourn CC. 1981. Hydroxyl radical production in body fluids. Biochem. J. 198: 125–131.
8. Gutteridge JMC, Paterson SK, Segal AW, and Halliwell B. 1981. Inhibition of lipid peroxidation by the iron-binding protein lactoferrin. Biochem. J. 199:259–261.
9. Bannister JV, Bannister WH, Hill HAO, Thornalley PJ. 1982. Enhanced production of hydroxyl radicals by the xanthine-xanthine oxidase reaction in the presence of lactoferrin. Biochim. Biophy. Acta. 715: 116–120.
10. Winterbourn CC. 1983. Lactoferrin-catalyzed hydroxyl radical production. Biochem. J. 210: 15–19.
11. Baldwin DA, Jenny ER, and Aisen P. 1984. The effect of human serum transferrin and milk lactoferrin on hydroxyl radical formation from superoxide and hydrogen peroxide. J. Biol. Chem. 259: 13391–13394.
12. Britigan BE, Hassett DJ, Rosen GM, Hamill DR, and Cohen MS. 1989. Neutrophil degranulation inhibits potential hydroxyl-radical formation. Biochem. J. 204: 447–455.
13. Nakamura M. 1990. Lactoferrin-mediated formation of oxygen radicals by NADPH-cytochrome P-450 reductase system. J Biochem 107: 395–399.
14. Britigan BE, and Edeker BL. 1991. Pseudomonas and Neutrophil products modify transferrin and lactoferrin to create conditions that favor hydroxyl radical formation. J Clin Invest 88: 1092–1102.
15. Britigan BE, Serody JS, Hayek MB, Charniga LM, Cohen MS. 1991. Uptake of lactoferrin by mononuclear phagocytes inhibits their ability to form hydroxyl radical and protects them from membrane autoperoxidation. J. of Immunology 147: 4271– 4277.
16. Cohen MS, Mao J. Rasmussen GT, Serody JS, and Britigan BE. 1992. Interaction of lactoferrin and lipopolysaccharide (LPS): effect on the antioxidant property of lactoferrin and the ability of LPS to prime human neutrophils for enhanced superoxide formation. J. of Infectious Dis. 166: 1375–1378.
17. Schultz PG, Taylor JS, Dervan PB. 1982. Design and synthesis of a sequence-specific DNA cleaving molecule. (Distamycin-EDTA) iron (II). J Am Chem Soc 104: 6861–6863.
18. Furmanski P, Li Z-P, Fortuna MB, Swamy CVB, Das MR. 1989. Multiple molecular forms of human lactoferrin - identification of a class of lactoferrin that possess ribonuclease activity and lack iron-binding capacity. J Exp Med 170: 415–429.
19. Hutchens TW, and Yip T-T. 1991. Metal ligand-induced alterations in the surface structures of lactoferrin and transferrin probed by interaction with immobilized copper(II) ions. J. Chromatogr 536: 1–15.
20. Miller DM, Buettner GR, and Aust SD. 1990. Transition metals as catalysts of "autoxidation" reactions. Free Radical Biol Med 8:95–108.
21. Bennett RM, Davis J, Campbell S, Portnoff S. 1983. Lactoferrin binds to cell membrane DNA. Association of surface DNA with an enriched population of B cells and monocytes. J Clin Invest 71: 611–618.
22. Sudar F, Csaba G, and Robenek H. 1986. Detection of localization and internalization of membrane DNA in the tetrahymena by the lactoferrin-colloidal gold technique. Acta Biologica Hangarica 37: 101–107.

23. Bennett RM, Merritt MM, and Gabor G. 1986. Lactoferrin binds to neutrophilic membrane DNA. British J. of Haematology. 63: 105–117.

24. Hutchens TW, Maguson JS, Yip T-T. 1989. Interaction of human lactoferrin with DNA: one-step purification by affinity chromatography on single-stranded DNA-agarose. Pediatr Res 26: 618–622.

25. Maniatis T, Fritsch EF, and Sambrook J. 1982. Molecular cloning: a laboratory manual. Cold Spring Harbor Laboratory, Cold Spring Harbor, NY.

26. Teuwissen B., Masson PL., Osinski P., and Heremans JF. 1972. Metal-combining properties of human lactoferrin. Eur. J. Biochem. 31: 230–245.

27. Ainscough EW, Brodie,AM, and plowman JE. 1979. The chromium, manganese, cobalt and copper complexes of human lactoferrin. Inorganic Chimica Acta, 33: 149–153.

28. Ainscough EW, Brodie AM, McLachlan SJ, and Ritchie VS. 1983. Spectroscopic studies on copper(II) complexes of human lactoferrin. J. Inorganic Biochem. 18: 103–112.

29. Smith CA, Anderson BF, Baker HM, and Baker EN. 1992. Metal substitution in transferrins: the crystal structure of human copper-lactoferrin at 2.1-Å resolution. Bioch. 31: 4527–4523.

30. Chevion M. 1988. A site-specific mechanism for free radical induced biological damage: the essential role of redox-active transition metals. Free Radical Biol. & Med. 5: 27–37.

31. Joshi RR and Ganesh KN. 1992. Chemical cleavage of plasmid DNA by Cu(II), Ni(II) and Co(III) desferal complexes. Biochem. Biophys. Res. Commun. 182: 588–592.

32. Sagripanti J and Kraemer. 1989. Site-specific oxidative DNA damage at polyguanosines produced by copper plus hydrogen peroxide. J. Biol. Chem. 264: 1729–1734.

33. Fusi P, Tedeschi G, Aliverti A, Ronchi S, Tortora P, and Guerritore A. 1993. Ribonucleases from the extreme thermophilic archaebacterium S. solfataricus. Eur. J. Biochem. 211: 305–310.

# LACTOFERRIN PROMOTES NERVE GROWTH FACTOR SYNTHESIS/SECRETION IN MOUSE FIBROBLAST L-M CELLS

Ichizo Shinoda, Mitsunori Takase, Yasuo Fukuwatari, and Seiichi Shimamura

Nutritional Science Laboratory
Morinaga Milk Industry Co.,
LTD. 1-83
5-chome
Higashihara, Zama-City
Kanagawa-Pref, 228, Japan

## SUMMARY

Fibroblast cells are known to have an ability to synthesize and secrete nerve growth factor (NGF). To investigate the mechanism of action of the iron-binding protein, lactoferrin (Lf), on cultured animal cells, the effect of bovine Lf (bLf) on NGF synthesis/secretion in mouse fibroblast cells was examined. Both apo- and holo-bLf induced an increase in NGF content in the cell-conditioned medium(CM) of mouse L-M cells, a line derived from L929 fibroblast cells, with similar effectiveness. The increase in NGF content in the CM of L-M cells cultured with bLf was not dependent on the induction of increase in cell numbers, but was due to induction of *de novo* synthesis of NGF in individual cells by bLf. Human Lf(hLf) also increased NGF content. However, apo- and holo- bovine transferrin (bTf) failed to stimulate the NGF synthesis.

The time-dependent induction of NGF in L-M cells by bLf was different from that induced by basic fibroblast growth factor(bFGF) and bLf showed an additive effect with bFGF.

These results suggest that the induction of NFG synthesis depends on a mechanism different from iron transport or bFGF.

## INTRODUCTION

Fibroblast cells produce nerve growth factor (NGF) (Furukawa et al., 1984, 1986). NGF is a protein required for the development and maintainance of sympathetic and sensory neurons in the peripheral nervous system (Thoenen and Barde, 1980). NGF also functions as a neurotrophic molecule for the magnocellular cholinergic neurons in basal forebrain nuclei (Whittermore and Seiger, 1987). Some aspects of the stimulatory mechanisms of NGF

*Lactoferrin: Structure and Function*
Edited by T.W. Hutchens *et al.*, Plenum Press, New York, 1994

synthesis in cultured cells have been elucidated recently (Furukawa et al., 1986, 1987, 1989; Furukawa and Furukawa, 1990; Lindholm et al., 1988; Hergerer et al., 1990).

Lactoferrin (Lf) is an iron-binding glycoprotein that appears to have a role in iron-adsorption and in the host defense system against microbial infections (Aisen and Listowsky, 1980; Arnold et al., 1980). This protein has been detected in most of the physiological fluids of mammals such as plasma, milk, saliva, tears, and secretions that bathe mucosal surfaces as well as in neutrophilic leukocytes (Manson et al., 1966, 1969). Recently, Lf was found to promote the growth of animal cells (Hashizume et al., 1983; Nichols et al., 1987; Azuma et al., 1989).

To investigate the mechanism of action of Lf on cultured animal cells, we examined the effect of bovine Lf (bLf) on NGF synthesis/secretion in mouse L-M cells, a line derived from L929 fibroblast cells. Since L-M cells can grow in a serum-free chemically defined medium, it is possible to strictly examine the effect of Lf on NGF synthesis in the absence of calf serum which contains both Lf and Tf.

## EXPERIMENTAL

### Materials

bLf was prepared from the whey protein fraction of cow's milk. Iron-free bLf (apo-bLf) and iron-saturated bLf (holo-bLf) were prepared according to the procedures described by Masson and Heremans (1968) and Suzuki et al. (1989), respectively. Bovine apo-transferrin (apo-bTf) and bovine holo-transferrin (holo-bTf) were purchased from Yagai Research Center; recombinant human basic fibroblast growth factor (rhbFGF) from Genzyme; Medium 199 from Flow Laboratories; Bacto-peptone from Difco; tissue culture flasks and 24- and 96-well culture plates from Corning; bovine serum albumin (BSA) from Armour.

### Cell Culture

Mouse L-M cells obtained from the American Type Culture Collection were routinely cultured in Medium 199 containing 0.5% Bacto-peptone. To examine the effect of bLf and other materials, L-M cells were cultured in 96-well plates ($5–10 \times 10^3$ cells/well (well surface $0.32$ cm$^2$)) and cultured for 24 h in peptone-containing medium. Then, the medium was changed to Medium 199 containing 0.5% BSA, with or without the test samples, and the cells were cultured for 24 h.

### Measurement of NGF Content in the Cell-Conditioned Medium (CM)

NGF content was determined by a two-site enzyme immunoassay (EIA) developed for mouse submaxillary gland β-NGF as described by Furukawa et al. (1983) with some modifications. Since bLf and other materials used in this study, as well as unconditioned medium, did not affect the EIA system, the CM was directly applied to the EIA.

## RESULT

Figure 1 shows the stimulatory effect of bLf on NGF synthesis in L-M cells. Apo- and holo-bLf were similarly effective for induction of an increase in NGF content in the CM of L-M cells. NGF production was stimulated by apo- and holo-bLf in a dose-dependent manner at Lf concentrations ranging from 1 to 1000 μg/ml. On the other hand, neither apo- nor holo-bTf were effective (Fig. 1). Since NGF synthesized by L-M cells is secreted rapidly into the CM,

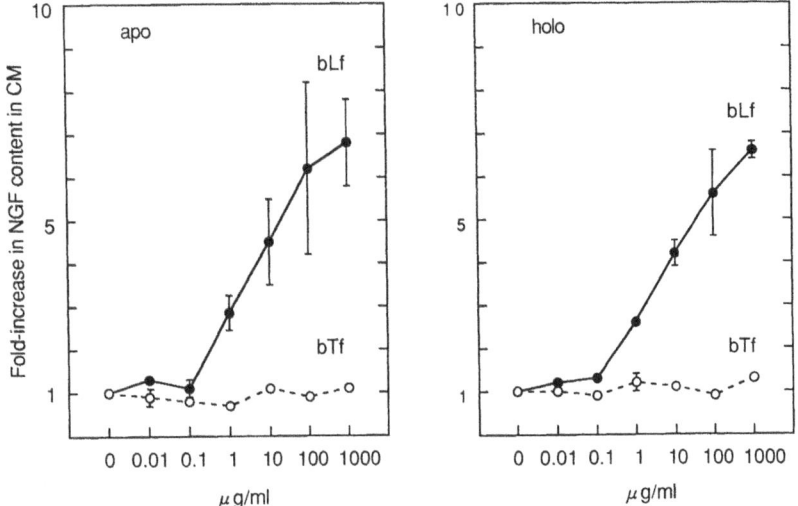

**Figure 1.** *Effects of bLf and bTf on NGF synthesis in L-M cells.* L-M cells were cultured with various concentrations of apo- and holo-bLf and apo- and holo-bTf in 96-well plates (0.1 ml) for 24 h. The NGF content in the medium was determined by EIA and is expressed as fold increase over that in the absence of samples. Each point is the mean ± S.E. of three determinations.

the measurement of NGF content in CM reflects the amount of NGF synthesized (Furukawa et al., 1986). To confirm that NGF synthesis in L-M cells was stimulated by bLf, the amount of NGF secreted per cell in cultures containing bLf was compared with that of control cultures after 24 h (Fig. 2). The results indicate that the increase in NGF content in the CM of L-M cells cultured with bLf was not dependent on induction of an increase in cell numbers by bLf during the culture period, but was due to induction of *de novo* synthesis of NGF in individual cells

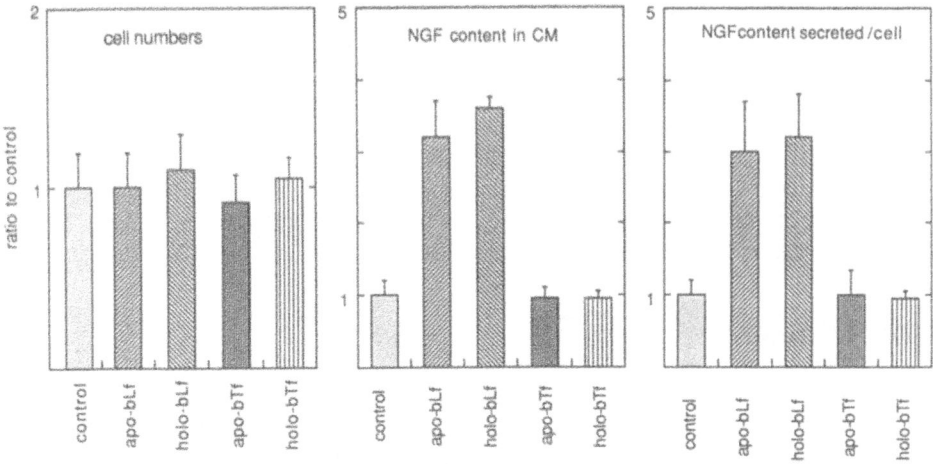

**Figure 2.** *Demonstration of de novo synthesis of NGF in L-M cells stimulated by bLf.* L-M cells (2.5 × 10⁴ cells/well) were cultured with apo- and holo-bLf and apo- and holo-bTf at 100 µg/ml in 24-well plates (1.0 ml) for 24 h. The NGF content in the medium was determined by EIA. The cell numbers and the NGF content in the medium as well as the NGF content per cell numbers are expressed as fold increase over that in the absence of samples. Each value is the mean ± S.E. of three determinations.

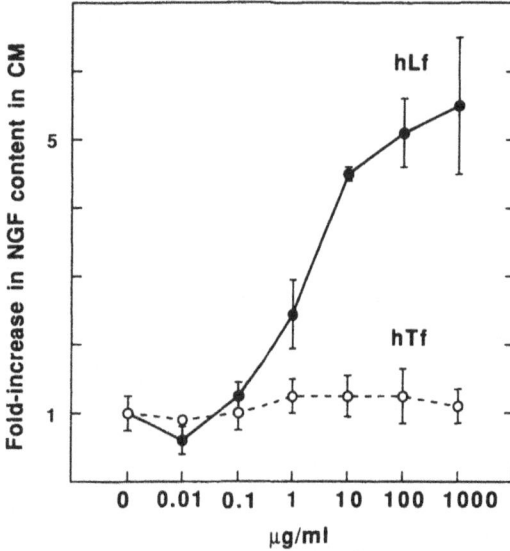

**Figure 3.** *Effect of hLf on NGF synthesis in L-M cells.* L-M cells were cultured with various concentrations of hLf (●) and hTf (○) in 96-well plates (0.1 ml) for 24 h. The NGF content in the medium was determined by EIA and is expressed as fold increase over that in the absence of samples. Each point is the mean ± S.E. of three determinations.

by bLf. Figure 2 shows the ineffectiveness of bTf on stimulation of NGF synthesis in L-M cells. Using human Lf and Tf, we obtained the same results as observed with bovine Lf and Tf (Fig.3).

It is interesting that different results were obtained between Lf and Tf although Lf is homologous to Tf in structure and has similar iron-binding properties (Metz-Boutigue et al., 1984). One of the physiological roles attributed to Lf is the transport of $Fe^{3+}$ ions into cells as Tf does (Azuma et al., 1989).

It has been found that basic fibroblast growth factor (bFGF) is a potent stimulator of NGF synthesis in cultured astroglial cells (Shinoda et al., unpublished results). Therefore, the effect of bLf on NGF synthesis was compared with that of bFGF with respect to time-dependent production of NGF. The increase in NGF content in the CM induced by bLf was found to be maximal at the culture time of 4 h, and it dropped at 12 h (Fig. 4). Comparing this curve with that generated by recombinant human bFGF (rhbFGF) (Fig. 5), the effect of rhbFGF after 2 h of culture was rather weaker than that of bLf. After 4 h, their effects were almost the same. Thereafter, rhbFGF markedly induced a further increase in NGF content in the CM. This difference between the effects of bLf and rhbFGF indicates that the regulatory mechanism of Lf in stimulating NGF synthesis differs from that of bFGF. The effect of bLf was assessed in the presence of rhbFGF. As shown in Fig. 6, bLf and rhbFGF additively induced an increase in NGF content in the CM, as expected.

## CONCLUSION

Bovine Lf induces an increase of NGF synthesis in L-M cells, and apo-Lf and holo-Lf have similar effectiveness. In contrast, Tf does not promote NGF synthesis in these cells. We assume that actions distinct from promotion of iron-transport by interaction with a putative receptor is required for the stimulatory effect of bLf on NGF synthesis in L-M cells.

The transitional process of the cell growth cycle from $G_0$ to $G_1$ phase is considered to be mainly connected with stimulation of NGF synthesis in cultured cells (Furukawa et al., 1987). Judging from the fact that Lf was effective for stimulation of NGF synthesis, we assume that Lf has the ability to promote this transitional process.

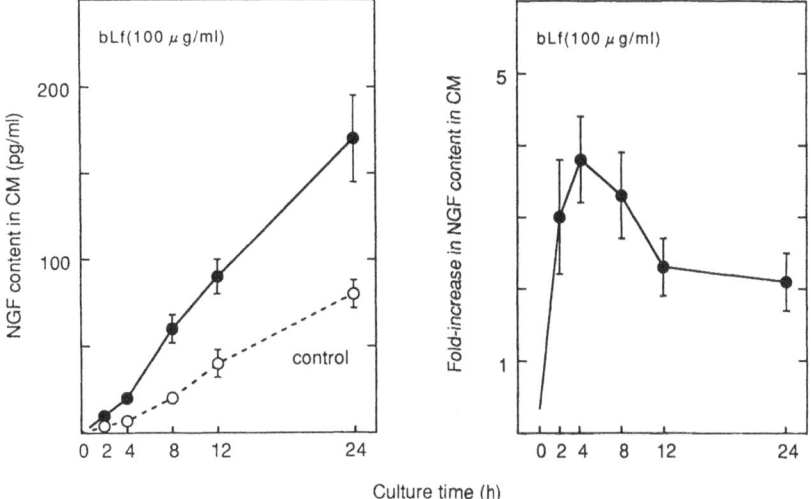

**Figure 4.** *Time course of NGF content in CM of L-M cells treated with bLf.* L-M cells were cultured with native bLf (100 μg/ml) in 96-well plates (0.1 ml), and the amounts of NGF in CM were measured at 2, 4, 8, 12, and 24 h. The NGF content is also expressed as fold increase over that in the absence of the bLf at the same culture time. Each point is the mean ± S.E. of four determinations.

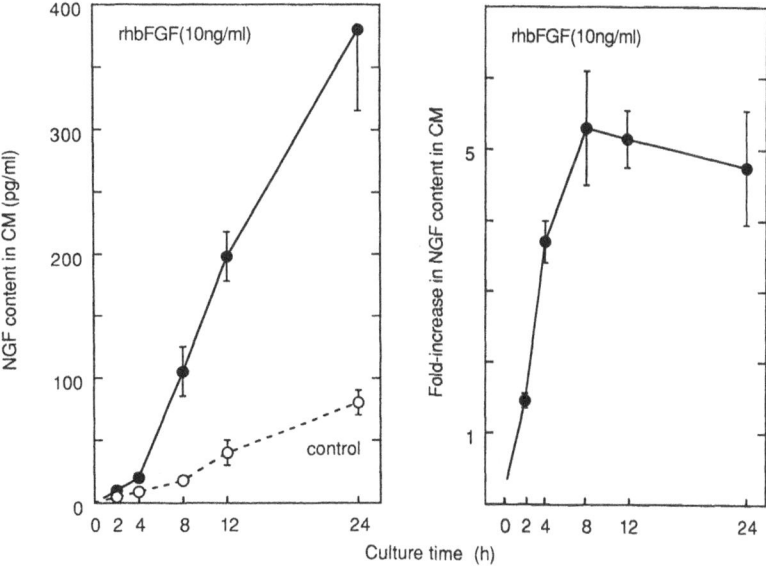

**Figure 5.** *Time course of NGF content in CM of L-M cells treated with rhbFGF.* L-M cells were cultured with rhbFGF (10 ng/ml) in 96-well plates (0.1 ml), and the amounts of NGF in CM were measured at 2, 4, 12, and 24 h. The NGF content is also expressed as fold increase over that in the absence of the rhbFGF at the same culture time. Each is the mean ± S.E. of four determinations tested.

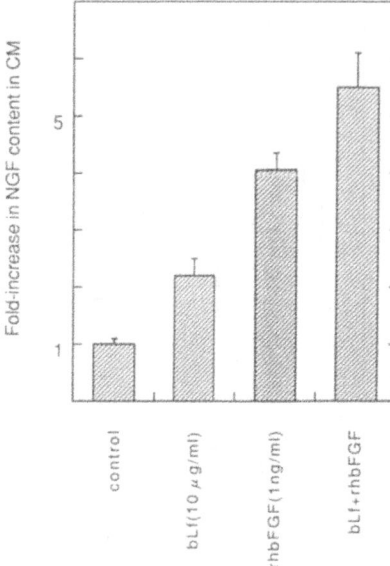

**Figure 6.** *Effect of bLf on NGF synthesis in L-M cells in the presence of rhbFGF.* L-M cells were cultured with native bLf (10 µg/ml) in the presence or absence of rhbFGF (1 ng/ml) in 96-well plates (0.1 ml) for 24 h. The NGF content in CM was determined by EIA and is expressed as fold increase over that of control. Each point is the mean ± S.E. of four determinations.

One of the early events of the transitional process is induction of the protooncogene c-fos and c-jun mRNAs, which encode the nuclear protein Fos and the Fos-associated protein Jun, respectively (Schönthal et al., 1988; Chiu et al., 1988). The complex of Fos and Jun (transcription factor AP-1) is known to activate the transcription of a set of genes which play an important role in cell proliferation (Chiu et al., 1988). According to Hengerer et al.(1990), an increase in c-fos mRNA levels preceded the increase in NGF mRNA levels in cultured rat sciatic nerve. Considering that an additive effect on NGF synthesis was observed with a combination of bLf and rhbFGF, which is reported to be a c-fos inducer (Müller et al, 1984), the possibility exists that the signals induced by Lf are terminally coupled with those induced by bFGF, resulting in an increase in c-fos mRNA levels and a concomitant increase in NGF-mRNA and NGF levels. It is very interesting to examine whether Lf induces c-fos mRNA.

## ACKNOWLEDGMENTS

We would like to thank Dr. Shoei Furukawa and Dr. Kyozo Hayashi of Gifu Pharmaceutical University for kindly providing rabbit anti-β-NGF antiserum and for helpful discussions.

## REFERENCES

Aisen P and Listowsky I. (1980) Iron transport and storage proteins. Annual Review of Biochemistry 49: 357–393.

Arnold RR, Brewer M, Gauthier JJ. (1980) Bactericidal activity of human lactoferrin: sensitivity of a variety of microorganisms. Infection and Immunity 28: 893–898.

Azuma N, Mori H, Kaminogawa S, Yamauchi K. (1989) Stimulatory effect of human lactoferrin on DNA synthesis in Balb/c 3T3 cells. Agricultural Biological Chemistry 53: 31–35.

Chiu R, Boyle WJ, Meek J, Smeal T, Hunter JB, Karin M. (1988) The c-Fos protein interacts with c-Jun/AP-1 to stimalate transcription of AP-1 responsive genes. Cell 54: 541–552.

Furukawa S and Furukawa Y. (1990) Nerve growth factor synthesis and its regulatory mechanisms: an approach to therapeutic induction of nerve growth factor synthesis. Cerebrovascular and Brain Metabolism Review 2, 328–344.

Furukawa Y, Furukawa S, Satoyoshi E, Hayashi K. (1984) Nerve growth factor secreted by mouse heart cells in culture. Journal of Biological Chemistry 259: 1259–1264.

Furukawa Y, Furukawa S, Satoyoshi E, Hayashi K. (1986) Catecholamines induce an increase in nerve growth factor content in the medium of mouse L-M cells. Journal of Biological Chemistry 261: 6039–6047.

Furukawa S, Furukawa Y, Satoyoshi E, Hayashi K. (1987) Synthesis/secretion of nerve growth factor is associated with cell growth in cultured mouse astroglial cells. Biochemical and Biophysical Research Communications 142: 395–402.

Furukawa S, Kamo I, Furukawa Y, Akazawa S, Satoyoshi E, Itoh K, Hayashi K. (1983) A highly sensitive enzyme immunoassay for mouse β-nerve growth factor. Journal of Neurochemistry 40: 734–744.

Furukawa Y, Tomioka N, Sato W, Satoyoshi E, Hayashi K, Furukawa S. (1989) Catecholamines increase nerve growth factor mRNA content in both mouse astroglial cells and fibroblast cells. FEBS Letters 247: 463–467.

Hashizume S, Kuroda K, Murakami H. (1983) Identification lactoferrin as an essential growth factor for human lymphocytic cell lines in serum-free medium. Biochimica Biophysica Acta 763: 377–382.

Hengerer B, Lindholm D, Heumann R, Rüther U, Wagner EF, Thoenen H. (1990) Lesion-induced increase in nerve growth factor mRNA is mediated by c-fos. Proceedings of The National Academy of Sciences of The United States of America 87: 3899–3902.

Lindholm D, Heumann R, Hengerer B, Thoenen H. (1988) Interleukin 1 increase stability and transcription of mRNA encoding nerve growth factor in cultured rat fibroblast. Journal of Biological Chemistry 263: 16348–16351.

Masson PL and Heremans JF. (1968) Metal-combining properties of human lactoferrin (red milk protein). The involvement of bicarbonate in the reaction. European Journal of Biochemistry 6: 579–584.

Masson PL, Heremans JF, Dive C. (1966) An iron-binding protein common to many external secretions. Clinica Chimica Acta 14: 735–739.

Masson PL, Heremans JF, Schonne E. (1969) Lactoferrin, an iron-binding protein in neutrophilic leukocytes. Journal Experimental Medicine 130: 643–658.

Metz-Boutigue MH, Jollês J, Mazurier J, Schoentgen F, Legrand D, Spik G, Montreuil J, Jollês P. (1984) Human lactoferrin: amino acid sequence and structal comparisons with other transferrins. European Journal of Biochemistry 145: 659–676.

Müller R, Bravo R, Burckhardt J, Curran T. (1984) Induction of c-fos gene and protein by growth factors precedes activation of c-myc. Nature 312: 716–720.

Nichols BL, Mckee KS, Henry JF, Putman M. (1987) Human lactoferrin stimulates thymidine incorporation into DNA of rat crypt cells. Pediatrics Research 21: 563–567.

Schönthal A, Herrlich P, Rahmsdorf HJ, Ponta H. (1988) Requirement for fos gene expression in the transcriptional activation of collagenase by other oncogenes and phorbol esters. Cell 54: 325–334.

Suzuki T, Yamauchi K, Kawase K, Tomita M, Kiyosawa I, Okonogi S. (1989) Collaborative bacteriostatic activity of bovine lactoferrin with lysozyme against Escherichia coli 0111. Agricultural and Biological Chemistry 53: 1705–1706.

Thoenen H and Barde YA. (1980) Physiology of nerve growth factor. Physiological Reviews 60: 1284–1335.

Whittermore SR and Seiger A. (1987) The expression, localization and functional significance of β-nerve growth factor in the central nervous system. Brain Research Reviews 12: 439–464.

# SUMMARY CHAPTER

**Lactoferrin Structure and Function: Remaining Questions, Methodological Considerations and Future Directions**

T. William Hutchens,[*] Bo Lönnerdal,[†] and Sylvia Rumball[‡]

[*]Department of Food Science and Technology
University of California
Davis, California 95616

[†]Department of Nutrition
University of California
Davis, California 95616

[‡]Department of Chemistry and Biochemistry
Massey University
Palmerston North, New Zealand

## SESSION I. STRUCTURE OF LACTOFERRIN

A goal expressed in this session was to increase attempts by all researchers in this field to correlate known structural features of the lactoferrin molecule with the mechanism(s) of physiological functions, e.g., the interaction with cells and membrane structures. Factors to consider regarding the three-dimensional structure would include the degree of domain movement (lobe flexibility) within both the N-lobe and the C-lobe, and the role of bound metal ions as structural determinants of these dynamics. The structural dynamics of the lactoferrin molecule in solution (i.e., possible mobility of loops and other domain structures) are more difficult to address and remain to be verified. While some individual amino acid residues can be very mobile, secondary structural domains appear to be relatively stable. The three-dimensional structure of the glycan moieties (different in number and sequence for bovine and human milk lactoferrins) and the effects of these posttranslational modifications on lactoferrin structure and function remain to be explored. Future priorities in the structural analysis of lactoferrin is co-crystallization of the lactoferrin-receptor complex and its structural analysis.

Given that we finally have some good detailed information about the structure of lactoferrin, in both its apo (metal-free) and holo configuration (state), it is imperative that we strive to perform all future experiments with lactoferrin of sufficient purity and structural integrity to allow interpretations that are relatively unaffected by the ambiguities otherwise associated with the use of lesser defined "lactoferrin preparations." In particular, metal-binding status, possible contamination of lactoferrin with LPS (bound), proteolytic cleavage (without dissociation), and structural changes induced by exposure to low/high pH should be understood and identified. As an example of this, the need to establish well-defined approaches to the removal of bound iron (or other metal ions) without inducing irreversible alterations in structure is quite difficult but an absolute neccessity. Buffers frequently contain small amounts of cations, Chelex treatment is rarely totally efficient in removing contaminants, EDTA dialysis often leaves some EDTA bound, citrate dialysis frequently alters the protein structure. Verification of apolactoferrin content is difficult. Iron-titration of apo-lactoferrin was suggested as one possible assay; however, this method has the caveat that iron salts are frequently contaminated and therefore the presence of other metal ions (especially those of lower affinity) may be underestimated. The use of immobilized chelators and short exposure to mildly acidic conditions was proposed. It may be better to physically separate apo- and hololactoferrin, and not attempt to remove bound metal ions for some types of experiments. In recognition of the above, it was recommended that the Methods section of papers on lactoferrin should include more detailed descriptions and characterizations of the protein.

## SESSION II. METAL AND ANION BINDING PROPERTIES OF LACTOFERRIN

The presence and the effect of other metal ions than iron should be carefully considered. This may be particularly important for the surface where metal ions like Cu, Zn, and Mn may become more or less non-specifically bound to structural features of the lactoferrin molecule or via associated artions. The conditions under which the assays are being performed may have a profound effect on the results. Changes in kinetic properties may not be the same as changes in spectroscopic properties. However, spectroscopic measurements may help to sort out many of these effects, e.g. salt effects. The potential consequences of changing concentrations of K, Na and Cl should be included in binding assays. It was cautioned that commercial lactoferrin contains polymers (up to 10 %) and that these forms should be removed prior to spectroscopic studies because of possible interference. The possibility of protein-protein interactions should also be considered, particularly as basicity and hydrophobicity can affect metal binding.

It was concluded that a high priority would be to remove the metal ion under non-disruptive conditions and then to study the effects of adding cations and anions back separately and in combination, using strictly defined stoichiometric ratios. More emphasis should be put on defining surface binding sites, although this will be difficult due to the much weaker $K_a$s for these sites. Further, a need for studies on the physiological release of iron from Lf was identified.

## SESSION III. BACTERIOSTASIS AND BACTERICIDAL EFFECTS OF LACTOFERRIN

There is clear evidence *in vitro* (at least under select conditions) that both iron chelation and bactericidal effects are exerted by lactoferrin. However, there are also "superpathogens" like Nesseria gonorrhea that use lactoferrin and transferrin as growth factors. It is important to sort out iron effects and membrane effects particularly as these effects differ between Gram positive, Gram negative bacteria and yeast. It must be emphasized that lactoferrin does (may) not act alone and postulated synergisms between sIgA and lysozyme need to examined in more detail. The precise mechanism(s) accounting for these putative effects, specifically the direct versus indirect (e.g., effects secondary to the interaction of lactoferrin with macrophages) role of lactoferrin, remain uncertain. Furthermore, the *in vivo* significance of these observations are in need of verification and further evaluation. In this regard there is a recognized need for the development of suitable animal model systems for the investigation of these individual and collective or linked events.

Because the various effects of lactoferrin are seemingly influenced by the presence and/or absence of other factors, presented at several concentrations, these effects must be interpreted within the context of the food or vehicle of ingestion. In this context, molecular biological approaches, perhaps tissue specific lactoferrin gene knockout experiments or hyperexpression, will allow these issues to be investigated with otherwise minimal perturbation of the system.

## SESSION IV.  MECHANISMS OF LACTOFERRIN INTERACTIONS WITH CELLS

There were several needs identified with regard to the interaction of lactoferrin with putative receptors on target cells. To date there is no information available on the primary sequence. Also, any type of signal induced in the cell by the binding of lactoferrin to the surface should be investigated. The possibility of the lactoferrin–lactoferrin receptor complex being internalized should also be explored and the site for the dissociation between the lactoferrin molecule and iron determined. The potential role of lactoferrin glycans in receptor recognition and other biological events needs to be clarified. It was emphasized that structural features of the lactoferrin molecule should be considered more carefully in studies on physiological functions and also that the purity of the lactoferrin used is clearly defined.

### Other Research Priorities Identified

Modification of structures within the lactoferrin molecule that will abolish the binding to the receptor.

Biological significance of receptor occupancy needs to be demonstrated and explored in detail. Despite the apparent differences in putative receptor properties (e.g., enterocytes, lymphocytes, monocytes) is there a receptor that is common to each of the tissue and cell types presently thought to have such receptors? A priority is to characterize and link the binding data with biochemical characterizations and biological effects. To this extent, there is a clear need to develop specific antibodies directed against the lactoferrin receptor. There is additional need to clarify the metabolic fate of plasma lactoferrin after release from specific granules (or introduction by other means). This would include clearance by liver uptake and possible degradation after receptor binding and possible internalization.

Receptor binding data may be flawed if the assumptions about added versus available lactoferrin concentrations differ significantly (i.e., the relative amounts of total and nonspecific binding). The term receptor protein, as opposed to binding sites, should be used with caution. Why are there so many "receptors" on the surface of presumed target cells? Should cultured cells be used as a source for lactoferrin receptor protein purification?

## SESSION V.  LACTOFERRIN EFFECTS ON CELL PROLIFERATION AND DEVELOPMENT

Most studies presented and discussed in this session were focused on the effects of lactoferrin on differentiated cell functions and not necessarily the initiation (i.e., directly) of cell growth *per se*. A strong need for collaborative research between basic scientists (immunologists, bacteriologists, biochemists) and clinicians was emphasized especially with respect to the need to define objective outcome variables. Close work between these categories of researchers coupled with a break-down of classical interdisciplinary barriers was considered likely to answer some of the fundamental questions on cell growth and development. It was recognized that there are both real and apparent disparities in the literature with respect to lactoferrin effects on myelopoisis; some of these differences, perhaps most, are likely to be the result of differences in the detail of experimental design and due to differences in the quality of the lactoferrin used in these experiments. A rigorous approach of phase 1 studies exploring the possibility of toxicity (considered unlikely) of lactoferrin followed by preclinical trials in animal models and clinical studies in various populations would yield valuable data if the nature of the lactoferrin was carefully defined, i.e., Fe saturation, LPS and/or lysozyme contamination, potential denaturation or structural derangement etc was closely examined. If a suppressive effect on infection was found, lactoferrin might be used in hyperproliferative disorders such as leukemia. The possibility of lactoferrin acting synergistically with other compounds such as gamma interferon may also be explored.

A pledge was made that the hypothesis of lactoferrin causing hypoferremia should be put to rest because of lack of evidence in support.

## SESSION VI. LACTOFERRIN METABOLISM AND IRON ABSORPTION

There were strong beliefs by some that lactoferrin is not directly involved in iron absorption *per se*. Is lactoferrin more of a buffer to help prevent the negative responses initiated by iron excess? Is lactoferrin a vehicle to help make iron accessible to the mucosal surface for absorption by a variety of possible mechanisms? So far there is very limited evidence to support a direct role of lactoferrin in the transmucosal transport of iron from the gut to storage proteins in the circulation. However, at least in preterm infants fed human milk at six weeks of age, lactoferrin is clearly absorbed intact. Furthermore, this lactoferrin is iron-saturated when it appears in the urine of these infants. It is not yet known to what extent these quantities are significant to the overall balance of iron metabolism.

## SESSION VII. MOLECULAR BIOLOGY OF LACTOFERRIN

Lactoferrin is not a tissue- or cell-specific protein; however, the regulation of lactoferrin gene expression is cell-specific. The upstream regulatory sequences identified for the lactoferrin gene are similar to the regulatory sequences identified for the estrogen receptor and related proteins in the erb gene family. The significance of this observation is not readily apparent.

The possible occurance of natural genetic variants (e.g., 3 versus 4 arginine residues in the N-terminus) was discussed. No genomic sequences have been presented to date.

The construction, expression, and use of recombinant human milk lactoferrin mutants is progressing. How do we use lactoferrin mutants to better understand lactoferrin function and specific structure/function relationships? Some of the more obvious recombinant lactoferrin mutants to produce involve changes in 1) specific residues in the two specific iron(metal ion)-binding sites, 2) alterations in the sites of glycosylation, and 3) alterations in residues thought to be involved in receptor binding (e.g., specific surface-exposed Lys and/or His residues). Although the N-lobe has been expressed, C-lobe expression has not been accomplished to date. The possible use of chimeric lactoferrin constructs was discussed, but there was no consensus or relative priority on this matter. Certainly there will be large amounts of recombinant lactoferrin produced in the near future, both in recombinant microorganisms, eukaryotic cells, and, hopefully, in transgenic cows (transgenic mice expressing human lactoferrin are now being characterized). Although the glycosylation of expressed recombinant lactoferrin appears to be different from human milk lactoferrin, it is not yet known to what extent glycosylation affects lactoferrin structure or its known functions (e.g., iron binding and receptor binding).

## WORKSHOP I. APPLIED AND COMMERCIAL APPLICATIONS

A very constructive and stimulating workshop was centered around the exploitation of lactoferrin as a natural reagent useful in many different contexts; these discussions were driven largely by the specific information presented in previous sessions but were influenced greatly by practical questions surrounding issues of lactoferrin processing as a food component. Although bovine milk lactoferrin is presently the only commercially viable source of lactoferrin for large scale applications (e.g., infant formulas), recombinant human lactoferrin is expected to become more available on the scale required for such applications. As apparent from the above accounts, there are numerous potential applications of lactoferrin, both bovine and human, once these proteins are available in large supply. Furthermore, specific domains and fragments derived from lactoferrin are potentially very important.

It can be anticipated that new and different sources of both bovine and human lactoferrin will facilitate greatly the identification of a more definitive biological role(s) for lactoferrin as well as a more definitive understanding of specific structure/function relationships and may be exploited in the future in a way which is different from the way in which intact lactoferrin is presently used.

# INDEX

The manufacturer's authorised representative in the EU is Springer
Nature Customer Service Centre GmbH, Europaplatz 3, 69115 Heidelberg,
Germany. If you have any concerns regarding our products, please
contact ProductSafety@springernature.com

Printed and bound by CPI Group (UK) Ltd, Croydon, CR0 4YY
23/04/2026
02095628-0011